ENVIRONMENTAL CHANGE IN ICELAND: PAST AND PRESENT

Glaciology and Quaternary Geology

VOLUME 7

Series Editor:

C. R. BENTLEY

University of Wisconsin-Madison,
Department of Geology and Geophysics,
Madison, Wisconsin, U.S.A.

The titles published in this series are listed at the end of this volume.

ENVIRONMENTAL CHANGE IN ICELAND: PAST AND PRESENT

Edited by

JUDITH K. MAIZELS

Department of Geography,
University of Aberdeen, Aberdeen, U.K.

and

CHRIS CASELDINE

Department of Geography,
University of Exeter, Exeter, U.K.

1991

KLUWER ACADEMIC PUBLISHERS

DORDRECHT / BOSTON / LONDON

Library of Congress Cataloging-in-Publication Data

```
Environmental change in Iceland : past and present / edited by Judith
  K. Maizels, Chris Caseldine.
        p.   cm. -- (Glaciology and quaternary geology ; v. 7)
    ISBN 0-7923-1209-0 (alk. paper)
    1. Geology, Stratigraphic--Holocene. 2. Glacial epoch.
  3. Paleoclimatology--Iceland. 4. Glacial landforms--Iceland.
  5. Landscape changes--Iceland.   I. Maizels, Judith K., 1948-   .
  II. Caseldine, Chris, 1951-   . III. Series.
  QE698.E58   1991
  551.7'93'094912--dc20                                     91-2726
```

ISBN 0-7923-1209-0

Published by Kluwer Academic Publishers,
P.O. Box 17, 3300 AA Dordrecht, The Netherlands.

Kluwer Academic Publishers incorporates
the publishing programmes of
D. Reidel, Martinus Nijhoff, Dr W. Junk and MTP Press.

Sold and distributed in the U.S.A. and Canada
by Kluwer Academic Publishers,
101 Philip Drive, Norwell, MA 02061, U.S.A.

In all other countries, sold and distributed
by Kluwer Academic Publishers Group,
P.O. Box 322, 3300 AH Dordrecht, The Netherlands.

Printed on acid-free paper

Printed in the Netherlands

DEDICATION

This volume is dedicated to the memory of Liz Martin (Thomas) who
tragically died before completing her paper. Her enthusiasm,
insight and spirit will be sadly missed but will live on in
the memory of all who knew her and worked with her.

ACKNOWLEDGEMENTS

This volume of papers arose originally from a meeting held in the Department of Geography at the University of Aberdeen in April 1989 sponsored by the Quaternary Research Association and the Geologists Association. Thanks to their help it was possible to bring together a wide range of scientists working on environmental change in Iceland and following a stimulating conference there has been a welcome increase in co-operation between those involved in this area of study. The editors are grateful to all the contributors to the volume for their valiant efforts to make deadlines and to the referees who completed their work so efficiently and to such effect. The eventual delay in publication was due to a printer breakdown, very much to the frustration of those involved in the production in the Department of Geography at the University of Exeter, especially Terry Bacon and Andrew Teed. Above all, for all those non-Icelandic individuals, we would like to thank the people of Iceland who, through their Research Council, have been so encouraging and welcoming and thus allowed such a wide range of nationalities to contribute to the understanding of one of the most interesting landscapes anywhere in the world.

CONTENTS

ENVIRONMENTAL CHANGE IN ICELAND: PAST AND PRESENT. AN INTRODUCTION

Judith Maizels
Department of Geography
University of Aberdeen
Elphinstone Road
Aberdeen AB9 2UF
UK

Chris Caseldine
Department of Geography
University of Exeter
Amory Building
Rennes Drive
Exeter EX4 4RJ
UK

1. INTRODUCTION

This volume brings together for the first time a collection of papers which explore the patterns of climatic and environmental changes in Iceland since the end of the last glaciation, through the 'Little Ice Age', to some of the processes modifying the present-day landscape. The book is particularly timely, for, while urgent interest is now being directed towards our understanding of global climate changes, Iceland is emerging as a crucial laboratory for advancing this understanding. Its position in the middle of the North Atlantic means that it is, and has been for most of its 3 million year Quaternary history, highly sensitive to north-south oscillations in the aerodynamic boundary that separates polar and tropical air streams. Hence, climatic variations and associated changes in oceanic circulation are, and have been, reflected in large-scale environmental changes in Iceland. The most dramatic example of this relationship was felt at the close of the last glaciation, when the marine polar front migrated thousands of kilometres between 13,000 and 11,000 BP, and again between 11,000 and 10,000 BP, associated with the complex climatic perturbations occurring during the final decay of the ice sheets in the northern hemisphere, and particularly those bounding the North Atlantic.

In order to develop a better understanding of the rates and mechanisms of climatic change, and of the feedback processes that control climatic change, there is clearly a need

1

J. K. Maizels and C. Caseldine (eds.), Environmental Change in Iceland: Past and Present, 1–9.

to obtain accurate, high resolution data on past climatic changes. Much of our knowledge of the Late Weichselian and Holocene environmental history of Iceland rests heavily on the work of two scholars, Thorleifur Einarsson and the late Sigurður Thórarinsson. Their pioneering and exceptionally all-encompassing work has laid the foundation for much of what appears in the following pages and Quaternary scientists owe them a considerable debt for providing models and methods to re-examine, refine and develop. It is only with the increase in the number of scientists now studying the past and present Icelandic environment that their ideas can be fully extended and evaluated. Despite the abundant morphological and stratigraphical evidence in Iceland there has been relatively little dating control, especially in comparison with other Scandinavian countries, and the overall Late Weichselian and Holocene chronology has rested heavily on long-distance analogies with Europe. There is no full pollen record for the Icelandic Holocene, nor definitive growth curves for lichenometric dating. Little is known of the differential and regional rates of isostatic uplift and the patterns of shoreline displacement during and following the last period of deglaciation. The existence and significance of possible refugia during the glacial maxima remain controversial, together with the origin and rates of expansion of its biota.

The lack of a comprehensive data base means that Quaternary scientists remain uncertain about the pattern and timing of the major environmental changes which occurred in Iceland during the Lateglacial and early Holocene. In particular, the limits and dates of the Weichselian ice maximum are uncertain, the age of the marine limit is in dispute, and the dates, rates and patterns of deglaciation are not resolved. Many uncertainties also remain in our understanding of environmental changes in Iceland during the mid and late Holocene. Of these, the most important questions address, first, the nature of the Holocene climatic record and timing of both the period of optimal postglacial climate and of maximum postglacial glacier expansion (i.e. the timing of the Neoglacial); and, second, the impact of human settlement from the late 9th century A.D. onwards on landscape change.

In terms of the Holocene climatic record, much detailed information has been gathered from historical records and accounts since the time of settlement. In addition, the soils of Iceland contain a detailed stratigraphic record of the numerous volcanic eruptions that have punctuated Iceland's history. The eruptions produced ash-falls, or tephra, deposited as a mantle across an entire landscape, providing an ideal chronozone marker in Iceland's soils. Many of the tephra layers are well dated, thanks to the work of S. Thórarinsson, allowing many Holocene landforms in turn to be relatively well dated. Much of our information on Holocene climatic change is largely derived from studies of moraine sequences and stratigraphic relations, but many of these have still not been accurately dated. The use of moraine sequences for inferring climatic changes has implied that glaciers tend to respond at a similar time to a given climatic change, regardless of local environmental conditions or the internal dynamics of the glacier system.

The second important question regarding the record of environmental change during the late Holocene is the role of human settlement on the landscape compared with that of natural climatic changes. Major changes have occurred in the physical environmental since Landnám (settlement) times. Widespread removal of native birch woodland, expansion of new biota and extinction of others, and extensive soil erosion reaching almost catastrophic proportions have led to desertification of many upland areas and abandonment by local populations. The role of climatic change as opposed to deforestation and sheep grazing in creating these new environments has proved a further issue of great controversy.

While our understanding of historic environmental changes remains inadequate, our knowledge of processes that are modifying the present-day landscape is also sparse and selective. Little is known of active periglacial processes, slope instabilities, and rates of soil erosion by slope wash and aeolian transport. Coastal processes of erosion and beach formation have been studied only locally. Most of our information on recent or active processes comprises records of glacier fluctuations, volcanic eruptions and jökulhlaup events, but still little is known of the mechanisms and processes of landscape change effected by these events.

This volume of papers, based on a conference sponsored by the Quaternary Research Association and the Geologists Association and held at the University of Aberdeen in April 1989, addresses many of these crucial uncertainties regarding environmental changes in Iceland from the Lateglacial onwards. The papers make a major contribution to dispelling many earlier uncertainties and clarifying areas of controversy. Many of the papers challenge traditional and poorly supported ideas, replacing them with hypotheses based on new data and new insights derived from the expansion of wider scientific expertise and theory. The volume focuses on three major areas of research in particular. First, the Lateglacial-Holocene glacial chronology, and associated changes in climate, sea-level, and isostatic uplift; second, the nature of environmental changes during the Postglacial period (including the mid Holocene, the historic period and the so-called 'Little Ice Age'), particularly in terms of (a) changes in the biological record, and their relation to patterns of human settlement and climatic change; and (b) the patterns of glacier fluctuation, as a means of identifying long-term climatic oscillations and periods of maximum and minimum postglacial temperatures; and finally, some of the processes that affect present-day landscape formation, and specifically, slope processes, and the impact of fluvial systems on deposition and removal of sediment in proglacial areas.

The research studies presented in this volume also draw on a wide range of field and analytical techniques to support the many new interpretations and theories regarding environmental change in Iceland. The studies of Lateglacial and early Holocene chronology depend on such techniques as radiocarbon dating of marine shells found in raised shoreline and till deposits, precise levelling of shorelines, detailed stratigraphic analysis, off-shore seismic profiling, and analysis of pollen and microfaunal assemblages.

The biological studies have been based on analysis or review of pollen assemblages, trap collections of beetle and insect fauna from a wide variety of natural and human habitats, including archaeological sites, and analysis of soil sections using tephrochronology as a dating control. Analyses of historic and Neoglacial glacier fluctuations (including rock glaciers of glacial origin), are widely based on dating of moraine sequences using new and more accurately derived lichen growth curves, combined with tephrochronology, radiocarbon dating, and interpretation of moraine sequences.

Finally, the analyses of recent slope and fluvial processes draw on a combination of detailed field monitoring of slope movements, and of suspended and solute load transport in a glacial meltwater river; geomorphic mapping, palaeohydrologic interpretation of sediment sequences and stratigraphic relations, and assessment of landscape sensitivity to change in areas subject to high erosion rates.

2. SIGNIFICANCE OF RESEARCH CONTRIBUTIONS TO UNDERSTANDING ENVIRONMENTAL CHANGE IN ICELAND

2.1 Lateglacial-early Holocene glacial chronology

The traditional Lateglacial-early Holocene chronology of Iceland has been developed over many years by the extensive research and syntheses of a number of Quaternary geologists, especially Th. Einarsson. It was Einarsson who interpreted major, distinctive moraines as representing successive stages of retreat of the Late Weichselian ice sheet in Iceland; the stages and their dates were correlated with those in mainland Europe. His work suggested that moraines representing the Older Dryas (Álftanes moraine; ca 12,000 BP) and the Younger Dryas (Búði moraine; ca 11-10,000 BP) could be identified, while the Alleröd interstadial was represented by a high marine limit. This conventional chronology is now being challenged, in the light of recently acquired radiocarbon dates, re-interpretation of former 'type sites', and analysis of new shoreline, off-shore and geological evidence.

The papers presented here on the Lateglacial-early Holocene period consistently demonstrate that the Late Weichselian ice sheet was far more extensive, and remained in existence until much later than was previously acknowledged. **Ingólfsson**, in his review of the Late Weichselian and early Holocene, demonstrates from extensive dating evidence that, at Borgárfjörður (in western Iceland), for example, the oldest moraine (Álftanes) dates from the Younger Dryas, not the Older Dryas; the marine limit occurred at ca 10,300 BP, and the Búði readvance dates from ca 9,700 BP, i.e. the Preboreal and not the Lateglacial. This new chronology is strongly supported by recent evidence both from north and south Iceland. **Norðdahl**, working in the Eyjafjörður area in north Iceland, confirms that the marine limit occurred during the late Younger Dryas-early Preboreal, and that final deglaciation was not complete until ca 9,650 BP. Norðdahl has also extended the chronology back to the Weichselian maximum. Using evidence from raised shoreline sequences and from the Skógar tephra found in ice-lake sediments, he concludes that the glacial maximum occurred ca 18,000 BP, rather than in the early Weichselian as some workers had previously thought. He also concluded, through modelling of the thickness of the last ice sheet, that extensive ice-free areas existed in north Iceland during the last glacial maximum.

Pétursson's paper further confirms that this new chronology is applicable in northern Iceland. Using data on raised shoreline gradients, stratigraphic and facies interpretation, and radiocarbon dates on shells from west Melrakkaslétta in northeast Iceland, Pétursson identified a high sea-level at ca 12,700 BP (Bölling), associated with glacio-isostatic loading and ice expansion, followed by a glaciomarine-marine transgression that occurred ca 12,100 BP, corresponding to the Older Dryas-Alleröd transition, and followed in turn by a younger Dryas marine limit at ca 10,200 BP. The new chronology suggested for the Lateglacial is also confirmed from south Iceland, based on the evidence presented by **Hjartarson** using a series of new radiocarbon dates on shells. Hjartarson additionally argues that the widespread absence of Alleröd dates from south Iceland indicates that this whole area was in fact still ice-covered at least until the Younger Dryas.

In addition to the growing evidence of more extensive and later deglaciation than previously accepted for Iceland, much of the shoreline evidence presented in this volume also throws significant light on patterns of isostatic readjustment and sea-level fluctuations during deglaciation. A high sea-level during early deglaciation is supported by evidence not only from Pétursson's work in northeast Iceland, but also from parts of northwest Iceland where **Hansom and Briggs** present evidence for an almost continuous shoreline sequence extending from over 135m a.s.l. to 1 m a.s.l. Hansom and Briggs regard this sequence as representing an exceptionally rapid fall in relative sea-level between ca 12,000 and 9,000 BP. An unusually rapid rate of postglacial isostatic rebound is also reported for southwest Iceland in the paper by **Thors and Helgadóttir**. The Alleröd shoreline, which they identified at ca 65 m a.s.l. and dated to ca 10,300 BP, was succeeded by a sea-level at -30 to -35 m. This low sea-level was identified by a series of seismic profiles distinguishing drowned coastal features, and was dated from overlying peats to ca 9,030 BP. Thors and Helgadóttir attribute this remarkably rapid rate of uplift to the low viscosity of the crustal rocks. However, as Hansom and Briggs emphasize in their paper, isostatic uplift rates are highly variable from one region to another, reflecting not only substrate geology but also proximity to regional centres of glaciation.

The evidence presented in these papers clearly indicates that the traditional deglaciation chronology for Iceland is no longer acceptable. The new chronologic framework that is now being developed for Iceland provides a sound, well-supported alternative model for understanding and modelling long-term environmental changes. Much still needs to be accomplished, especially in terms of acquiring more radiocarbon dates from a wide range of sites and environments, and particularly from lake and mire sequences; and in terms of linking the chronology with evidence of other environmental changes based on biological, morphological, and lithostratigraphical studies.

2.2 Environmental changes in Iceland during the Postglacial

2.2.1 Biotic changes, climatic change and human settlement

Many of the papers addressing the problems of Postglacial environmental change in Iceland also challenge conventional schools of thought. One major point of controversy lies in identifying the origin of the Iceland biotic populations. The traditional view, proposed by such workers as Lindroth, Dahl and Einarsson, considers that the Icelandic flora and fauna are derived from species that survived during the Quaternary glaciations in ice-free areas, or refugia. A more recently introduced view is that biota reached Iceland at some time in the late Tertiary or Quaternary interglacials via a land-bridge. The paper by **Buckland and Dugmore** provides persuasive arguments for an alternative origin which accommodates a number of features about Iceland's history and the characteristics of its biota. For example, Iceland has been an island for at least 15 million years, and probably glaciated numerous times over the past 2 million years; the Icelandic biota are of European affinity, contain no endemic species, and include a number of wingless species of fauna with low aerial dispersal potential. Buckland and Dugmore therefore propose that the Icelandic biota did not survive in refugia, nor arrived across a land-bridge, nor solely by aerial dispersal. Instead, they present evidence to support the view that plant and animal species arrived through ice-rafting associated with a short-lived period of massive melting of the continental ice sheets around the eastern North Atlantic, probably around 10,000 BP.

However, the natural biotic assemblages of Iceland appear to have been subject to considerable modification since the beginning of the Postglacial. In their paper, **Buckland, Dugmore and Sadler** review the evidence for major changes that occurred in insect populations (particularly the Coleoptera) following the arrival of the Norse settlers. Buckland et al summarize the species evidence from pre-Landnám and post-Landnám archaeological sites and from a range of 'natural' habitats. They demonstrate that the major faunal changes identified over the historic period are the result of human activity. Extensive loss of birch woodland has led irrevocably to changes in available habitats, while more recently, drainage and reseeding of pasture lands, has led to loss of species diversity and the introduction of new species associated with pastoral agriculture, such as the dung beetle. Buckland et al also explore the many problems associated with analysis of biotic changes during the Holocene, including those of sampling, quantification, identification of species, and data interpretation.

Throughout Iceland there is evidence of long-term, widespread soil erosion, with wind being the dominant eroding agent. Indeed, about half of Iceland's original soil cover, some 20,000 km², has now been stripped away. **Dugmore and Buckland**, in their paper, use a detailed tephrochronologic framework to demonstrate that the timing and location of erosion, and associated sediment accumulation rates, exhibit significant local variations throughout the Holocene. The authors found that major soil erosion first commenced with the Norse settlement, starting in upland areas, and migrating to lower lying areas. The greatest period of soil erosion during the Holocene occurred during the ' Little Ice Age', when grazing pressures from large sheep populations increased at lower levels, eventually leading to complete loss of vegetation cover, soils and farm abandonment. **Sveinbjarnardóttir's** paper focuses on identifying some of the causes of more widespread farm abandonment in Iceland. Traditionally farm abandonment was believed to have been triggered by climatic deterioration and/or by epidemics. Sveinbjarnardóttir tests this view in the light of the many sources of geological, archaeological and historical evidence, together with examination of two areas of farm abandonment located in highly distinctive physical environments. Her findings primarily support those of Dugmore and Buckland, in that the overwhelming evidence demonstrates that soil erosion, induced by severe overgrazing, led to land deterioration and farm abandonment. This process led to farm abandonments as early as the

12th century in many inland areas (e.g. parts of Þórsmörk). In coastal areas farm abandonment was triggered by episodes of coastal erosion, and in other areas local river erosion, climatic deterioration, disease and, more recently, economic and social isolation, have all played a role in patterns of farm abandonment.

2.2.2 Glacier fluctuations: the Holocene climatic optimum and the 'Little Ice Age'

A second particular point of contention in studies of the Postglacial in Iceland is the traditional view that there was a major period of glacier expansion in the ' Sub-Atlantic' , around ca 2500 BP. This view was largely based on analogy rather than stratigraphically secure ^{14}C dates, and only now are accurate dates emerging from different parts of Iceland which suggest an alternative model of Postglacial climatic and glacial change. **Stötter** demonstrates from evidence of fluctuations of climatically sensitive, alpine glaciers in the Tröllaskagi area of northern Iceland, that the warmest Postglacial period may have occurred between 4,800 and 3, 400 BP. He considers that temperatures may have been sufficiently high not only to have raised the treeline by at least 150m, but also to have caused complete disappearance of local glaciers at that time. Stötter's evidence throws severe doubts on the traditional model of Postglacial climatic chronology in Iceland. Further, **Häberle**, working in the same part of northern Iceland, draws on a combination of radiocarbon dating, tephrochronology and lichenometry to show that glacier advances occurred at different times in different valleys periodically throughout the late Holocene.

Some of the problems in using lichenometry for dating purposes are discussed in the papers by Kugelmann and Caseldine. **Kugelmann** demonstrates that many of the growth curves established in Iceland for *Rhizocarpon geographicum* agg. are erroneous, since the calibration surfaces themselves have been inaccurately dated and the curves have been plotted from too few data points. Using a revised dating curve for part of northern Iceland, based on 19 dated surfaces, Kugelmann produces a chronology of glacier fluctuations for the latter part of the 'Little Ice Age' . He confirms that glacier advances in the late 19th and early 20th centuries correspond to periods of low temperature and increased incidence of sea-ice, and hence that records of glacier fluctuations may be used as indicators of climatic oscillations.

Caseldine similarly emphasizes the need to link glacial chronology derived from lichenometric dating of moraines, to climatic variations, particularly as a means of evaluating present-day glacier-climate relations. This approach clearly requires that lichenometric methods are as accurate as possible. Since many lichen populations include components of both older populations and/or of disturbed populations Caseldine proposes that, rather than using just the single largest lichen thallus for dating, the gradient of the whole population distribution curve should be used. From this and traditional lichenometric techniques Caseldine found that in the Tröllaskagi area, the ' Little Ice Age' represents the Holocene glacial maximum for most glaciers. Here, as at many other sites in Iceland, it appears that there are no lichens more than about 200 years old. Both Kugelmann and Caseldine surmise that the absence of older lichens may be 'an intriguing coincidence' (Caseldine), perhaps representing a response to the major eruption of Lakagígar in 1783.

2.3 Processes affecting recent landscape change

Landscapes in Iceland continue to be modified by a wide range of active geomorphic processes, often producing dramatic changes in the environment. **Gerrard** argues that much of Iceland is highly sensitive to landscape change, largely in response to its steep slopes, unstable substrate materials and high energy processes. Landscape instability can be promoted by volcanic activity and climatic events, such as intense rainfall and seasonal snowmelt, as well as from increased settlement or grazing pressures. Gerrard suggests that this instability is in turn reflected in the wide range of slope movements and episodes of large-scale soil erosion. He describes sites where slopes have become increasingly unstable through time, with slips, debris slides, and earth flows developing through spring sapping into gullies and pipes. By contrast, he also identifies sites which have become stabilized after periods of slope movement . These contrasting results reflect the complexity of episodic slope development and of the controls on landscape instability in Iceland.

Whilst a great deal of research has been published on glacier fluctuations in Iceland, and on the palaeoclimatic and climatic significance of these fluctuations, as well as on the geomorphic impact within the proglacial environment, little is known of glacier systems lying in marginally-glacierized areas where glaciers may be at the threshold of growth or extinction. **Martin, Whalley and Caseldine** report on a rock glacier of glacial origin in the Tröllaskagi area of north Iceland. The rock glacier is shown to be a relict feature whose response rate to wider environmental changes is subject to prolonged lags, possibly exceeeding 80 years (i.e. since the end of the ' Little Ice Age'), because of the thick, insulating cover of debris mantling the ice core. Indeed, whilst some glaciers in the area have retreated by over 300m since 1946, in response to overall climatic warming, the rock glacier exhibited virtually no change in activity or morphology. The present-day landscape comprises both actively retreating or fluctuating glaciers, and ice masses that have remained as stagnant relicts of larger glaciers active in the past. The rock glaciers therefore provide potential evidence both of palaeoclimatic change and of the rates of landform response to environmental change.

Maizels' paper is also concerned with the impact of episodic events on landscape development. This paper focuses on the rôle of jökulhlaup events on sandur formation, in areas subject to jökulhlaups generated by either sudden drainage of ice-dammed or sub-glacial lakes, or from subglacial volcanic eruptions. She demonstrates that the stratigraphy and sedimentology of the main sandar of south Iceland are dominated by a long history of jökulhlaup events, while evidence of normal seasonal meltwater flows is rarely preserved in the recent geological record. Maizels also demonstrates that the nature of the sedimentological sequence provides a key to the type of jökulhlaup event, the nature of the flood hydrograph, and the relative concentrations of sediment transported by the flood. Because of the highly friable substrates, the production of fresh tephra during volcanic eruptions, and high runoff amounts, Icelandic jökulhlaups characteristically contain extremely high sediment concentrations, resulting in hyperconcentrated flood flows, and extensive but episodic accumulation of sandur sediments.

A number of papers in this volume have demonstrated that Icelandic catchments have notoriously high soil erosion rates, and Maizels has shown that jökulhlaups can transport huge volumes of sediment. **Lawler'** s paper provides an analysis of a series of detailed measurements of annual, seasonal and diurnal variations in the concentrations of suspended and solute loads in a glacial meltwater river in southern Iceland. His records confirm that at this site, at least, the 15-year mean sediment yield of 14,500 t km^{-2} a^{-1} (equivalent to a catchment erosion rate of ca 5.4 mm a^{-1}) is one of the highest recorded anywhere in the world. However, sediment transport rates are also shown to be highly variable through time, with maximum annual rates possibly linked to an initial period of ice readvance, while on a short-term timescale pulses of high sediment transport were found to occur over periods of under 1 hour. Lawler argues that our understanding of proglacial sediment transfer systems could prove vital in analysis of off-shore sediment sequences, prediction of subglacial geothermal activity, and assessment of the role of normal as opposed to jökulhlaup events in modifying proglacial landscapes in Iceland.

3. CONCLUSIONS

This volume of papers provides a framework for future research into environmental change in Iceland. The papers present the reader with a range of new and challenging ideas, data, models and directions for future research. We believe that this volume represents an important stage in the development of ideas about environmental change in Iceland, marking a new confidence in challenging traditional or poorly founded ideas, and drawing on a new and sound store of scientific evidence. The papers here indicate that future work needs to be directed particularly towards establishing a full Lateglacial-Holocene pollen sequence, based on analysis of lake and mire sediments, and correlated with climatic, biotic and geomorphological evidence. The research presented here represents the first step towards developing this new, detailed chronologic framework for understanding climatic change in Iceland and the numerous and complex environmental changes that have taken place over Iceland's recent geological history. The models of environmental change emerging from research in Iceland will provide a valuable input to our understanding of climatic perturbations over the North Atlantic and, in turn, to models of global climate systems.

PART 1

LATEGLACIAL AND EARLY HOLOCENE

ENVIRONMENTAL CHANGES

A REVIEW OF THE LATE WEICHSELIAN AND EARLY HOLOCENE GLACIAL AND ENVIRONMENTAL HISTORY OF ICELAND

Ólafur Ingólfsson
Department of Quaternary Geology
Lund University
Sölvegatan 13
S-223 62 Lund
Sweden

ABSTRACT. The pattern and timing of the last deglaciation of Iceland is critically reviewed and discussed. The interpretations of the glacial history are controversial, but new data suggest a relatively late deglaciation, with larger glaciers during the Younger Dryas and Preboreal periods than earlier assumed. Interpretations of the early Holocene environmental development are hampered by a scarcity of data and poor resolution in the biostratigraphical and chronostratigraphical records. The Icelandic evidence on the Late Weichselian/early Holocene climatic and environmental development is examined in the light of data from adjacent North Atlantic areas. It is concluded that the contribution of Icelandic data to the discussion of climatic changes in the North Atlantic region has been limited. The strategy and aims of continued research on Iceland are briefly discussed.

1. INTRODUCTION

Iceland lies geographically in the middle of the North Atlantic (Fig.1), and large scale environmental changes in Iceland reflect climatic oscillations and changes in the atmospheric and oceanic circulation. High resolution data for the glacial and environmental history of Iceland can thus contribute to our understanding of climatic development around the North Atlantic. Iceland is an especially good reference area for studying climatic development, as the late colonisation by man (9th century A.D.) excludes human influence on the environment for most of the Holocene. The purpose of this paper is to critically review the present status of research on the Late Weichselian and early Holocene glacial and environmental history of Iceland, and to point out future research aims.

J. K. Maizels and C. Caseldine (eds.), Environmental Change in Iceland: Past and Present, 13–29.

2. DEGLACIATION HISTORY AND RELATIVE SEA-LEVEL CHANGES

Our present knowledge of the Weichselian glacial history of Iceland is limited to the latest part of the Late Weichselian, ca 13,000-10,000 B.P., when glaciers in parts of the country had already retreated to positions inside the present coast (Einarsson, 1968, 1978; Ashwell, 1975; Andersen 1981). The Weichselian glacial maximum has neither been properly mapped nor dated (Hoppe, 1968, 1982; Norðdahl, 1981, 1983; Hjort et al., 1985). On the basis of shelf morphological features it is assumed that the maximum Weichselian ice front probably extended to the edges of the shelf (Einarsson, 1978, 1985). It is assumed that glaciers had their maximum volume and extent around 18,000 B.P. (Einarsson, 1971, 1985; Einarsson, Tr., 1966). The existence of ice-free areas (Fig.2) which could have served as refugia for terrestrial fauna and flora during the Weichselian glaciation (Einarsson, 1961, 1963, 1967; Steindórsson, 1962, 1963; Hoppe, 1968, 1982; Sigbjarnarsson, 1983; Lindroth et al., 1988) is controversial and has not been resolved.

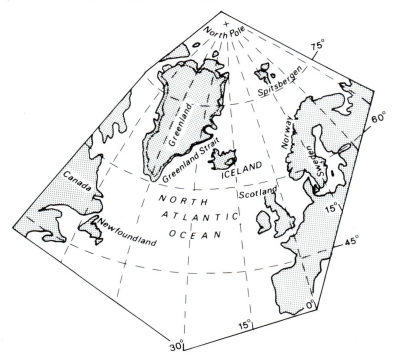

Fig.1 The position of Iceland in the North Atlantic Ocean.

The deglaciation history of Iceland is mainly based on geomorphological and litho-stratigraphical studies of moraines, raised deltas and outwash deposits, which can be successively followed inland from the present coastal areas. In the coastal areas, raised beaches have been correlated with radiocarbon-dated fossiliferous glaciomarine sediments, to obtain a chronology for the relative sea level changes. The glacial chronology is mainly based on time-space correlations between dated glaciomarine sediments, raised beaches and ice front deposits. The marine limit (Fig.6) is highest in southern Iceland, at ca 110m a.s.l. In the Borgarfjörður area, western Iceland, it lies at 60-80m, but elsewhere in Iceland usually at 40-50m. There is a gradient in raised beaches, which rise in altitude from the outer coast inland. The age of the marine limit is a matter of dispute (Einarsson, 1978; Ingólfsson, 1987,1988; Hjartarson and Ingólfsson, 1988), and probably it is not synchronous around the island. The varying altitude and age of the marine limit could be due to either of two main reasons, or a combination of them: a) differential downwarping of Iceland caused by differential glacial load during the Weichselian glaciation and as a consequence of that a subsequent differential isostatic rebound; and, b) metachronous age of the marine limit due to regionally different deglaciation patterns. The rate of isostatic uplift may complicate the picture. It has been suggested that glacial isostatic responses are very rapid in Iceland (Einarsson, Tr., 1966; Imsland, 1982), but detailed stratigraphical control of relative sea-level changes does not exist for any part of Iceland.

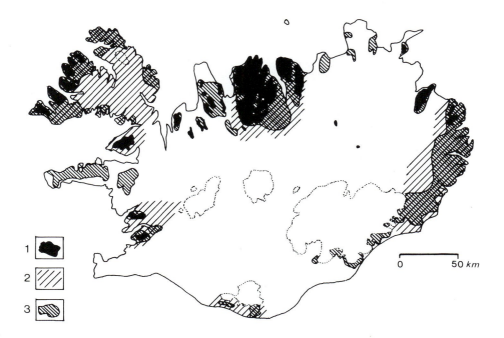

Fig. 2 Areas of proposed limited glaciation during the Weichselian period.
Legend: (1) probable ice-free areas (Einarsson (1963); (2) possible plant refugia areas (Steindórsson, 1963); (3) areas of alpine landscape development, including a relatively limited glaciation (Sigbjarnarson, 1983). Major present-day glaciers are dotted.

Two basically different reconstructions of the history of glacial retreat in Iceland have been presented:

Model 1: Einarsson (1961, 1967, 1968, 1971, 1973, 1978, 1979, 1985) presented a morphostratigraphical synthesis for the deglaciation based on a broad summary of the results of earlier workers (e.g. Keilhack, 1884; Thoroddsen, 1906; Askellsson, 1934; Kjartansson, 1940, 1943, 1964; Thórarinsson, 1951), which has for a long time influenced geological thinking in Iceland. He recognised two interstadials and two stadials during the Late Weichselian deglaciation, which he correlated with the Late Weichselian chronozones of NW-Europe (Einarsson, 1979; Mangerud et al., 1974; Mangerud and Berglund, 1978). According to Einarsson, the ice front in parts of the country had retreated to positions inside the present coast shortly after 13,000 B.P. Glaciomarine sediments recognised in western Iceland (Borgarfjörður) and northeastern Iceland (Kópasker) were related to the Kópasker interstadial of Einarsson, which he correlated with the Bölling interstadial of Scandinavia.

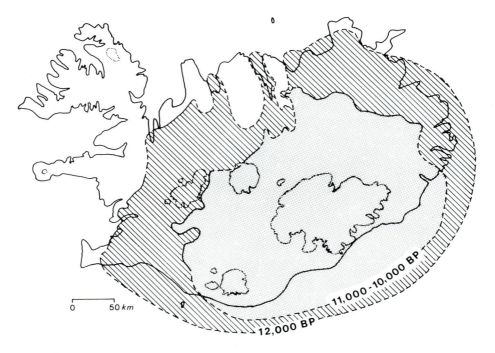

Fig.3 Reconstruction of the 12,000 B.P. and 11,000-10,000 B.P.
ice-frontal isochrones by Einarsson (1978).

A climatic deterioration around 12,000 B.P. caused the glaciers to readvance. The Álftanes moraines (Figs. 3 and 4), a chain of terminal moraines in the coastal areas of southwestern Iceland, which Einarsson also correlated with moraines in northern and eastern Iceland (Fig.4), were formed during this readvance. Einarsson called this event the

Álftanes stadial, and broadly correlated it with the Older Dryas of Scandinavia. The type locality for the stadial is at Álftanes in the greater Reykjavík area (Fig.4). According to Einarsson's reconstruction the marine limit was reached around most of Iceland during the following interstadial, the Saurbær interstadial, broadly correlated with the Alleröd interstadial of Scandinavia. Hardly anything is known about the extent of ice during the Saurbær interstadial, other than that the coastal areas were probably ice-free around most of Iceland. The type locality is at Saurbær in western Iceland (Fig.4), where fossiliferous glaciomarine sediments have been reported (Bárðarson, 1921; Andrésdóttir, 1987).

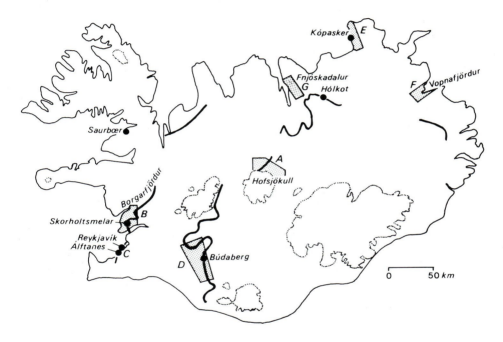

Fig.4 Conspicuous terminal moraines, as mapped by Einarsson (1978).
Localities A-G are discussed in the text.

The Saurbær interstadial was followed by the Búði stadial, during which conspicuous moraines were formed (Figs.3 and 4), both in southern Iceland (the Búði moraine complex) and northern Iceland (the Hólkot moraines). The type locality for the Búði stadial is at Búðaberg in southern Iceland (Fig.4). Einarsson (1979) correlated the Búði stadial with the Younger Dryas of Scandinavia. According to Einarsson, the Preboreal chronozone was a time of rapid glacial retreat and wasting of ice in the central highlands, without any significant advance or glacial episode causing prominent moraine formation.

Recent work by Einarsson and Albertsson (1988) and Norðdahl and Einarsson (1988) largely affirm the deglaciation synthesis of Einarsson, although some modifications are suggested for the 12,000 B.P. ice-frontal isochrone for northeast Iceland.

Model 2: Einarsson's reconstruction has been increasingly challenged during the past few years, both with respect to the morpho-stratigraphical correlations and the chronology of glacial events.

Kaldal (1978) mapped a number of ice-marginal positions north of Hofsjökull, central Iceland (Fig.4, locality A), and discussed the possibility of a Preboreal age for the deposits. She stated that in view of the lack of absolute chronology for the deglaciation of the area, the age of the moraines should be considered an open question for future studies. Einarsson (1978) took the moraines to represent the position of a Younger Dryas ice margin.

In western Iceland Ingólfsson (1985, 1987, 1988) concluded that glaciers had extended to or beyond the present coastline in the Borgarfjörður region (Fig.4, locality B) more or less continuously from 12,000 B.P. to 10,300 B.P. Ingólfsson recognised two glacial advance episodes, the Skípanes event and the Skorholtsmelar event which he suggested were Older Dryas and Younger Dryas events, respectively. He proposed that glaciation in central western Iceland was considerably more extensive during the Younger Dryas than Einarsson's reconstruction recognises. Ingólfsson (1988) suggested that the marine limit at 60-80m a.s.l. was reached around 10,300 B.P., and that moraines in the Borgarfjörður tributary valleys were of Preboreal age, which suggests a relatively heavy early Holocene glaciation. Ingólfsson's (1987, 1988) deglaciation chronology is based on 32 radiocarbon dates, and is by far the best dated deglaciation sequence in Iceland.

Hjartarson (1987, 1988) proposed a radical revision of the deglaciation chronology for the Reykjavík area (Fig.4, locality C) in southwestern Iceland. Marine sediments, found below till and previously thought to be of last interglacial (Eemian) age, were radiocarbon dated to around 11,500 B.P. (Hjartarson 1989; Anderson et al. in press). He also suggested that the deposits at the type locality for the Álftanes stadial derived from a Younger Dryas glacial event, and that the marine limit in the Reykjavík area was reached around 10,200 B.P., rather than during the Saurbær (Alleröd) interstadial as follows from the deglaciation synthesis of Einarsson.

Hjartarson and Ingólfsson (1988) re-investigated the type locality for the Búði stadial and a number of new sections relating to the formation of the Búði moraines (Fig.4, locality D). Based on four new radiocarbon dates they came to the conclusion that the moraine was formed after 9,700 B.P. This indicates a considerably later deglaciation of southern Iceland than does Einarsson's synthesis. A Preboreal age for the Búði moraines implies that the history of relative sea-level changes and the age of the marine limit in southern Iceland has to be re-evaluated. This conclusion also has important implications for the interpretation of the biostratigraphical record from southern Iceland (see below).

In northeastern Iceland, Pétursson (1986, 1987) reinvestigated the type locality of the Kópasker interstadial (Fig.4, locality E), correlated by Einarsson (1978, 1979) with the Bölling of Scandinavia. Einarsson (1978, 1979) suggested that a till of Older Dryas age overlies glaciomarine deposits of Kópasker (Bölling) age. Pétursson suggests that the till is of Younger Dryas age and radiocarbon dates the glacial retreat from the area and age of the marine limit to about 10,100 B.P. Norðdahl and Hjort (1987), working in the Vopna-fjörður area, northeastern Iceland (Fig.4, locality F), concur with Pétursson on an extensive Younger Dryas glaciation as their results indicate that glaciers in Vopnafjörður reached the present coast around 10,000 B.P.

Norðdahl (1981, 1983) studied the glacial history of the Fnjóskadalur area, northern Iceland (Fig.4, locality G). He described an important marker unit in the Fnjóskadalur sequence, the Skógar tephra, to which he could relate stages in the glacial history of Fnjóskadalur. Norðdahl (1983) estimated the age of the tephra to be around 17,000 B.P. In the absence of dateable material, Norðdahl (1983) related his reconstruction of the deglacia-tion to Einarsson's (1971, 1978) ice-marginal positions and chronology. In a recent paper, Norðdahl and Haflidason (1990) suggest a radical revision of the deglaciation chronology for the Fnjóskadalur area. They correlate the Skógar tephra to the Vedde ash, described and dated to about 10,600 B.P. in Norway (Mangerud et al. 1984). Norðdahl and Haflidason (1990) conclude that the new deglaciation chronology implies a much heavier Younger Dryas and Preboreal glaciation in central northern Iceland than previously suggested.

3. GLACIAL READVANCES AND CHRONOLOGY

One problem with the last deglaciation is that although significant ice-margin positions have been recognised (Fig.4), usually the stratigraphy is too poorly known to disprove a readvance. The Skorholtsmelar moraines in western Iceland have been shown to relate to a glacial advance (Ingólfsson, 1987, 1988). Other parts of the Álftanes ice-margin, as mapped by Einarsson (1968, 1978), are defined by geomorphological criteria, without any stratigraphic evidence for a readvance. The Hólkot moraine in northern Iceland (Thórarinsson, 1951), correlated with Búði by Einarsson (1968), is not dated by any absolute means and has not been proven to relate to a glacial advance. The Búði moraines in southern Iceland, assigned a Younger Dryas age by Einarsson (1968) but a Preboreal age by Hjartarson and Ingólfsson (1988), is probably complex with regard to ice-marginal positions during its formation. At the type locality at Búðaberg, Hjartarson and Ingólfsson (1988) found structural and stratigraphical evidence for a readvance, whilst further to the southeast the stratigraphic evidence rather suggested a prolonged period of ice-marginal sedimentation in front of a grounded glacier.

The deglaciation chronology, regardless of which reconstruction one chooses, is based on only a few tens of radiocarbon dates. The stratigraphic significance of dates, i.e. how they can be related to a specific geological event, is often unclear, and may explain some of the differences between the discussed reconstructions. Model 2 above is largely based on new and stratigraphically more significant radiocarbon dates compared to Model 1.

4. HOLOCENE VEGETATION AND CLIMATIC HISTORY OF ICELAND

A search of the geological literature for references on the Holocene vegetation and climate history of Iceland resulted in a list of twenty papers. Of these, 11 papers discussed early Holocene (Preboreal and Boreal) development, using data from 15 sites in Iceland (Fig.5). Additional studies are presently being prepared for publication, and valuable information can also be found in a few unpublished undergraduate theses from the University of Iceland (Hallsdóttir, pers. comm. 1989).

Palaeoenvironmental studies from lake and mire sediments in Iceland have focused on reconstructing the vegetational history by means of pollen analysis. Other methods (Berglund, 1986) have as yet been only rarely applied in Iceland (Bradshaw and Thompson, 1985; Thompson et al., 1986). The present Icelandic flora comprises about 440 species of vascular plants. The only native trees in Iceland are birch (*Betula pubescens*), rowan (*Sorbus aucuparia*) and aspen (*Populus tremula*). The pollen production in the vegetated parts is low (Hallsdóttir, 1987), which is reflected as a relative scarcity of pollen grains in the lake and mire sediments (Einarsson, 1963).

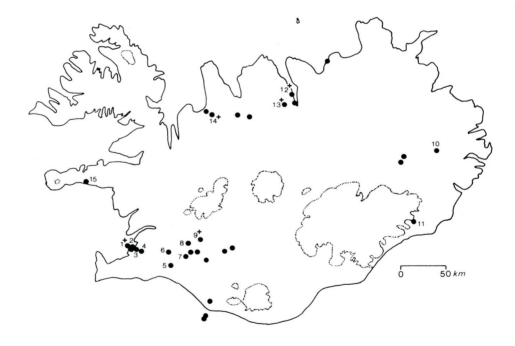

Fig.5 Pollen analysed sites in Iceland. Sites 1-15 probably reach early
Holocene strata. Sites where early Holocene age has been
established by means of radiocarbon dates are marked by a plus
sign. Modified after Hallsdóttir (1987, pers. comm. 1989).

Pollen-analytical work on lake and mire sediments was started by Thórarinsson (1944, 1955), Okko (1956) and Straka (1956). Einarsson (1961, 1963) published results from 20 different localities in Iceland, and made a generalised pollen diagram for the Holocene (Einarsson, 1968, 1975), where five periods or pollen zones were recognised:

The Upper Mire Period	0 - 2500 B.P.
The Upper Birch Period	2500 - 5000 B.P.
The Lower Mire Period	5000 - 7000 B.P.
The Lower Birch Period	7000 - 9000 B.P.
The Birch-Free Period	9000 - 10000 B.P.

The naming of the periods reflects the vegetation development, and hints at the Holocene climatic oscillations. A treeless pioneer phase occurred in the Preboreal Chronozone, during which tundra-like vegetation prevailed, followed by a birch woodland phase in a dry and warm Boreal type of climate. Around 7,000 B.P. mires expanded as a cooler and more humid Atlantic type of climate developed, which again turned towards a drier climate around 5,000 B.P., with birch woodlands expanding onto the mires. Einarsson (1968, 1985) suggests that the hypsithermal in Iceland falls within the Upper Birch Period, with an annual mean temperature 2-3°C higher than today. Around 2,500 B.P. the mires expanded again, as a consequence of a more humid and cooler climate. After 1,100 B.P. human influence on the vegetation takes over the dominance.

The results of Okko (1956), Straka (1956), Vasari (1972, 1973), Bartley (1968), Schwaar (1978), Påhlsson (1981) and Hallsdóttir (1982, 1987) fit reasonably well with the general pattern of the results of Einarsson. There are, however, different opinions on the spread of birch woodland in the early Holocene; while Einarsson assumes birch to have spread in southwestern Iceland around 9,000 B.P., Vasari (1972) suggests a little before 8,000 B.P. (dating in Jungner 1979). Hallsdóttir (1987) also suggested that birch spread in southwestern Iceland close to 8,000 B.P. A complication when interpreting and correlating the biostratigraphically studied sites is that nowhere is the whole Holocene represented. Also, radiocarbon dates for the lowest parts of the sequences are few (Fig.5), and perhaps somewhat unreliable due to the low organic content of the sediments.

As yet biostratigraphical studies on the Holocene development of Iceland as a whole are supported by only about 35 radiocarbon dates, spanning 10,000 years and spread out on 11 different sites (Hallsdóttir pers. comm.). One reason for the scarcity of radiocarbon dates in biostratigraphic sequences in Iceland is the extensive use of tephrochronology for dating and correlation. However, the Icelandic tephrochronology does not at present extend back to the early Holocene.

5. CONCLUSIONS

Einarsson's morphostratigraphical division for deglaciation, based on the recognition and correlation of terminal moraines and associated ice-marginal features from different parts of Iceland, has been questioned with regard to correlation of ice-frontal positions and the chronolgy of glacial events. The new concept of a more extensive glaciation in Iceland during Younger Dryas and Preboreal times calls for a re-evaluation of the relative sea level history.

There seems to be an agreement on the general outline of the biostratigraphic zonation of the Holocene, but opinions differ as to the chronology of the vegetation history during the early Holocene, i.e. the Preboreal and Boreal chronozones.

In summary, the period ca 12,000-8,000 B.P. is controversial with regard to the interpretation of the morphological, lithostratigraphical and biostratigraphical records, i.e. the timing and pattern of deglaciation and the spread of vegetation after deglaciation. The scarcity and often unclear stratigraphic significance of radiocarbon dates make chronological reconstructions for glacial and climatic events somewhat ambiguous.

6. THE ICELANDIC DATA IN A NORTH ATLANTIC PERSPECTIVE

From a regional (North Atlantic) perspective on climatic change, the deglaciation period is very interesting as it represents a highly dynamic section of the environmental history. Interpretations of stratigraphical data in terms of climatological signals, suggest that the more maritime the climate has been, the larger the fluctuation in environment and in ice volume.

In the Alps, the boundaries between the Alleröd-Younger Dryas biozones and the Younger Dryas-Preboreal biozones are reflected by relatively small scale fluctuations (e.g. Amman and Lotter, 1989). In continental NW Europe and southern Scandinavia there is a clear indication of a climatic deterioration at the end of the Alleröd interstadial, and the Younger Dryas chronozone is characterised by arctic conditions, before a rapid warming during the Preboreal (e.g. Björck and Möller, 1987; Lemdahl, 1985; Lundqvist, 1988; Nyberg, 1988).

In the British Isles Atkinson et al. (1987) reported a similar pattern in temperature fluctuations to that recognised in southern Scandinavia, and in Scotland there was a significant glacial advance during the second half (10,500-10,000 B.P.) of the Younger Dryas or Loch Lomond stadial (Peacock et al., 1989). The evidence of the Loch Lomond stadial as indicating a severe climatic deterioration was discussed by Sutherland (1984) and Ballantyne (1984). Data from maritime western and northern Norway indicate that the ice fronts retained similar positions during the Older Dryas and Younger Dryas stadials, and in parts of southwest Norway a major readvance occurred during the Younger Dryas (e.g. Anundsen, 1985; Mangerud, 1980; Mangerud et al., 1979). The deglaciation of Svalbard has been debated (Troitsky et al., 1979; Boulton, 1979; Miller,1982; Boulton et al., 1982; Mangerud and Salvigsen, 1984), but Landvik et al. (1987) presented evidence for a Younger Dryas readvance in eastern Svalbard, and suggested that the glacial histories of Svalbard and Scandinavia showed greater similarities than earlier assumed.

The North Atlantic deep-sea record suggests strong fluctuations of the marine polar front during the deglaciation (Ruddiman and McIntyre, 1981; Bard et al., 1987), i.e. in the period ca 13,000-10,000 B.P., with a strong southward advance of the front between 11,000 and 10,000 B.P. The same pattern is recognised in the Greenland ice cores (Dansgaard et al., 1989), though chronostratigraphical correlations with terrestrial and marine data are somewhat circumstantial. The glacial history of East Greenland includes a Late Weichselian Younger Dryas-Preboreal glacial advance, called the Milne Land stade (Funder and Hjort, 1973; Funder, 1978; Funder, in press) or the Nanok (Nanok II) advance (Hjort, 1981; Hjort and Björck, 1984). Moraines belonging to this advance were formed in the period ca 10,300-9,500 B.P. Funder and Hjort (1973) suggested that increases in precipitation over the region in connection with the return of climate to interglacial conditions caused the readvance. The break-up of sea-ice cover and opening of the Greenland (Denmark) Strait is necessary for cyclones to be able to bring precipitation to East Greenland, north of 69°N.

The contribution of Icelandic data to the discussion of climatic changes in the North Atlantic region during and after the last deglaciation has been limited. Scarcity or lack of data and poor stratigraphical control of available evidence, resulting in controversial interpretations of the environmental history, are surely to blame.

7. DISCUSSION AND PROSPECTS FOR CONTINUED RESEARCH

A problem in reconstructing the deglaciation history is that although landforms such as moraines and raised beaches can be successfully relatively dated, it is often difficult to relate them to stratigraphic sequences and absolutely date their origin. The Late Weichselian strata of Iceland are often incomplete and lacking dateable material. An illustration of this is that although raised beaches occur around the whole island, late- and postglacial fossil shells have mainly been discovered in southwestern, western and northwestern Iceland (Fridriksdóttir, 1978)(Fig.6). Fossil shell localities have neither been found in central northern nor central southern Iceland. Contributions based on lithostratigraphical and/or morphological investigations from areas without dateable material (e.g. Víkingsson, 1978; Kaldal, 1978; Norðdahl, 1983; Norðdahl and Einarsson, 1988) are speculative when it comes to the chronology of glacial events.

In order to develop research on the deglaciation and climate history of Iceland, it is necessary to parallel lithological/biological studies of lake sediments and mires with the morphological/lithostratigraphical effort and the search for shell-bearing deglaciation sequences in the coastal areas. The first step is to look for lake and peat basins where sequences with a potentially high stratigraphic resolution are preserved. The study of palaeoenvironmental changes from lake sediments and mires is well established in Quaternary geology, with clearly defined research strategy, sophisticated sampling and mapping techniques, well developed laboratory procedures and advanced methods of data treatment (for a review see Berglund, 1986). As outlined above, only a few studies of this kind have been carried out in Iceland. The aim of continued research should be to gather concise information on the timing and extent of glacial oscillations and associated shoreline displacements in relation to the last deglaciation, as well as studying the Holocene vegetational and environmental histories and assessing the climatological background.

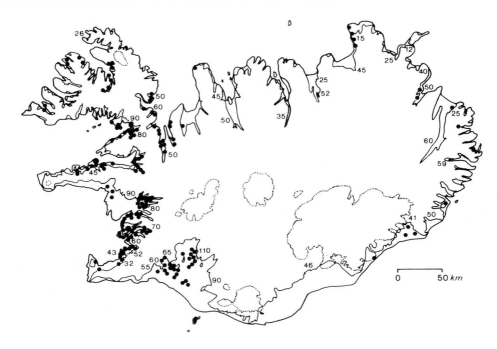

Fig.6 Sites with subfossil shells in Iceland (dots). The altitude of the
marine limit is given in metres, and areas that probably have been
submerged some time during or after the deglaciation are outlined
(full line outside the coast). Based on Einarsson (1963),
Sæmundsson (1977), Fridriksdóttir (1978), Hjort et al. (1985),
Pétursson (1986), Norðdahl and Hjort (1987), Ingólfsson (1988) and
Norðdahl and Einarsson (1988). Major present day glaciers are
dotted.

ACKNOWLEDGEMENTS

My research on the deglaciation history of Iceland has been supported by the Swedish
Natural Science Research Council and the Science Foundation of Iceland, which is
gratefully acknowledged. The manuscript benefitted greatly from critical reviews by
Christian Hjort and Jan Mangerud. Margret Hallsdóttir is thanked for giving me information
on unpublished biostratigraphical data, and for critically reading a draft version on the
Holocene vegetational history.

REFERENCES

Amman, B. and Lotter, A.F. (1989) 'Late-Glacial radiocarbon- and palynostratigraphy on the Swiss Plateau', Boreas 18, 109-126.

Andersen, B.G. (1981) 'Late Weichselian Ice Sheets in Eurasia and Greenland', in G.H.Denton and T.J.Hughes (eds.) The Last Great Ice Sheets, Wiley, 1-65.

Andersen, B.G., Heinemeier, J., Nielsen, H.L., Rud, N., Thomsen, M.S., Johnsen, S.J., Sveinbjörnsdóttir, A.E. and Hjartarson, Á. (in press) 'AMS 14-C dating on the Fossvogur sediments, Iceland', Radiocarbon.

Andrésdóttir, A. (1987) 'Glacial geomorphology and raised shorelines in the Skarðsstrond-Saurbær area, West Iceland', Examensarbeten i Geologi vid Lunds Universitet, Geologiska Institutionen, Lund, 25pp.

Anundsen, K. (1985) 'Changes in shore-level and ice-front position in Late Weichsel and Holocene, southern Norway', Norsk Geografisk Tidsskrift 39, 205-225.

Ashwell, I.Y. (1975) 'Glacial and Late Glacial processes in Western Iceland', Geografiska Annaler 57, 225-245.

Askellsson, J. (1934) 'Quartärgeologische Studien von Island I.', Geologiska Föreningens i Stockholm Förhandlingar 56, 596-618.

Atkinson, T.C., Briffa, K.R. and Coope, G.R. (1987) 'Seasonal temperatures in Britain during the past 22,000 years, reconstructed using beetle remains', Nature 325, 587-592.

Ballantyne, C.K. (1984) 'The Late Devensian periglaciation of upland Scotland', Quaternary Science Reviews 3, 311-343.

Bárðarson, G.G. (1921) 'Fossile Skalaflejringer ved Breidafjörður i Vest-Island', Geologiska Föreningens i Stockholm Förhandlingar 43, 323-380.

Bard, E., Arnold, M., Duprat, J., Moyes, J. and Duplessy, J.-C. (1987) 'Reconstruction of the last deglaciation: deconvolved records of ^{18}O profiles, micropaleontological variations and accelerator mass spectrometric ^{14}C dating', Climatic Dynamics 1, 101-112.

Bartley, D.D. (1973) 'The stratigraphy and pollen analysis of peat deposits at Ytri Bægisá near Akureyri, Iceland', Geologiska Föreningens i Stockholms Förhandlingar 95, 410-414.

Berglund, B.E. (1986) Handbook of Holocene Palaeoecology and Palaeohydrology, Wiley, Chichester.

Björck, G.S. and Möller, P. (1987) 'Late Weichselian environmental history in south-eastern Sweden during the deglaciation of the Scandinavian ice sheet', Quaternary Research 28, 1-37.

Boulton, G.S. (1979) 'Glacial history of the Spitsbergen archipelago and the problem of a Barents Shelf ice sheet', Boreas 8, 31-57.

Boulton, G.S., Baldwin, C.T., Peacock, J.D., McCabe, A.M., Miller, G., Jarvis, J., Horsefield, B., Worsley, P., Eyles, N., Chroston, P.N., Day, T.E., Gibbard, P., Hare, P.E. and von Bruun, V. (1982) 'A glacio-isostatic facies model and amino acid stratigraphy for late Quaternary events in Spitsbergen and the Arctic', Nature 298, 437-441.

Bradshaw, R. and Thompson, R. (1985) 'The use of magnetic measurements to investigate the mineralogy of Icelandic lake sediments and to study catchment processes', Boreas 14, 203-215.

Dansgaard, W., White, J.W.C. and Johnsen, S. (1989) 'The abrupt termination of the Younger Dryas climatic event', Nature 339, 532-533.

Einarsson, Th. (1961) 'Pollenanalytische Untersuchungen zur spät- und postglazialen Klimageschichte Islands', Sönderveröffentlichungen des Geologischen Institutes der Universität Köln 6, 52pp.

Einarsson, Th. (1963) 'Pollen-analytical studies on vegetation and climate history of Iceland in late and post-glacial times', in A.Löve and D.Löve (eds.) North Atlantic Biota and their History, Pergamon, 355-365.

Einarsson, Th. (1967) 'Zu der Ausdehnung der weichselzeitlichen Vereisung Nordislands', Sönderöffentlichungen des Geologischen Institutes der Universität Köln 13, 167-173.

Einarsson, Þ. (1968) Jarðfræði, saga bergs og lands. Mál og Menning, Reykjavík.

Einarsson, Þ. (1971) Jarðfræði. Heimskringla, Reykjavík.

Einarsson, Th. (1973) 'Geology of Iceland', in M.G.Pitcher (ed.) Arctic Geology, Memoirs of the American Association of Petroleum Geologists 19, 171-175.

Einarsson, Þ. (1975) 'Um myndunarsögu íslensks mýrlendis', in A.Garðarson (ed.) Votlendi, Landvernd, Reykjavík, pp.15-21.

Einarsson, Þ. (1978) Jarðfræði. Mál og Menning, Reykjavík.

Einarsson, Th. (1979) 'The deglaciation of Iceland', Norsk Geologisk Förening Medlemsblad 13, 18.

Einarsson, Þ. (1985) Jarðfræði. Mál og menning, Reykjavík.

Einarsson, Th. and Albertsson, K.E. (1988) 'The glacial history of Iceland during the past three million years', Philosophical Transactions of the Royal Society of London B318, 637-644.

Einarsson, Tr. (1966) 'Late and Post-glacial rise in Iceland and sub-crustal viscosity', Jökull 16, 157-166.

Fridriksdóttir, S. (1978) 'Fundarstaðir skelja frá síðjökultíma', Náttúrufræðingurinn 48, 75-85.

Funder, S. (1978) 'Holocene stratigraphy and vegetation history in the Scoresby Sound area, East Greenland', Bulletin Grönlands Geologiske Undersögelse 129, 66pp.

Funder, S. (in press) 'Quaternary geology of Greenland', in R.J.Fulton and J.A.Heginbottom (eds.) Quaternary Geology of Canada and Greenland, Geological Society of Canada and Geological Society of Greenland.

Funder, S. and Hjort, C. (1973) 'Aspects of the Weichselian chronology in central East Greenland', Boreas 2, 69-84.

Hallsdóttir, M. (1982) 'Frjógreining tveggja jarðvegssnida úr Hrafnkellsdal', in H.Þórarinsdóttir, O.Oskarsson, S.Steinþórsson and Þ.Einarsson (eds.) Eldur er í Norðri, Sögufélag, Reykjavík, 253-266.

Hallsdóttir, M. (1987) 'Pollen analytical studies of human influence on vegetation in relation to the Landnám tephra layer in southwest Iceland', Lundqua Thesis 18, Lund, 45pp.

Hjartarson, Á. (1987) 'Tímatal í ísaldarsögu Reykjavíkur', in Ísaldarlok á Íslandi, Jarðfræðifélag Íslands, Reykjavík, 4-5.

Hjartarson, Á. (1989) 'The ages of the Fossvogur layers and the Álftanes end-moraine, SW-Iceland', Jökull 39, 21-31.

Hjartarson, Á. and Ingólfsson, Ó. (1988) 'Preboreal glaciation of Southern Iceland', Jökull 38, 1-16.

Hjort, C. (1981) 'A glacial chronology for northern East Greenland', Boreas 10, 259-274.

Hjort, C. and Björck, S. (1984) 'A re-evaluated chronology for northern East Greenland', Geologiska Föreningens i Stockholms Förhandlingar 105, 235-243.

Hjort, C., Ingólfsson, Ó. and Norðdahl, H. (1985) 'Late Quaternary geology and glacial history of Hornstrandir, Northwest Iceland: a reconnaissance study', Jökull 35, 9-29.

Hoppe, G. (1968) 'Grimsey and the maximum extent of the last glaciation of Iceland', Geografiska Annaler 50, 16-24.

Hoppe, G. (1982) 'The extent of the last inland ice sheet of Iceland', Jökull 32, 3-11.

Imsland, P. (1982) 'Um flotjafnvægi, tengsl þess við eldvirkni, gerð jarðskorpu og áhrif á landnytingu', in H.Þórarinsdóttir, O.Oskarsson, S.Steinþórsson and T.Einarsson (eds.) Eldur er í Norðri, Sögufélag, Reykajvík, pp.311-317.

Ingólfsson, Ó. (1985) 'Late Weichselian glacial geology of the lower Borgafjörður region, western Iceland: a preliminary report', Arctic 38, 210-213.

Ingólfsson, Ó. (1987) 'Investigations of the Late Weichselian glacial geology of the lower Borgarfjörður region, western Iceland', Lundqua Thesis 19, Lund.

Ingólfsson, Ó. (1988) 'Glacial history of the lower Borgarfjörður area, western Iceland', Geologiska Föreningens i Stockholms Förhandlingar 110, 293-309.

Jungner, H. (1979) Radiocarbon dates I - report no.1. Radiocarbon Dating Laboratory, University of Helsinki.

Kaldal, I. (1978) 'The deglaciation of the area north and northwest of Hofsjökull, central Iceland', Jökull 28, 18-31.

Keilhack, K. (1884) 'Über postglaziale Meeresablagerungen in Island', Zeitschrift Deutscher Geologische Geselleschaft 35, 145-160.

Kjartansson, G. (1940) 'Stadier i isens tilbagerykning fra det sydvest -islandske lavland', Meddelelser fra Dansk Geologisk Forening 9, 426-458.

Kjartansson, G. (1964) 'Ísaldarlok og eldfjöll á Kili', Náttúrufræðingurinn 34, 9-38.

Landvik, J.Y., Mangerud, J. and Salvigsen, O. (1987) 'The Late Weichselian and Holocene shoreline displacement on the west-central coast of Svalbard', Polar Research 5, 29-44.

Lemdahl, G. (1985) 'Fossil insect fauna from Late-Glacial deposits in Scania (South Sweden)', Ecologica Mediterranea XI, 185-191.

Lindroth, C.H., Bengtson, S.-A. and Enckell, P.H. (1988) 'Terrestrial faunas of four isolated areas: a study in tracing old fauna centres', Entomologica Scandinavia Supplement 32, 31-65.

Lundqvist, J. (1988) 'Younger Dryas-Preboreal moraines and deglaciation in southwestern Varmland, Sweden', Boreas 17, 301-316.

Mangerud, J. (1980) 'Ice-front variations of diferent parts of the Scandinavian ice sheet, 13,000 - 10,000 B.P.', in J.J.Lowe, J.M.Gray and J.E.Robinson (eds.) Studies in the Lateglacial of North West Europe, Pergamon, Oxford, pp.23-30.

Mangerud, J. and Berglund, B.E. (1978) 'The subdivision of the Quaternary of Norden: a discussion', Boreas 7, 179-181.

Mangerud, J. and Salvigsen, O. (1984) 'The Kapp Ekholm section, Billefjorden, Spitsbergen: a discussion', Boreas 13, 155-158.

Mangerud, J., Andersen, S.T., Berglund, B.E. and Donner, J. (1974) 'Quaternary stratigraphy of Norden, a proposal for terminology and classification', Boreas 3, 109-128.

Mangerud, J., Larsen, E., Longva, O. and Sonstegard, E. (1979) 'Glacial history of western Norway 15,000 - 10,000 B.P.', Boreas 8, 179-187.

Mangerud, J., Lie, S.E., Furnes, H., Kristiansen, I.L. and Lomo, L. (1984) 'A Younger Dryas ash bed in western Norway and its possible correlations with tephra in cores from the Norwegian Sea and the North Atlantic', Quaternary Research 21, 85-104.

Miller, G.H. (1982) 'Quaternary depositional episodes, western Spitsbergen, Norway: aminostratigraphy and glacial history', Arctic and Alpine Research 14, 321-340.

Norðdahl, H. (1981) 'A prediction of minimum age for the Weichselian maximum glaciation in North Iceland', Boreas 10, 471-476.

Norðdahl, H. (1983) 'Late Quaternary stratigraphy of Fnjóskadalur, central north Iceland', Lundqua Thesis 12, Lund.

Norðdahl, H. and Hjort, C. (1987) 'Aldur jökulhörfunar í Vopnafirði', in Ísaldarlok á Íslandi, Jarðafræðifélag Íslands, Reykjavík, pp. 18-19.

Norðdahl, H. and Einarsson, T. (1988) 'Hörfun jökla og sjávarstöðubreytingar í ísaldarlok á Íslandi', Náttúrufræðingurinn 58, 59-80.

Norðdahl, H. and Hafliðason, H. (1990) 'Skógar tefran, en senglacial kronostratigrafisk marker på Nordisland', Geolognytt 17, 84.

Nyberg, R. (1988) 'Moraine-like ridges at Skaralið canyon, southern Sweden, and their relationships to lateglacial cold periods', Boreas 17, 333-345.

Okko, V. (1956) 'Glacial drift in Iceland, its origin and morphology', Acta Geographica 15, 1-133.

Påhlsson, I. (1981) 'A pollen analytical study on a peat deposit at Lágafell, southern Iceland', Striae 15, 60-64.

Peacock, J.D., Harkness, D.D., Housley, R.A., Little J.A. and Paul, M.A. (1989) 'Radiocarbon ages for a glaciomarine bed associated with the maximum of the Loch Lomond Readvance in west Benderloch, Argyll', Scottish Journal of Geology 25, 69-79.

Pétursson, H.G. (1986) Kvartærgeologiske undersögelser på Vest-Melrakkaslétta, Nordost Island, Unpublished Cand. Real Thesis, Universitetet i Tromsö, 157pp.

Ruddiman, W.F. and McIntyre, A. (1981) 'The North Atlantic during the last deglaciation', Palaeogeography, Palaeoclimatology and Palaeoecology 35, 145-214.

Schwaar, J. (1978) 'Moorkundliche Untersuchungen am Laugarvatn (Südvest Island)', Berichte aus der Forschungstelle Neðri-Ás, Hverageröi, Ísland 29, 1-29.

Sigbjarnarson, G. (1983) 'The Quaternary alpine glaciation and marine erosion in Iceland', Jökull 33, 87-98.

Steindórsson, S. (1962) 'On the age and immigration of the Icelandic flora', Rit Vísindafélags Íslendinga 35, 157pp.

Steindórsson, S. (1963) 'Ice age refugia in Iceland as indicated by the present distribution of plant species', in A.Löve and D.Löve (eds.) North Atlantic Biota and their History, Pergamon, Oxford, 303-320.

Straka, H. (1956) 'Pollenanalytische Untersuchungen eines Moorprofiles aus Nord-Island', Neue Jahrbuch Geologische Pälaontologische 6, 262-272.

Sutherland, D.G. (1984) 'Modern glacier characteristics as a basis for inferring former climates with special reference to the Loch Lomond stadial', Quaternary Science Reviews 3, 291-309.

Sæmundsson, K. (1977) Jarðfræðikort af Íslandi, blað 7. Náttúrufræðistofnun Íslands og Landmælingar Íslands, Reykjavík.

Thompson, R., Bradshaw, R.H.W. and Whitley, J.E. (1986) 'The distribution of ash in Icelandic lake sediments and the relative importance of mixing and erosion processes', Journal of Quaternary Science 1, 3-11.

Thórarinsson, S. (1944) 'Tephrokronologiska studier på Island', Geografiska Annaler 19, 1-217.

Thórarinsson, S. (1951) 'Laxárgljúfur and Laxárhraun, a tephrochronological study', Geografiska Annaler 33, 1-90.

Thoroddsen, Th. (1906) Island. Grundriß der Geographie und Geologie, Justus Perthes, Gotha, 358pp.

Troitsky, L. Punning, J.-M., Hutt, G. and Rajamae, R. (1979) 'Pleistocene chronology of Spitsbergen', Boreas 8, 401-407.

Vasari, Y. (1972) 'The history of the vegetation of Iceland during the Holocene', in Y.Vasari, H.Hyvärinen and S.Hicks (eds.), Climatic Changes in the Arctic Areas During the Last Ten Thousand Years, Acta Universitatis Ouluensis A 3, Geologica 1, 239-251.

Vasari, Y. (1973) 'Post-glacial plant succession in Iceland before the period of human interference', Proceedings of the III International Palynological Conference, Nauka, Moscow, 7-14.

Víkingsson, S. (1978) 'The deglaciation of the southern part of the Skagafjörður district, Northern Iceland', Jökull 28, 1-17.

Þórarinsson, S. (1955) 'Nákudungslögin við Húnaflóa í ljósi nyrra aldursákvardanna', Náttúrufræðingurinn 25, 172-186.

A REVIEW OF THE GLACIATION MAXIMUM CONCEPT AND THE DEGLACIATION OF EYJAFJÖRÐUR, NORTH ICELAND

Hreggviður Norðdahl
Science Institute
Dunhaga 3
IS-107 Reykjavík
Iceland

ABSTRACT. The Weichselian glaciation in North Iceland is divided into three main phases: (1) A maximum phase, characterized by merging outlet glaciers from local ice domes in Tröllaskagi with major ice streams from an inland ice sheet. These combined glaciers reached out to the shelf off North Iceland and overrode the island of Grímsey. (2) A phase characterized by repeated advances of the glaciers and the formation of several ice-dammed lakes in Fnjóskadalur. (3) A final phase which was denoted by an Alpine type of glaciation and the final departure of glaciers from the North Icelandic lowlands. A tephro-chronological correlation has afforded a possible Younger Dryas age for the youngest ice-lakes in Fnjóskadalur. A ^{14}C date in Flateyjardalur has provided a minimum Preboreal age for the final deglaciation of the area and for the marine limit in Eyjafjörður.

1. INTRODUCTION

This paper is a review of the present state of knowledge on the Weichselian maximum glaciation, the deglaciation and marine limit in North Iceland. This review is based on the work of geoscientists working in the area from the beginning of this century and until present. The area under investigation is central North Iceland between Skagafjörður in the west and Skjálfandi in the east (Fig. 1). The main topographical features of this area are the deep and long north-south orientated valleys; Skagafjörður, Eyjafjörður, Fnjóskadalur and Bárðardalur, which are separated by the peninsulas Tröllaskagi and Flateyjarskagi. The mountains in Tröllaskagi reach 900-1500 m a.s.l. and the mountains in Flateyjarskagi and south thereof reach up to about 1200 m a.s.l.. A dendritic arrangement of valleys is characteristic for the landscape of Tröllaskagi and the northwestern part of Flateyjarskagi (Fig. 1). Another conspicuous and important topographical feature in central North Iceland is that Eyjafjörður continues topographically across Leirdalsheiði (about 300 m a.s.l.) and

31

J. K. Maizels and C. Caseldine (eds.), Environmental Change in Iceland: Past and Present, 31–47.
© 1991 Kluwer Academic Publishers. Printed in the Netherlands.

along Hvalvatnsfjörður. Fnjóskadalur extends topographically to the north across Flateyjardalsheiði (about 200 m a.s.l.) and along Flateyjardalur, which faces the north-western part of Skjálfandi (Fig. 1).

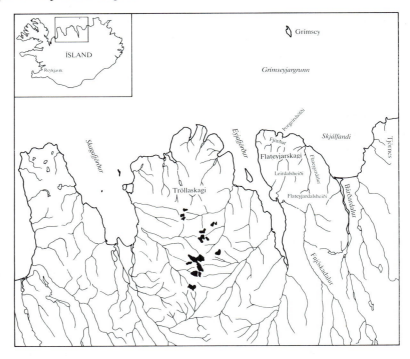

Fig. 1 Central North Iceland, with names of localities mentioned in the text. A broad outline of the topography is given by the coastline and the rivers in the approximately 1000 m high mountains. The black areas are present day glaciers in Tröllaskagi.

2. THE WEICHSELIAN GLACIATION IN NORTH ICELAND

A general concept for the Weichselian glacial history of North Iceland accounts for three main phases or periods of glaciation (Norðdahl 1979, 1981):

A. The maximum extent of glaciation was reached when the glaciers reached as far beyond the present coast of North Iceland as the island of Grímsey .

B. Subsequently to the maximum extent the glaciers retreated and both Flateyjardalur and Flateyjardalsheiði and later Fnjóskadalur were deglaciated. This period was characterized by the formation and drainage of several ice-lakes in Fnjóskadalur, most likely due to climatologically controlled variations in the configuration of the inland ice sheet and the extent of the major glaciers in North Iceland.

C. The third phase comprises the final recession of the ice sheet and withdrawal of glaciers from Skagafjörður, Eyjafjörður and Bárðardalur and, hence, the final drainage of the youngest ice-lake in Fnjóskadalur. Furthermore, this period was characterized by an advance of cirque and valley glaciers in the northern part of the area, and by an enlargement of the inland ice sheet which advanced but to a more southerly position than before.

2.1. The maximum extent

At the turn of the century Thoroddsen (1905-06), who was a monoglacialist, concluded that Iceland had been covered by a continuous ice sheet which reached beyond the present coastline during a glaciation maximum. This conclusion was based on the fact that he had found glacial striae at sea level on all major peninsulas around Iceland.

Trausti Einarsson (1959) was of the opinion that Þorgeirshöfði in Fjörður (211 m a.s.l.) had only been glaciated up to about 40 m a.s.l. and that Grímseyjargrunn and Grímsey itself had never been overridden by a glacier. Thorleifur Einarsson (1961) was also of the opinion that the shelf off central North Iceland had been ice-free during the maximum glaciation, until he for the first time found SE to SSE orientated glacial striae in three different places in Grímsey (Thorleifur Einarsson, 1967). Consequently, he concluded that the island had been overridden by a glacier that flowed from an ice-divide above Melrakkaslétta in Northeast Iceland and the mountains south thereof and across the present coastline between Tjörnes and Melrakkaslétta. Later Hoppe (1968, 1982) confirmed the findings of Einarsson of SSE orientated striae and reported new findings of WNW orientated striae in the southernmost part of the island.

According to the author's observations on Þorgeirshöfði in Fjörður, the top of the promontory has been overridden and sculptured by an actively eroding glacier. Recently, a few SSE orientated glacial striae have been recorded at about 200 m a.s.l. on the west side of Þorgeirshöfði (Pétursson, pers. comm.). Measurements of glacial striae in Grímsey revealed the SSE orientated striae in the southern part of the island but failed to recognize Hoppe's WNW orientated striae. Instead, older ESE orientated striae were found there on a NW facing facet and they might be the same as found by Hoppe (Fig. 2). According to these observations on the orientation of glacial striae the island was overridden by a glacier that moved from Skjálfandi or Flateyjardalur and across Grímseyjargrunn. This maximum state of glaciation ended when the glaciers retreated towards the south and into the mouth of the major north-south orientated valleys of North Iceland (Norðdahl, 1979, 1981).

2.2. The deglaciation of Eyjafjörður

Glacial striae in Eyjafjörður are mainly orientated parallel to the direction of the fjord (Thoroddsen, 1905-06). On the island of Hrísey Trausti Einarsson (1959) observed two sets of striae; an older one SSE orientated and an younger set which is SE orientated. Hjartarson

(1973) reported SE orientated striae on Hámundarstaðaháls in outer Eyjafjörður (Fig. 3). Above Akureyri the direction of glacial striae was successively changed during the de-

Fig. 2 Glacial striae in the southern part of the island of Grímsey. Photo: Hreggviður Norðdahl.

glaciation from a southerly direction to a southeasterly direction due to diminishing influence of a tributary glacier in Glerárdalur on the direction of the glacier in Eyjafjörður (Hallsdóttir, 1973, 1984). The very same development is seen in the orientation of lateral channels, both at Glerárdalur above Akureyri (Hallsdóttir, 1984) and at the mouth of Hálsdalur and Þorvaldsdalur in the northern part of Eyjafjörður (Hjartarson, 1973; Þóroddsson, 1982).

The deglaciation of Eyjafjörður was repeatedly interrupted by glacier advances which closed the natural drainage path of Fnjóskadalur through Dalsmynni and caused the formation of at least eight ice-lakes in Fnjóskadalur with their outlet north across the Flateyjardalsheiði threshold (Norðdahl, 1983). A seismic study of submarine sediments in Eyjafjörður (Hafliðason, 1983) has revealed up to 10 or 12 glacier advance stages and terminal moraines which all postdate the maximum glaciation in North Iceland (Fig. 3). At least 7 of these moraines were formed by glacier advances that reached north of Dalsmynni and two moraines are situated on a height with Dalsmynni. The number of moraines north of Dalsmynni is in accordance with the number of high-level ice-lakes in Fnjóskadalur, which must support the conclusion that all the ice-lakes were formed during repeated advances of the glacier in Eyjafjörður (Norðdahl, 1983).

Norðdahl (1982) shows two glacier termini just south of Hrísey concurrently with the Fnjóskadalur ice-lake, the youngest high-level ice-lake in the valley. Hjartarson (1973) described a 2 km long lateral moraine feature on eastern Hámundarstaðaháls and lateral drainage features on Látraströnd north of Grenivík apparently marking the youngest terminal position of a glacier in the northern part of Eyjafjörður. Þoroddsson and Hafstað (1975) also noted these conspicuous lateral features at Grenivík (Fig. 3). The Fnjóskadalur ice-lake was finally drained when the glacier in Eyjafjörður retreated south of Dalsmynni and lost its damming capacity (Norðdahl, 1983).

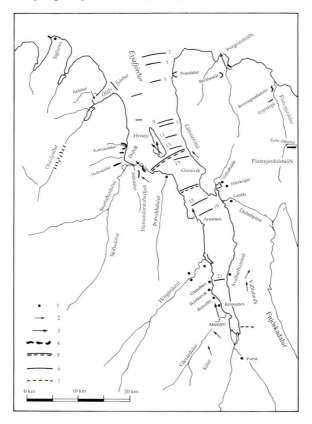

Fig. 3 Map of the Eyjafjörður fjord and its surroundings with names of localities mentioned in the text. A broad outline of the topography is given by the coastline and the rivers. (1) Raised marine features. (2) Lateral channels. (3) Glacial striae. (4) Moraines. (5) Terminal moraines in Eyjafjörður (cf. Norðdahl 1982). (6) Terminal moraines in Eyjafjörður (cf.Hafliðason 1983). (7) Terminal moraine in Eyjafjörður (cf. Þorkelsson 1922, 1935).

Hafliðason (1983) recognized two younger terminal moraines in the marine sediments south of Dalsmynni: one opposite Arnarnes and a diffuse one just north of Akureyri. Þorkelsson (1922, 1935) interpreted a till accumulation reaching about 200 m a.s.l. opposite Akureyri as a morainic feature formed by the glacier in Eyjafjörður during an advance and a stillstand. He referred to it as the Akureyri-stage. According to Thorleifur Einarsson (1967) a set of lateral terraces and a terminal moraine at Hólar in Eyjafjörður were formed during a glacier advance into the southernmost part of the Eyjafjörður valley.

2.3. The deglaciation of the tributary valleys

The deglaciation of the tributary valleys of Eyjafjörður was also frequently interrupted by glacier advances. In Ólafsfjörður (Fig. 3) the glacier retreated towards the southeast and into the main valley. The arrangement of lateral moraines and drainage features in the valley indicates that the glaciers in the tributary valleys retreated somewhat earlier than the main Ólafsfjörður glacier except in Árdalur where lateral channels indicate the opposite. In the inner half of the Ólafsfjörður valley, where it narrows and ascends above the 20-40 m level, a series of transverse morainic ridges probably indicate changes in the rate and mode of deglaciation (Norðdahl, 1978).

Venzke and Meyer (1986) are of the opinion that since the glacier in Svarfaðardalur retreated from a terminal moraine and dead ice features near Dalvík, the deglaciation of Svarfaðardalur-Skíðadalur was interrupted by two advances. The first left terminal moraines just inside the junction between Svarfaðardalur and Skíðadalur and the latter at the mouths of their tributary valleys (Fig. 3).

Fig. 4 The lateral and terminal moraines in Fossdalur on northern Látraströnd. Photo: Hreggviður Norðdahl.

Subsequent to the retreat of the glaciers in Eyjafjörður and the tributary valleys, small cirque and valley glaciers in the northern part of Tröllaskagi and Flateyjarskagi advanced and reached out of their valleys and into the sea (Þorleifur Einarsson, 1967; Hjartarson, 1973; Norðdahl, 1979). During this advance a number of lateral and terminal moraines were formed by valley glaciers in Flateyjarskagi (Fig. 3), and the terminal moraine in front of Brettingsstaðadalir has been dated to a minimum age of 9650±120 years B.P. (Norðdahl, 1979). Lateral moraines in Grjótárgil, at Syðri-Jökulsá further south in Flateyjardalur, and the very conspicuous moraines in Fossdalur on Látraströnd (Fig. 4) and in Breiðaskál in Fjörður are probably of the same age as the moraine in front of Brettingsstaðadalir (Norðdahl, 1983). Lateral and terminal moraines in Sauðadalur, Karlsárdalur, Holtsdalur, and Hálsdalur on the west side of Eyjafjörður (Hjartarson, 1973), were probably formed at the same time. A comparable glacier advance has been demonstrated in Glerárdalur above Akureyri (Hallsdóttir, 1973, 1984).

3. SEA-LEVEL CHANGES

During the deglaciation of Eyjafjörður and the tributaries the sea successively transgressed the area. According to Trausti Einarsson (1959) raised marine beaches and terraces in Eyjafjörður are now e.g. found at successively higher altitude, from present sea level on Siglunes in the outermost part of the fjord, 20 m a.s.l. in lowermost Hörgárdalur, 25 m a.s.l. at Þvera', to about 35 m a.s.l. at Melgerðismelar in the southernmost part of the Eyjafjörður valley (Trausti Einarsson, 1959). These marine features are found over a distance of about 80 km, (erroneously put to about 40 km by Trausti Einarsson (1959, Fig. 3)) and were interpreted to have been formed at a synchronous sea-level in Eyjafjörður. According to Trausti Einarsson (1959) the marine limit in Eyjafjörður is at about 35 m a.s.l. at Melgerðismelar. According to the author's studies of raised marine features in Eyjafjörður these are not synchronous but heterochronous features successively formed along with the glacier retreat in Eyjafjörður.

Numerous marginal deltas and terraces were formed at, or in close connection with, a temporarily stationary glacier terminus at Glæsibær (20 m a.s.l.), Skjaldarvík (20 m a.s.l.), Brávellir (20 m a.s.l.), and Krossanes (25 m a.s.l.) on the west coast of Eyjafjörður (Hallsdóttir, 1973, 1984; Norðdahl, 1974a). At Grýtubakki, Hléskógar, Helguhóll, and Laufás near Dalsmynni on the east side of Eyjafjörður, marine terraces are found at about 20 m a.s.l., and further south on Svalbarðsströnd these features are found between 20 m and 30 m a.s.l. (Norðdahl, 1974b). Hjartarson (1973) and Venzke and Meyer (1986) have reported marine terraces in the outer part of Svarfaðardalur at altitudes between 16 and 20 m a.s.l. On the west side of Eyjafjörður and south of Hrísey Þóroddsson (1982) described a raised beach feature at 10-15 m a.s.l.. In lower Hörgárdalur the marine limit is between 20 m and 30 m a.s.l. (Fig. 3) (Norðdahl, 1974a).

Unpublished data from investigations at Arnarnes on the west side of Eyjafjörður (Fig. 3) have revealed marine sediments up to about 60 m a.s.l.. This high sea-level was reached prior to deposition of a superposed till bed and the formation of marine terraces at about 20 m a.s.l.

4. THE CONCEPT OF MAXIMUM GLACIATION

Thoroddsen (1905-06) introduced the idea that the highest parts of the coastal mountains in Iceland remained ice-free during the maximum extent of glaciation. On the basis of entomological studies in Southeast Iceland, Lindroth (1931) concluded that ice-free areas must have existed there during the maximum glaciation. On the basis of botanical studies in North Iceland Gelting (1934) and Steindórsson (1962) concluded that ice-free areas and plant refugia must have existed there during the maximum glaciation. The ice-free areas were, according to the botanical studies, mainly located in the peripheral and mountainous parts of Northwest, North, and East Iceland.

The thickness of the major glaciers in North Iceland is an important factor when discussing the possible existence and extent of nunataks in North Iceland during the maximum glaciation. On the basis of the morphology of the area between Skagafjörður and Skjálfandi Thórarinsson (1937) pointed out the general possibility of nunataks in the area and concluded that considerable areas in Tröllaskagi and Flateyjarskagi had been above the surface of the glaciers, and only local valley and cirque glaciers existed in the area. Sigbjarnarson (1983) agreed with this when he classified the area as being that of a 'typical Alpine glaciation'. Trausti Einarsson (1959) studied and mapped the uppermost limit of glaciation in Eyjafjörður. Concentrating his work on features such as lateral moraines,

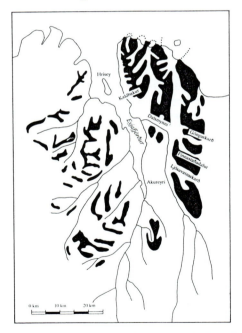

Fig. 5 The maximum extent of glaciers and ice-free areas in Eyjafjörður. Modified after Trausti Einarsson (1959).

glacial erosion and weathering, he performed a consistent three dimensional reconstruction of the Eyjafjörður glacier, and the glaciers in the western part of Tröllaskagi, and in Flateyjarskagi. According to his findings, the glacier in Eyjafjörður reached up to about 200 m a.s.l. at the mouth of the fjord, about 500 m a.s.l. at Kaldbakur, and about 1000 m a.s.l. at Akureyri (Fig. 5). Lateral drainage channels, which show the minimum thickness of the glaciers, have been described up to 870 m a.s.l. on Súlur near Akureyri (Hallsdótir, 1973, 1984) and up to 550 m a.s.l. on Hámundarstaðafjall near Dalvík (Fig. 3) (Hjartarson, 1973).

The uppermost limit of glaciation in northern Flateyjardalur has been determined to about 400 m a.s.l. (Norðdahl, 1979), and a relatively thick moraine cover and lateral channels on Vaðlaheiði east of Akureyri shows that it was definitely overridden by glaciers (Trausti Einarsson 1959; Norðdahl, 1983). The area between Gönguskarð and Finnastaðadalur in Kinnarfjöll was ice-covered up to about 700 m a.s.l. This limit is characterized by a seemingly matrix-free talus primarily made of about 50 cm long and sharp edged boulders above the limit, but below the talus is made of 20-30 cm long and less sharp edged boulders in a sandy silty matrix. This estimated upper limit of glaciation shows the possibility of relatively large ice-free areas north of Ljósavatnsskarð and Dalsmynni. The mountains south thereof were apparently covered and overridden by actively eroding glaciers, seen in the presence of ground moraines, glacial striae, erratics and other glacial geological features (Norðdahl, 1983).

4.1. The model of reconstruction

As information concerning the location and extent of ice-free areas in North Iceland has been fragmented, it was considered feasible to try and reconstruct the uppermost limits of glaciation in North Iceland. Such a reconstruction of the maximum situation of glaciation and thus of the minimum extent of ice-free areas was performed by Norðdahl (1983). He stated that the method Nye (1952) used for calculating the thickness of ice sheets is not suitable for North Iceland as it does not account for glaciers flowing from ice sheets. A method for calculating the overall gradient of glaciers that extended from an inland ice sheet to any part of a fjord or ice shelf was developed by Buckley (1969). This method was applied for North Iceland, where the main boundary conditions for the reconstruction are the following: first, large glaciers are presumed to have flowed from a main inland ice sheet and into Bárðardalur, Fnjóskadalur, Eyjafjörður, and Skagafjörður, the major north-south orientated valleys of North Iceland. Secondly, these glaciers are supposed to have reached as far beyond the present coast of North Iceland as the island of Grímsey, and probably to the edge of the shelf (Norðdahl, 1983).

The glacier type fitting best to Trausti Einarsson's (1959) uppermost limit of glaciation in Eyjafjörður is a combination of (1) the type (B) overall gradient of the Antarctic group, which contains glaciers flowing from ice-sheets into fjords, including those continuing as ice-shelves, and (2) the surface profile of the 60-100 km long Greenland profile type (Fig. 6 A, B) (cf. Buckley 1969). This model of a coherent and consistent three dimensional reconstruction of the glaciation maximum in North Iceland constitutes the basis for (1) establishing of the configuration of the glacier surface, (2) the principal direction of ice-flow in the area and (3) the approximate location of ice-free areas in the Tröllaskagi and Flateyjarskagi mountains (Fig. 7). This condition of maximum extent and thickness of glaciers may have coincided with the Weichselian maximum glaciation in North Iceland and the maximum concept of Thorleifur Einarsson (1967) and Hoppe (1968).

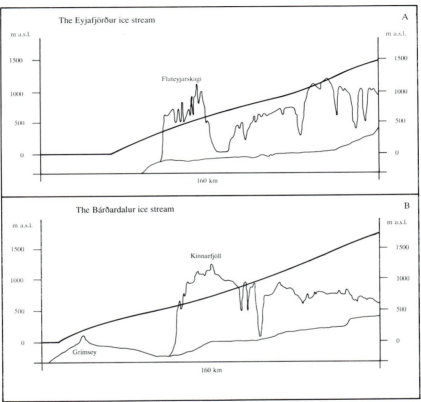

Fig. 6 A reconstruction of two major Weichselian ice streams in North Iceland. A. Profile along Eyjafjörður and the mountains east of the fjord, and the surface profile of the maximum Eyjafjörður ice stream and ice shelf. B. Profile along Bárðardalur and the mountains west of the valley, and the surface profile of the maximum Bárðardalur ice stream and ice shelf extending across the island of Grímsey.

4.2. The flow of glaciers

One of the main topographical features in North Iceland, apart from the north-south orientated valleys, is the dendritic arrangement of valleys in Tröllaskagi and in the northwestern part of Flateyjarskagi (Fig. 1). The north-south orientated valleys were eroded by ice streams that flowed from an inland ice sheet and towards the north. The valleys in Tröllaskagi, displaying a typical U shaped cross section, are obviously a part of a glacially eroded and moulded valley system, and the dendritic pattern of this valley system indicates that it was formed by glaciers originating within Tröllaskagi itself (Sugden and John, 1976). The flow of glaciers through these valleys was controlled by the mass balance of independent ice domes in Tröllaskagi (Fig. 7).

Given the boundary conditions for the model of glacier reconstruction, the dendritic Tröllaskagi valley system could not be occupied by glaciers from an inland ice sheet. Instead it was occupied by glaciers from independent ice domes in Tröllaskagi. On the basis of this model, the flow of ice in North Iceland, at the time of maximum glaciation, is divided into three different groups, the first one being the four major ice streams in Skagafjörður, Eyja-fjörður, Fnjóskadalur, and Bárðardalur (Fig. 7).

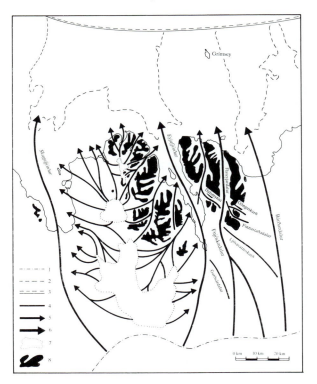

Fig. 7 The concept of the Weichselian maximum glaciation in North Iceland. (1) The boundary between the main inland ice sheet and the major ice streams. (2) The 200 m depth curve and a probable grounding line for the glaciers. (3) The edge of the ice shelf. (4) Interflowing ice streams. (5) Outlet glaciers. (6) Major ice streams. (7) Ice domes. (8) Ice-free areas with small valley and cirque glaciers.

The second group contains outlet glaciers which originated from ice domes in Tröllaskagi and occupied the dendritic Tröllaskagi valley system. The location of these former ice domes coincides with areas where small glaciers are today. Furthermore, also belonging to this group of glaciers, were the NW and NNW moving glaciers in Gönguskarð, Finnastaðadalur, and Dalsmynni in addition to those above Ljósavatnsskarð and Garðsárdalur. During the maximum glaciation an ice stream diverged from the Eyjafjörður ice stream and flowed towards the north across Leirdalsheiði (Fig. 7).

The third group of glaciers comprised small cirque and valley glaciers originating below the edge of the ice-free area in the northeastern part of Tröllaskagi and the northwestern part of Flateyjarskagi. These glaciers flowed along their enclosing valleys and coalesced with glaciers occupying valleys such as Ólafsfjörður, Svarfaðardalur, Skíðadalur, and Hörgárdalur on the west side of Eyjafjörður. East of Eyjafjörður valley glaciers in the northwestern part of Flateyjarskagi coalesced with the glacier above Leirdalsheiði, a branch of the Eyjafjörður ice stream. In the northernmost part of Flateyjardalur valley glaciers coalesced with the Fnjóskadalur ice stream reaching the shelf north thereof (Fig. 7).

A generalized picture of the maximum glaciation in North Iceland displays a successive merging of glaciers flowing from an inland ice sheet with glaciers from ice domes in Tröllaskagi, which subsequently reached onto the shelf and as far to the north as the island of Grímsey (Fig. 7). Due to the topography of the sea floor off North Iceland, the ice moving across the shelf constituted a floating ice shelf in areas which today are at greater depth than - 200 m, the grounding/floating boundary of the model of glacier reconstruction (Norðdahl, 1983). An intensive calving of the glaciers at the edge of the shelf is supposed to have prohibited any further northward extent of the ice shelf there.

4.3. The ice-free areas

Thórarinsson (1937) pointed out the general possibility of ice-free areas in North Iceland and Steindórsson (1962) argued that a part of the Icelandic flora had survived the Weichselian glaciation. A more coherent image of ice-free areas in North Iceland was presented by Trausti Einarsson and his map shows the thickness of a glacier in Eyjafjörður and the extent and location of ice-free areas (Fig. 5).

An interesting result of the present model of glacier reconstruction is that it accounts for the possibility of considerable ice-free areas both in Tröllaskagi and Flateyjarskagi concurrently with the maximum glacier extent, and that ice-free areas were mainly met in the northeastern part of Tröllaskagi and the northern part of Flateyjarskagi. The main reason for this uneven distribution of ice-free areas is the absolute height and configuration of the mountains in the eastern part of Tröllaskagi, which on average are 200-500 m higher than the mountains in the western part of Tröllaskagi. The 700-1200 m high mountains in the northern part of Flateyjarskagi protruded above the glaciers occupying the relatively closely spaced Eyjafjörður, Fnjóskadalur and Bárðardalur. The mountains west and east of Eyjafjörður constituted an 'Alpine' scenery of ice-free horns and arêtes. The mountains in the eastern part of Flateyjarskagi constituted a scenery of plateaux at about 700-900 m a.s.l., and most likely held ice-fields and perennial snow-fields which reduced the really ice-free areas there to ice-free enclaves below the edge of the plateaux (cf. Hjort et al. 1985).

5. DATING THE DEGLACIATION OF EYJAFJÖRÐUR

Þorkelsson (1935) compared the Akureyri-stage in Eyjafjörður with the Ra-stage in Norway, and on the basis of the average rate of deposition at the head of the fjord he estimated the date of deglaciation at Akureyri to 10-12,000 years ago.

As very few absolute dates concerning the Weichselian glacial history of North Iceland are available, stratigraphical correlations with other areas have most frequently been used in order to estimate the age of Weichselian events in the Eyjafjörður area. According to Trausti Einarsson (1959) the sediments in northern Fnjóskadalur are older than the Weichselian glaciation and represent; 'the whole duration of the classical Pleistocene which begins with the Günz glaciation'. This great age of the sediments in Fnjóskadalur was later contradicted by Norðdahl (1983) as he demonstrated that all the sediments in Fnjóskadalur are of Weichselian age and postdate the Weichselian maximum glaciation in North Iceland. Through the years two different opinions on the timing of the Weichselian maximum glaciation in Iceland have been put forward. The first one states that the maximum glaciation occurred at about 18,000 years B.P. or concurrently with the European Weichselian maximum (Kjartansson, 1962; Trausti Einarsson, 1966). The second one, which accounts for a somewhat different point of view, puts forwards the possibility that the maximum extent was reached in Early Weichselian time (Norðdahl, 1981, 1983), concurrently with the glaciers in Arctic Canada (Miller et al. 1977) and in Northeast Greenland (Funder and Hjort 1973; Hjort 1981). The estimated great age of the maximum glaciation in Iceland was first and foremost based on the calculated age estimates arrived upon through combination of strandline gradients and time in Fnjóskadalur (Norðdahl, 1981, 1983).

Thorleifur Einarsson (1967) attempted an age estimate for an ice-lake in Fnjóskadalur as he correlated its formation with the Álftanes advance, which was dated to about 12,000 years B.P. in West Iceland (Ashwell, 1967; Þorleifur Einarsson, 1968). Furthermore, Thorleifur Einarsson (1973) also estimated the age of the marine limit in Eyjafjörður and North Iceland to about 11,000 years B.P.

One of the most important findings in Fnjóskadalur are four separated ice-lake strandlines, which have been related to the four youngest ice-lakes in the valley (Norðdahl, 1983). The strandlines display a positive gradient between 1.59 and 2.65 m km[-1] toward the south, and the formation of these different gradients has been attributed to four separated periods of glacial isostatic recovery (Norðdahl, 1981, 1983). The magnitude of strandline gradients usually decreases with time towards the present or the end of uplift, a development that is preferably described by an exponential expression (cf. Norrman 1964; Andrews and Dugdale 1970; Norðdahl, 1981). It was thus possible on the basis of such an expression for the Fnjóskadalur area, the assumed age of the youngest ice-lake and that of the marine limit in Eyjafjörður, to calculate the time of the duration for the different periods of uplift. Given the duration, the estimated ages of about 12,000; 17,000; 20,800 and 23,900 years B.P. were attributed to the four youngest ice-lakes in Fnjóskadalur (Norðdahl, 1983).

The occurrence of the Skógar Tephra in the ice-lake sediments in Fnjóskadalur offers an unique opportunity for stratigraphical correlations far beyond that area and e.g. with sedimentary sequences from the North Atlantic Ocean. Norðdahl (1983) mapped the distribution of the tephra in Fnjóskadalur and found it most abundant near Skógar and named it the Skógar Tephra. The Skógar Tephra was basically transported by running water into the Fnjóskadalur basin and it is found in numerous deltas formed at the strandline of the Austari-Krókar ice-lake, the second youngest ice-lake in Fnjóskadalur and also in ice-lake sediments in Hrossadalur. The age of the Austari-Krókar ice-lake was estimated to about 17,000 years B.P., which was also regarded as an age estimate for the Skógar Tephra (Norðdahl, 1983). Redeposited Skógar Tephra is found in successively higher stratigraphical position in Fnjóskadalur, Eyjafjörður and Bárðardalur. The establishment of a correlation between the Skógar Tephra in North Iceland and the Vedde Ash Bed in West Norway (Norðdahl and Hafliðason, 1990), and thus dating the Skógar Tephra and the Austari-Krókar ice-lake to 10,600 years B.P., will unavoidably change the current view concerning the chronological position of the Weichselian maximum glaciation, glacier advances and sea-level changes in North Iceland.

6. DISCUSSION AND CONCLUSIONS

During the maximum glaciation the flow regime of glaciers originating from the ice domes in Tröllaskagi was only indirectly affected by the inland ice sheet as its height must have reduced the amount of precipitation that reached North Iceland and the Tröllaskagi glaciers. The maximum glaciation was terminated by recession of the inland ice sheet, which caused the North Icelandic ice shelf and the major ice streams to retreat towards the south and into the mouth of Skagafjörður, Eyjafjörður, Flateyjardalur and Bárðardalur. This glacier recession was probably due to changes in the climate and in the configuration of the inland ice sheet.

At some time during the general glacier recession in North Iceland, the northernmost part of the Fnjóskadalur basin became ice-free and the first of at least eight high-level ice-lakes were formed (Norðdahl, 1983). The formation of such an ice-lake in Fnjóskadalur was dependent upon the morphology of the area and the extent of glaciers within it, and ice-lakes could only be formed when Dalsmynni and Ljósavatnsskarð were simultaneously blocked by glacier tongues from glaciers in Eyjafjörður and Bárðardalur, respectively (Hospers, 1954; Trausti Einarsson, 1959; Norðdahl, 1983). Repeated glacier advances and the formation of ice-lakes in Fnjóskadalur are attributed to climatic changes and short-term oscillations in the extent and volume of an inland ice sheet and preferably in its major glaciers in North Iceland (Norðdahl, 1983). This continued development of glacier recession in North Iceland, in the period between the maximum glaciation and the final departure of glaciers from the area, was mainly characterized by successive transformation from an inland ice sheet and ice streams dominated type of glaciation towards a more local and independent Alpine type of glaciation.

It can be concluded, on basis of the inferred age of Skógar Tephra, that the maximum extent of the Weichselian glaciation in North Iceland was reached about 18,000 years B.P., concurrently with the European Weichselian maximum glaciation. Furthermore, the retreat from the maximum position out on the shelf and into the major fjords in North Iceland and the repeated glacier advances recorded in Fnjóskadalur and Eyjafjörður occurred during the Late Weichselian Substage. The retreat of the glacier from the moraine in front of Brettingsstaðadalir in Flateyjardalur and the ^{14}C dated beginning of the deposition of limnic organic material provides a minimum Preboreal age of about 9650±120 years B.P. for the final deglaciation of the area. Furthermore, this date postdates the Langhóll glacier advance in Flateyjardalur and the Belgsá glacier advance in Fnjóskadalur, and the latest possible water-outflow from an ice-lake in Fnjóskadalur through Flateyjardalur (Norðdahl, 1979). The Langhóll advance in Flateyjardalur and the Ljósavatn advance in Ljósavatnsskarð are now correlated with a Preboreal glacial event. The youngest ice-lakes in Fnjóskadalur were consequently formed during the Younger Dryas and the marine limit in Eyjafjörður was reached in late Younger Dryas or early Preboreal time.

ACKNOWLEDGEMENTS

I am indebted to my colleague Docent Christian Hjort at the Lund University in Sweden for his cooperation on the island of Grímsey and to Halldór G. Pétursson at the Akureyri Museum of Natural History for allowing me to use his observations on glacial striae on Þorgeirshöfði, for reading the manuscript and making valuable comments upon it. Professor Gylfi Már Guðbergsson at the University of Iceland is greatly acknowledged for reading the manuscript and correcting my English.

REFERENCES

Andrews, J. T. and Dugdale, R. E. (1970) 'Age prediction of glacio-isostatic strandlines based on their gradients', Geological Society of America Bulletin 81, 3769-3772.
Ashwell, I. Y. (1967) 'Radiocarbon ages of shells in the glaciomarine deposits of western Iceland', Geographical Journal 133, 48-50.
Buckley, J. T. (1969) 'Gradients of past and present outlet glaciers', Geological Survey of Canada, Paper 69-29, 1-13.
Einarsson, Tr. (1959) 'Studies of the Pleistocene in Eyjafjörður', Societas Scientiarum Islandica, Rit 33.
Einarsson, Tr. (1966) 'Late and Post-glacial rise in Iceland and sub-crustal viscosity', Jökull 16, 157-166.
Einarsson, Th. (1961) 'Pollenanalytische Untersuchungen zur spät- und postglazialen Klimageschichte Islands', Sonderveröffentlichungen des Geologischen Institutes der Universität Köln 6, 52 pp.

Einarsson, Th. (1967) 'Zu der Ausdehnung der weichselzeitlichen Vereisung Nordislands', Sonderveröffentlichungen des Geologischen Institutes der Universität Köln 13, 167-173.

Einarsson, Þ. (1968) Jarðfræði, saga berg og lands, Mál og menning, Reykjavík.

Einarsson, Þ. (1973) 'Geology of Iceland', in M. G. Pitcher (ed.), Arctic Geology, American Association of Petroleum Geologists Memoir 19, 171-175.

Funder, S. and Hjort, C. (1973) 'Aspects on the Weichselian chronology in central East Greenland', Boreas 2, 69-84.

Gelting, P. (1934) 'Studies on the vascular plants of East Greenland between Franz Josephs Fjord and Dove Bay', Meddelelse om Grønland 101.

Hafliðason, H. (1983) The marine geology of Eyjafjörður, North Iceland: sedimentological, petrographical and stratigraphical studies, Unpublished M. Phil. Thesis, Grant Institute of Geology Faculty of Science, University of Edinburgh.

Hallsdóttir, M. (1973) Um ísaldarlok á Glerárdal og í nágrenni Akureyrar, Unpublished BS Thesis, University of Iceland.

Hallsdóttir, M. (1984) 'Um ísaldarlok á Glerárdal og í nágrenni Akureyrar', (English summary: On the deglaciation in Glerárdalur and at Akureyri, North Iceland), Museum of Natural History, Akureyri, Mimeograph No. 12.

Hjartarson, Á. (1973) Rof jarðlagastaflans milli Eyjafjarðar og Skagafjarðar og ísaldarmenjar við utanverðan Eyjafjörð, Unpublished BS Thesis, University of Iceland.

Hjort, C. (1981) 'A glacial chronology for northern East Greenland', Boreas 10, 259-274.

Hjort, C., Ingólfsson, Ó. and Norðdahl, H. (1985) 'Late Quaternary geology and glacial history of Hornstrandir, Northwest Iceland: A reconnaissance Study', Jökull 35, 9-29.

Hoppe, G. (1968) 'Grímsey and the maximum extent of the Last Glaciation of Iceland', Geografiska Annaler 50, 16-24.

Hoppe, G. (1982) 'The extent of the last inland ice sheet of Iceland', Jökull 32, 3-11.

Hospers, J. (1954) 'The geology of the country between Akureyri and Mývatn in northern Iceland', Geologie en Mijnbouw 16, 491-508.

Kjartansson, G. (1962) 'Jökulminjar á Hálsum milli Berufjörður og Hamarsfjarðar', Náttúrufræðingurinn 32, 83-92.

Lindroth, C.H. (1931) 'Die Insektfauna Islands und ihre Probleme', Zoologiska bidrag från Uppsala 13, 105-599.

Miller, G.H., Andrews, J.T. and Short, S.K. (1977) 'The last interglacial-glacial cycle, Clyde foreland, Baffin Island, N.W.T.: stratigraphy, biostratigraphy and chronology', Canadian Journal of Earth Sciences 14, 2824-2857.

Norðdahl, H. (1974a) 'Skrýsla um efniskönnun fyrir Vegagerð ríkisins á Akureyri, vegna fyrirhugaðrar vegagerðar í Kræklingahlíð við Eyjafjörð', Mimeographed report, Icelandic Public Road Administration.

Norðdahl, H (1974b) 'Skrýsla um efniskönnun fyrir Vegagerð ríkisins á Akureyri, vegna fyrirhugaðrar vegagerðar um Víkurskarð S.-Þing', Mimeographed report, Icelandic Public Road Adminstration.

Norðdahl, H. (1978) 'Landmótun svæðisins Lágheiði-Ólafsfjörður, frum könnun á möguleikum til malartekju í Ólafsfirði, Eyjafjarðarsýslu', Mimeographed report, Icelandic Public Road Administation.

Norðdahl, H. (1979) 'The Last Glaciation in Flateyardalur central North Iceland, a preliminary Report', University of Lund, Department of Quaternary Geology, Report 18.

Norðdahl, H. (1981) 'A prediction of minimum age for the Weichselian maximum glaciation in North Iceland', Boreas 10, 471-476.

Norðdahl, H. (1982) 'Ljós vikurlög frá seinni hluta síðasta jökulskeiðs í Fnjóskadal', in Þórarinsdóttir, H., Óskarsson, Ó., Steindórsson, S. and Einarsson, P. (eds.), Eldur er í Norðri, Sögufélag, Reykjavík, pp. 167-175.

Norðdahl, H. (1983) 'Late Quaternary Stratigraphy of Fnjóskadalur central North Iceland, a study of sediments, ice-lake strandlines, glacial isostasy and ice-free areas', Lundqua Thesis 12.

Norðdahl, H. and Hafliðason, H. (1990) 'Skógar tefran, en sen glacial kronostratigrafisk marker på Nordisland', Abstract, 19th Nordic Geological Winter Meeting, Norsk Geologisk Forening, Geolognytt 17-1, 84.

Norrman, J. O. (1964) 'Vätterbäckenets senkvartära strandlinjer', Geologiska Föreningens i Stockholm Förhandlingar 85, 391-413.

Nye, J. F. (1952) 'A method of calculating the thickness of the ice sheets', Nature 169, 529-530.

Sigbjarnarson, G. (1983) 'The Quaternary Alpine glaciation and marine erosion in Iceland', Jökull 33, 87-98.

Steindórsson, S. (1962) 'On the age and immigration of the Icelandic flora', Societas Scientiarum Islandica, Rit 35.

Sugden, D. E. and John, B. S. (1976) Glaciers and Landscape, Edward Arnold, London.

Thórarinsson, S. (1937) 'The main geological and topographical features of Iceland', Geografiska Annaler 19, 161-175.

Thorkelsson, Th. (1935) 'Old shore-lines in Iceland and isostasy', Societas Scientiarum Islandica, Greinar I.1.

Thoroddsen, Th. (1905-06) 'Island. Grundriß der Geographie und Geologie', Petermanns Mitteilungen, Ergänzungsheft no. 152 und 153. Justus Perthes, Gotha.

Venzke, J.-F. and Meyer, H.-H. (1986) 'Remarks on the Late Glacial and early Holocene deglaciation of the Svarfaðardalur and Skíðadalur valley system, Tröllaskagi, Northern Iceland', Research Institute Neðri-Ás, Hveragerði, Iceland Bulletin No. 46.

Þorkelsson, Þ. (1922) 'Um ísaldarmenjar og forn sjávarmál kringum Akureyri', Andvari 47, 44-65.

Þóroddsson, Þ. (1982) 'Jarðfræfi', in Hallgrímsson, H. (ed.), Vesturströnd Eyjafjarðar, náttúrufar og minjar, Ministry of Industry, Reykjavík, pp. 16-56.

Þóroddsson, Þ. and Hafstað, Þ. (1975) Neysluvatnsrannsókn fyrir Grenivík', Mimeographed report OSJHD 7509, Energy Authority of Iceland.

THE WEICHSELIAN GLACIAL HISTORY OF WEST MELRAKKASLÉTTA, NORTHEASTERN ICELAND

Halldór G. Pétursson
The Akureyri Museum of Natural History
P.O. Box 580
602 Akureyri
Iceland

ABSTRACT. Melrakkaslétta is part of the volcanic zone in Northeastern Iceland and in the area subglacial mountains, lava fields, faults and fissures are evidence of recent volcanic activity. On West Melrakkaslétta glacial striae and deglaciation landforms have been correlated with Weichselian sediments that have been divided into 13 different stratigraphical units. The oldest sediments show that an oscillating glacier overrode marine sediments of Eemian or Weichselian age. Finally the glacier retreated and at least northwest Melrakkaslétta became ice free. A lava flow represents this Weichselian interstadial in the stratigraphy. The intermediate sediments show that the glacier advanced, then retreated to a floating stagnating glacier and advanced once more. The youngest sediments show that a marine environment was transformed into a glaciomarine one as a floating glacier advanced at a rising sea level at 12,700 B.P. The environment developed into marine at 12,100 B.P. and deglaciation and isostatic recovery set in. Ice-rafted debris show that this coincided with major deglaciation around the North Atlantic Ocean. During the Younger Dryas the northernmost part of the Icelandic inland ice sheet constituted a glacial dome over Melrakkaslétta. The final deglaciation and formation of marine limit is dated to 10,200 B.P.

1. INTRODUCTION

Melrakkaslétta is a flat and low peninsula situated between two broad fjords, Öxarfjörður and Þistilfjörður in northeast Iceland (Fig. 1). West Melrakkaslétta is a part of the northern volcanic zone of Iceland and the topography of the area is consequently influenced by the associated volcanism. In this area recent volcanic and tectonic activity is common and is illustrated by postglacial and interglacial lavafields and craters, subglacial hyaloclastite ridges and table mountains, open fissures and faults. In the eastern part, where the bedrock

49

J. K. Maizels and C. Caseldine (eds.), Environmental Change in Iceland: Past and Present, 49–65.
© 1991 *Kluwer Academic Publishers. Printed in the Netherlands.*

is older and outside the volcanic zone, the topography is dominantly shaped by erosive processes.

The geology of the area has been mapped by Thoroddsen (1895, 1897, 1905-1906), Sæmundsson (1977) and Pétursson (1979). Áskelsson (1938) described fossiliferous sediments from the locality of Röndin at Kópasker. Kjartansson (1955) observed and reported on his measurements of glacial striae on Melrakkaslétta. According to his observations the glacier on Melrakkaslétta was a part of the major inland ice sheet that covered Iceland during the Weichselian. On the basis of Áskelsson's stratigraphy (1938) and a radiocarbon date (Olsson et al., 1972), Einarsson (1971, 1973, 1979) proposed the Röndin locality as a type site for the Bölling interstadial in Iceland.

The present paper is a summary of the author's research efforts under the supervision of Dr. T.O. Vorren at Tromsö University, Norway, that have been focused on the stratigraphy of the sediments in five different profiles and on the geomorphology of West Melrakkaslétta (Pétursson, 1986, 1987, 1988).

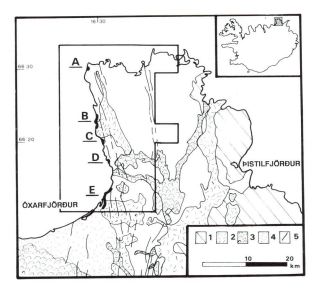

Fig.1 Location and simplified bedrock geology of Melrakkaslétta. The research area of West Melrakkaslétta is framed. (A) Rauðinúpur sediments; (B) Hvalvík sediments; (C) Háubakkavík sediments; (D) Röndin sediments; (E) Naustarvík sediments; 1: bedrock older than 0.7 Ma; 2: hyaloclastite mountains; 3: postglacial lavafields; 4: postglacial sandur; 5: fissures and faults.

2. GEOMORPHOLOGY

The Melrakkaslétta peninsula is mostly a flat and low plain, where vegetation covers glacially sculptured lava bedrock. On the western and eastern parts, long hyaloclastite ridges and table mountains are the remains of subglacial volcanism. On West Melrakkaslétta, tectonic movements in the forms of fissures and faults have occurred in postglacial time. The research on the geomorphology of the area has been concentrated on glacial striae and landforms formed during the last deglaciation of the area.

On Melrakkaslétta glacial striae form a radial pattern over the peninsula, which is in good agreement with Kjartansson's (1955) results. As the glacier of Melrakkaslétta was a part of the Weichselian inland ice sheet, changes in the regime of the ice sheet were sub-

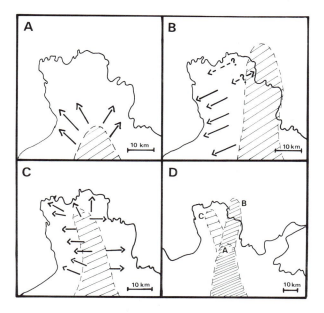

Fig.2 Direction of movement of the glaciers according to the glacial striae of Melrakkaslétta (Pétursson, 1986). (A) The glacier movement according to the oldest striae, phase A with ice divides south of Melrakkaslétta. (B) The glacier movement according to the most marked striae, phase B with ice divides over East Melrakkaslétta or on the shelf east of Melrakkaslétta. (C) The glacier movement according to the youngest striae, phase C of Younger Dryas age with ice divides in the centre of Melrakkaslétta. (D) The ice divides of the three different directions of glacier movement over Melrakkaslétta in relation to the ice divides of the inland ice sheet of Northeastern Iceland.

sequently reflected on Melrakkaslétta. The glacial striae on West Melrakkaslétta have been assigned to three different phases of glacier movements (Fig. 2):

Phase A, the oldest phase is explained by movement of glacier ice from the topographic higher area south of Melrakkaslétta (Fig 2A).

Phase B is explained by movement from an ice divide that was situated over the easternmost part of Melrakkaslétta or possibly over the shelf east of Melrakkaslétta (Fig 2B and 2D). This phase is the most marked on West Melrakkaslétta and bedrock forms such as whalebacks were formed by this direction of ice movement.

Phase C, the youngest direction of ice flow, produced a radial pattern of glacial striae away from the centre of Melrakkaslétta (Fig 2C). This ice flow is correlated with the stratigraphy

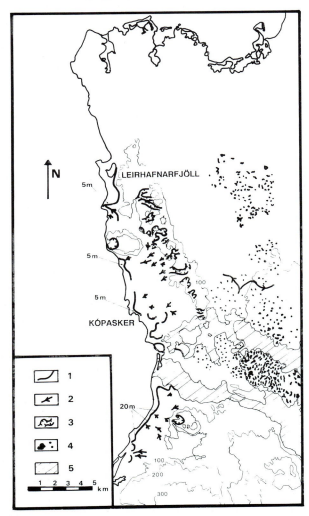

Fig.3 The landforms of the Younger Dryas deglaciation and the strandline of the marine limit on West Melrakkaslétta. 1: The strandline of the marine limit; 2: meltwater channels; 3: canyons in hyaloclastite; 4: hummocky terrain; 5: postglacial lavafield.

by similarities between the fabric in a till unit and the main direction of the glacial striae of phase C. The till was formed by an ice advance in Younger Dryas time.

3. THE STRATIGRAPHY OF THE SEDIMENTS ON WEST MELRAKKASLÉTTA

Landforms from the last deglaciation, which dates from the end of the Younger Dryas Chronozone (10,200 B.P.), are shown in Figure 3. The landforms are raised terraces, deltas, small sandurs and strandlines of the marine limit on West Melrakkaslétta. In the area of raised features, postglacial tectonic subsidence is evidenced by numerous open fissures and normal faults. The general trend is that the marine limit rises towards the south, probably due to a different ice load and isostatic recovery. Dry channels and small canyons as e.g. in Leirhafarfjöll date from the last deglaciation (Fig.3) and many of these landforms are correlated with strandlines of the marine limit. In the middle of Melrakkaslétta a considerable area of ridges and hummocky terrain (Fig.3) may mark a stillstand or a small scale advance of the glacier on Melrakkaslétta.

The location of the five sedimentary profiles of West Melrakkaslétta is shown on Figure 1. The sediments rest on a foundation of bedrock, either in the form of lava flows or hyaloclastite. The sediments are lithified and very resistant to e.g. glacial erosion. The high degree of diagenesis in such young sediments is probably due to the nearness of the volcanic zone. Unstable volcanic glass is produced in great quantities in subglacial eruptions and some of this material was deposited on the glacier and became part of its debris load. Unstable glass is stabilized at some time by the process of palagonitization, a process that includes hydration and migration of elements in the glass and results in the debris being lithified (Jakobsson, 1971, 1972, 1978). Sediments containing a lot of glass in the finer grain sizes quickly become lithified. Tills and glaciomarine sediments with silt and clay matrix are lithified, while sorted sediments are unlithified. The tills of West Melrakkaslétta are in fact so lithified, that they are classified as tillites.

3.1 The Háubakkavík profile, the oldest group of sediments

The oldest group of sediments is found in the coastal cliffs of Háubakkavík (Fig.4). In Naustárvík (Fig.5) two beds may possibly be compared with some of the sediments in Háubakkavík, but correlation can not be justified on the basis of the available material. The profile in Háubakkavík is complex, and in addition to the oldest group, sediments of the two younger groups are also represented.

The lowest unit in the group displays layers of sand and rounded gravel (unit 1, Fig. 4). A layer often starts as a thin lens of sand but thickens laterally as coarser material replaces the sand. Where the material is coarse, erosion of the underlying layers is often observed. A few moulds of dissolved shells and shell fragments are found in the lower part of unit. *Mya truncata, Hiatella arctica* and *Macoma calcarea* were identified, but for the most part they were unidentifiable. The unit is interpreted as to have been formed in a shallow marine to sublittoral coastal environment. The marine limit at the formation of this unit was higher than present.

An erosional unconformity separates units 1 and 2. Unit 2 (Fig. 4) is a diamictite, rich in rounded gravel, which is interpreted as tillite. The erosional unconformity probably represents a hiatus and the unit is evidence of an ice advance in the area. Another erosional unconformity separates units 2 and 3. The erosion has in some parts of the profile been so effective that it removed the till of unit 2, and unit 3 rests directly on the sublittoral gravel of unit 1. Unit 3 (Fig. 4) is composed of sorted, rounded gravel and boulders. The largest boulders are concentrated at the bottom of the unit, resting on the erosional surface. Upwards there is a decrease in grain size indicating that depositional energy fell with time. Unit 3 is interpreted as glaciofluvial gravel, the remains of channels of a braided river system. The fluvial erosion of the underlying sediments (unit 1 and 2) is considered to have occurred just prior to, or contemporary with, the deposition of unit 3.

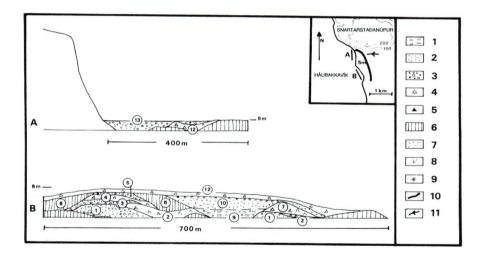

Fig.4 Schematic profile of Háubakkavík with different stratigraphical units marked. The location of the Háubakkavík sediments is given in Fig.1. Legend - 1: perlite; 2: sand; 3: gravel; 4: diamictite; 5: stratified drift; 6: lava; 7: hyaloclastite; 8: glaciotectonic structures; 9: location of radiocarbon-dated shells; 10: strandline of the marine limit; 11: meltwater channel.

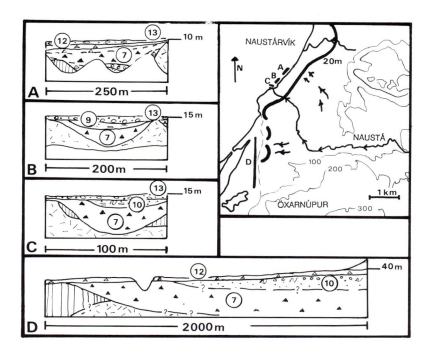

Fig.5 Schematic profile of Naustarvík with different stratigraphical units marked. For legend see Fig.4. The location of the Naustarvík sediments is given in Fig.1.

An erosional unconformity also separates units 3 and 4. Unit 4 (Fig.4) is a gravel- and boulder-rich diamictite, interpreted as a tillite. The contact between units 4 and 5 varies but is mostly sharp and erosional, and the units are interfingered. Unit 5 (Fig. 4) is composed of poorly sorted sand, gravel and boulders, interpreted as glaciofluvial sediments. The tillite of unit 4 is taken as evidence of an ice advance in the area and the glaciofluvial sediments of unit 5 were probably formed during the following deglaciation.

Unit 6 (Fig. 4) is a basaltic lava flow. In the profile it is found in two small basins, which were eroded down into the bulk of the underlying sediments. The erosion has in some parts of the profile been so intensive that the sediments have been completely removed. It seems likely that the lava flowed along pre-eroded channels, around and up onto the sides of a small hillock of older sediments. No sediments have been found in the channels. The lava flow indicates an eruption during ice-free conditions.

3.2 The Naustárvík profile, the intermediate group of sediments

The intermediate group of sediments is found in the coastal cliffs of Naustárvík (Fig. 5).

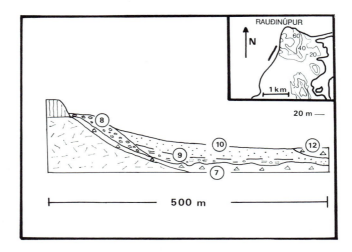

Fig.6 Schematic profile of Rauðinúpur with different stratigraphical units marked. For
legend see Fig.4. The location of the Rauðinúpur sediments is given in Fig.1.

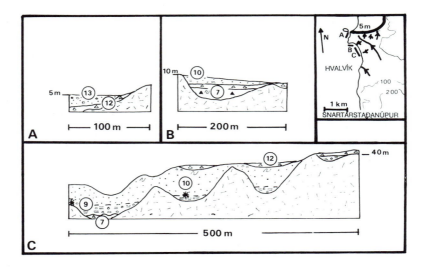

Fig.7 Schematic profile of Hvalvík with different stratigraphical units marked. For
legend see Fig.4. The location of the Hvalvík sediments is given in Fig.1.

Parts of the profiles in Háubakkavík (Fig. 4), Rauðinúpur (Fig. 6) and Hvalvík (Fig. 7) contain sediments of this group.

The intermediate group consists of one unit (unit 7), which is made up of beds of different appearance, but all were deposited in a glacial environment. The lower part of the unit is a diamictite, which is interpreted as a tillite and taken as evidence of an ice advance on West Melrakkaslétta. The middle part consists of thick, poorly sorted sand, gravel and boulders with thin irregular beds of diamictite. Some parts of the Naustárvík profile (Fig. 5) consist of a thick bed of diamictite with rounded gravel clasts. The pebbles are in bands in the fine material as the sediment has been liquefied at the time of deposition. These sediments are interpreted as flow tills in front of a stagnating floating glacier, where the input of material is high (Boulton, 1968, 1971, 1972; Marcussen, 1973, 1975; Dreimanis, 1976, 1979, 1982; Evenson et al., 1977).

During a subglacial eruption the confining glacier ice is melted and a lake is formed in the ice, and material from the eruption is deposited in the lake. Debris from the melted glacier and tephra from the eruption is also sedimented in the lake (Kjartansson, 1943, 1956, 1966; Jones, 1969, 1970). On West Melrakkaslétta four subglacial eruptions are considered to have occurred during the Weichselian and in the vicinity of Naustárvík is the table mountain Öxarnúpur, probably of late Weichselian age (Pétursson, 1979, 1986). The sediments of the Naustárvík profile are covered to the south by debris from the cliffs of Öxarnúpur (Fig. 4D). The cliffs are made up of a foreset breccia formed when the lava flowed into the glacial lake. It is therefore possible that this part of unit 7 in Naustárvík was formed in connection with a subglacial eruption. In other profiles of West Melrakkaslétta parts of unit 7 have the same appearance as the middle part of the unit in Naustárvík. On the basis of the present data, the middle part of unit 7 is considered to represent an environment of a stagnating floating glacier.

The upper part of the intermediate group is a diamictite, interpreted as tillite. The tillite

Table 1. Radiocarbon dates from West Melrakkaslétta. The base year is 1950. The value of 5570 years is used for the half life ^{14}C. Correction of 365 B.P. years for deviation from standard ^{13}C/^{12}C ratio and apparent age of living marine organisms (Håkansson, 1983) is applied. Samples marked (*) have been corrected with 440 B.P. years (Pétursson, 1986). Sample marked (**) is uncorrected (Olsson et al., 1972).

Sample name	Laboratory Number	Reported Age (Years B.P.)	Corrected Age (Years B.P.)
Kópasker	T-4467	10,130±80*	10,205±80
Hvalvík	T-4468	12,580±90*	12,655±80
Röndin	T-4469	11,880±110*	11,955±110
Hvalvík	T-4470	12,060±150*	12,135±150
Röndin	U-2225	12,830±170**	12,465±170

and an erosional contact below it are taken as evidence of an ice advance in the area. The fabric of this tillite is the same as the direction of glacial striae of phase B of glacier movement on West Melrakkaslétta (Fig. 2B).

3.3 The Rauðinúpur, Hvalvík and Röndin profiles, the youngest group of sediments

The youngest group of sediments is found in the cliffs and steep slops of Rauðinúpur (Fig. 6), Hvalvík (Fig. 7) and Röndin (Fig. 8). The youngest group is also represented in Háubakkavík and Naustárvík. The sediments of Röndin have previously been described by Áskelsson (1938) and Einarsson (1968, 1971, 1979).

The lowest unit (unit 8) in the youngest group is only found in the northernmost profile, Rauðinúpur (Fig. 6). It consists of clastic sediments of a local delta, deposited when relative sea level stood about 20 m a.s.l. The contact to the under- and overlying sediments

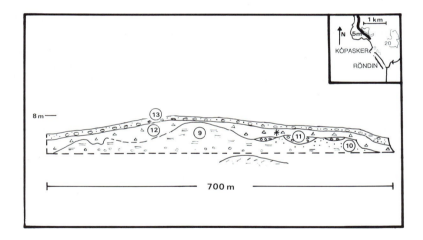

Fig.8 Schematic profile of Röndin with different stratigraphical units marked. For legend see Fig.4. The location of the Röndin sediments is given in Fig.1.

is always sharp and the lower one is probably erosional. In all the other profiles unit 9 follows the tillite of unit 7, but unit 9 is a perlite with stones and shells, formed in a glaciomarine environment with occasional dropstones from ice rafts. The unit is mostly found at 40 m a.s.l. in Hvalvík (Fig. 7), and was deposited when sea level was well above the 40 m level. Of the shells, the species *Mya truncata* and *Hiatella arctica* have been identified, and many of them were found in situ. Moulds of individuals of the same species are also common in the unit. Radiocarbon dating of a shell from the bottom of the unit in Hvalvík gave an age of 12,655±90 B.P. (Table 1).

Unit 9 is gradually taken over by unit 10 as fine grained material disappears. The lower part of unit 10 consists of layers of sand of different grain sizes, but the upper part is made of somewhat coarser sand and gravel. Shells and moulds of shells are found in the lower part of the unit. The sediments of the unit were deposited in a marine environment. In Hvalvík this unit is over 11 m thick, and it rests on top of a 15 m high bedrock cliff (Fig. 7). The sea level at the time of deposition was at least about 30 m a.s.l. At one locality in Hvalvík moulds of *Mya truncata* with especially well developed sipho display a fight to survive against a growing input of sediments. This has happened repeatedly and several thick beds display this type of structure. A radiocarbon date on a shell from the contact between unit 9 and 10 yielded an age of 12,135±150 B.P. (Table 1).

Unit 11 consists of well rounded gravel which was deposited in a sublittoral coastal environment. In Röndin, a few stones of foreign lithology are found in the gravel, and at the time of deposition sea level stood below 10 m a.s.l. In Röndin this unit is found in small basins eroded into the sediments of units 9 and 10 (Fig. 8). Unit 12 is found in all the described profiles on West Melrakkaslétta and is composed of a sand-rich diamicton with clasts of stones and boulders. It is interpreted as a till and much of the material in the till is redeposited sediments from the underlying units. The till of unit 12 is taken as evidence of a glacier advance which eroded and deformed the underlying sediments in units 9, 10 and 11 in all the studied profiles. Radiocarbon dating of resedimented fragments of shells from unit 10, collected from the till in unit 12 at Röndin (Fig. 8), gave an the age of 11,955±110 B.P. (Table 1).

Unit 13 is composed of beds of well rounded gravel which was deposited in a sublittoral or littoral environment. Landforms such as strandlines, meltwater channels and deltas are correlated with this unit. Radiocarbon dating of a *Balanus* sp. from a sublittoral gravel below the marine limit at 5 to 6 m a.s.l., just north of Kópasker, gave an age of 10,205±80 B.P. (Table 1).

4. CONCLUSION

The coastal cliffs of West Melrakkaslétta are made up of sediments resting on a foundation of lava flows or hyaloclastite rocks. The bedrock is considerably younger than 0.7 Ma (Sæmundsson 1977, Pétursson 1979), and a Weichselian age is inferred for the bulk of sediments on Melrakkaslétta.

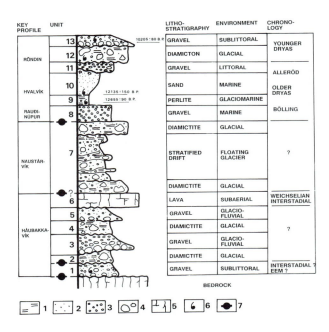

Fig.9 Lithostratigraphy, the depositional environments and chronology of the sediments of West Melrakkaslétta. 1: Perlite; 2: sand; 3: gravel; 4: stones and boulders; 5: lava; 6: shells or moulds of shells; 7: hiatus.

The Weichselian stratigraphy of West Melrakkaslétta (Fig. 9) starts with a gravel (unit 1) deposited in a coastal environment at a sea level higher than at present. The age of unit 1 is unknown, but it could date from a Weichselian interstadial or even from the Eemian. The gravel has been overridden by a glacier that deposited the tillite of unit 2. When the glacier retreated, glaciofluvial gravel of unit 3 was deposited on top of the tillite. During a later glacier advance another tillite (unit 4) was deposited. During a subsequent retreat of the glacier, glaciofluvial sediment (unit 5) was deposited.

The next unit is the lava flow of unit 6 which is correlated with an eruption in the tuff crater Rauðinúpur in the northwestern part of Melrakkaslétta. It is clear on the basis of the areal extent of the lava flow that parts of Melrakkaslétta were ice-free at that time (Pétursson 1986). In Háubakkavík the lava (unit 6) flowed along abandoned melt water channels eroded into the older sediments. This probably occurred during a Weichselian interstadial. The next unit (unit 7) is composed of beds of tillite and poorly sorted glaciofluvial sediments. The lower part of the unit is a tillite deposited during a glacier advance on Melrakkaslétta, the middle part is made of a flow till in an environment of a floating, stagnating glacier. The upper part of the unit is a tillite which must be evidence that the glacier of Melrakkaslétta had advanced again. The next unit (unit 8) is a gravel of a marine delta deposited when relative sea level was 20 m a.s.l. The perlite of unit 9 contains both shells and stones, but the unit was deposited in a glaciomarine environment when sea level was more than 40 m a.s.l. The different characteristics of unit 8 and unit 9 and the sharp contact that separates them, probably indicates some age difference and a hiatus between the formation of these two units. In that time sea level had risen and the depositional environment was changed from marine to glaciomarine. A glacier in the vicinity of Melrakkaslétta may have advanced to a floating position beyond the coastline, and radiocarbon dating of a shell from the lowest part of unit 9 gave an age of about 12,700 B.P. The environment gradually changed from glaciomarine to marine as demonstrated by the thick sand of unit 10. Radiocarbon dating of a shell from the contact between units 9 and 10 gave an age of about 12,100 B.P.

Unit 11, which is a gravel formed in a coastal environment is found in small eroded basins in units 9 and 10. The gravel was deposited when relative sea level was less than 10 m a.s.l. and the different character of units 9, 10 and 11 and the different sea level at deposition clearly indicates an age difference between units 9 and 10, and between units 10 and 11. In that time relative sea level changed from greater than 40 m to less than 10 m a.s.l. The stones of foreign lithology in the gravel are interpreted as debris from ice rafts that stranded on the beaches of Melrakkaslétta. Unit 12 is a till, rich in resedimented material from units 9, 10 and 11, which was deposited during a glacier advance that eroded and deformed the sediments of units 9, 10 and 11. Radiocarbon dating of fragmented shells that originated from the marine sand of unit 10 and collected from the till gave an age of about 12,000 B.P. for material eroded, but not for the advance itself. A Younger Dryas age is therefore inferred for the glacier advance.

Landforms of the final deglaciation and formation of the marine limit on West Melrakkaslétta are correlated with the sublittoral gravel of unit 13. Radiocarbon dating of a shell from a sublittoral gravel at a marine limit of 5 to 6 m a.s.l. gives an age of 10,200 B.P. for the deglaciation.

5. DISCUSSION

In the lower part of the stratigraphy of West Melrakkaslétta a glacier advanced and overrode older shallow marine sediments. The glacier retreated, advanced again and finally retreated from at least northwestern Melrakkaslétta as demonstrated by the lava from Rauðinúpur (Pétursson, 1986, 1988). In the Tjörnes area, where lava and sediment stratigraphy are interwoven as on Melrakkaslétta, Eiríksson (1981, 1985) has described and defined cyclic series of sedimentary environments from a glacial to an interglacial status. In that cycle the glacial environment is followed by the glaciofluvial, glaciomarine, marine, littoral and terrestrial environment, the latter represented by a lava flow. On Melrakkaslétta, the marine environment is replaced by series of glacial and glaciofluvial environments, and finally by lava flow. This is different from the glacial/interglacial cycle of Tjörnes and the sediments on West Melrakkaslétta are instead best explained as interstadial or interglacial sediments overlain by a series of sediments deposited by an oscillating glacier.

Finally the northwestern part of Melrakkaslétta became ice-free. A Weichselian interstadial age is suggested as the time of eruption for the Rauðinúpur lava flow, but its absolute age is still unknown. This point of view is supported by the environments of the intermediate group where the lower part was deposited during a glacial advance. During the formation of the middle part of the intermediate group the glacier had been changed from a grounded advancing glacier into a floating, stagnating glacier in the present coastal regions of West Melrakkaslétta. The upper part was on the other hand deposited by an advancing glacier and the clast fabric of the tillite in this upper part is the same as the main direction of the glacial striae of phase B of glacier movement on West Melrakkaslétta. Bedrock erosion from this glacial phase is the most marked in the area and the ice divide of the glacier on Melrakkaslétta was situated in a glacier dome over the eastern part of Melrakkaslétta or on the shelf east of it during a phase of a relative low sea level (Fig. 2D).

This part of the West Melrakkaslétta sediments and the interstadial status of some of them, may possibly be compared with some of the interstadials that Norðdahl (1983) has described and defined in Fnjóskadalur, central North Iceland, as the glaciers in both areas were a part of the Weichselian inland ice sheet that covered Iceland. A hiatus occurs between the sediments of the intermediate group and the sediments of Rauðinúpur, Hvalvík and Röndin. Melrakkaslétta may have been ice-free for some time before relative sea level rose to about 20 m a.s.l. and a marine delta at Rauðinúpur was formed. A subsequent glaciomarine environment that transgressed above the 40 m level has been dated to about 12,700 B.P. which is also the age of an advance of a floating glacier in the vicinity of Melrakkaslétta. This high sea level can be explained by increased glacio-isostatic load due to changes in the extent and thickness of the Icelandic inland ice sheet. This glaciomarine environment was later gradually changed into a marine environment, the change occurring at about 12,100 B.P. Later, when relative sea level had dropped to less than 10 m a.s.l., a coastal environment developed on West Melrakkaslétta and ice-rafted debris of foreign lithology indicates a time of great ice melting on the continents surrounding the North Atlantic Ocean.

The glacier of Melrakkaslétta advanced again and deposited a till, and the orientation of the clast fabric is the same as glacial striae of phase C of glacier movement in the area. It is clear from the distribution of the direction of the glacial striae of phase C in the coastal areas of North and West Melrakkaslétta that the glaciers advanced beyond the present coast of Melrakkaslétta. The radial pattern of striae of phase C around Melrakkaslétta, suggests that a glacier dome was built up over the central parts of the peninsula. This dome was the northernmost prolongation of the inland ice sheet in northeastern Iceland. The strong radial pattern of the striae of phase C also suggests that the glacier of Melrakkaslétta originated within the area itself, during a time when the glaciation limit suddenly fell and ice started to accumulate in the central part of the peninsula. Compared with the deglaciation environments of the sediments in units 9, 10 and 11, a considerable ice advance and a major cooling had occurred since the deposition of these sediments.

The advancing glacier eroded and deformed older sediments. Radiocarbon-dated fragments of shells from the till give an age of about 12,000 B.P. for resedimented shells from the marine sand but not the age of the ice advance itself. On the basis of stratigraphical evidence, the ice advance is considered to be younger and to coincide with the major cooling during the Younger Dryas. This view is supported by a shell from the final deglaciation which has been radiocarbon dated to 10,200 B.P. Sublittoral and littoral gravels, meltwater channels, deltas and strandlines of the marine limit are consequently dated to the end of the Younger Dryas Chronozone.

On the basis of Áskelsson's (1938) stratigraphy from a part of the Röndin profile (Fig. 10) and a shell radiocarbon dated to 12,465±170 B.P. (Olsson et al. 1972) (Table 1), Einarsson (1971, 1979) proposed the Röndin locality as a type site for the Bölling Interstadial in Iceland. The last ice advance on Melrakkaslétta consequently occurred in Older Dryas time and deglaciation and formation of marine limit during the Alleröd.

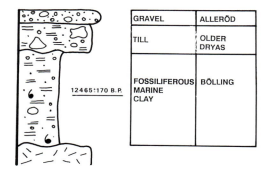

Fig.10 Einarsson's (1971,1979) interpretation of Áskelsson's (1938) stratigraphy of a section at Röndin.

The results of the present work on West Melrakkaslétta are as follows: an ice advance and increased isostatic load is registered in Bölling time (12,700 B.P.); the Older Dryas and Alleröd were times of deglaciation and isostatic recovery; the last ice advance occurred in Younger Dryas time and the following deglaciation and formation of marine limit are dated to the end of the Younger Dryas (10,200 B.P.). These results from West Melrakkaslétta suggest that the Younger Dryas glaciation of northeastern Iceland was more extensive than was earlier suggested by Einarsson (1968, 1971, 1973, 1979).

ACKNOWLEDGEMENTS

Þorleifur Einarsson at the University of Iceland proposed this investigation on Melrakkaslétta as early as 1976. This was an essential support in the beginning. Tore O. Vorren was my supervisor at Tromsö University. Hreggviður Norðdahl of the University of Iceland critically reviewed the manuscript and made valuable comments upon it. Chris Caseldine corrected the English. Erland T.H. Lebesbye, Ólafur Ingólfsson and Hörður Kristinsson are remembered for the years of fieldwork. Especially my old friend Sigmar Benjaminsson, who died in 1984. Grethe-Kristin, Anik, Inger og Liv, en klem.

REFERENCES.

Áskelsson, J. (1938) 'Um íslensk dýr og jurtir frá jökultíma', Náttúrufræðingurinn 8, 1-16.

Boulton, G.S. (1968) 'Flow tills and related deposits on Vestspitsbergen glaciers', Journal of Glaciology 9, 391-412.

Boulton, G.S. (1971) 'Till genesis and fabric in Svalbard, Spitsbergen', in R.P.Goldthwait (ed.) Till, a symposium, Ohio State University Press, Ohio, 41-72.

Boulton, G.S. (1972) 'Modern arctic glaciers as depositional models for former ice sheet', Quarterly Journal of the Geological Society of London 128, 361-393.

Dreimanis, A. (1976) 'Tills: their origin and properties', in R.F.Legg (ed.), Glacial Till. An interdisciplinary study, Royal Society of Canada Special Publication 12, 11-49.

Dreimanis, A. (1979) 'The problems of waterlain tills', in Ch. Schluchter (ed.), Moraines and Varves, A.A. Balkema, Rotterdam, 167-177.

Dreimanis, A. (1982) 'The origin of stratified Catfish Creek till at Plum Point, Ontario Canada', Boreas 11, 173-180.

Einarsson, Þ. (1968) 'Jarðfræði, saga bergs og lands', Mál og Menning, Reykjavík.

Einarsson, Þ. (1971) 'Jarðfræði', Heimskringla, Reykjavík.

Einarsson, Th. (1973) 'Geology of Iceland', in M.G.Pitcher (ed.), Arctic Geology, American Association of Petroleum Geologists Memoir 19, 171-179.

Einarsson, Th. (1979) 'On the deglaciation of Iceland', abstract in 14. Nordiske geologiske vintermøte, Bergen University 1980, Norsk geologiske forening, Geolognytt 13, 18.

Eiríksson, J. (1981) 'Lithostratigraphy of the upper Tjörnes sequence, North Iceland: the Breiðavík group', Acta Naturalia Islandica 29, 31 pp.

Eiríksson, J. (1985) 'Facies analysis of the Breiðavík group sediments on Tjörnes, North Iceland', Acta Naturalia Islandica 31, 56 pp.

Evenson, E.D., Dreimanis, A. & Newsome J.W. (1977) 'Subaquatic flow tills: a new interpretation for the genesis of some laminated till deposits', Boreas 6, 115-133.

Håkansson, S. (1983) 'A reservoir age for the coastal waters of Iceland', Geologiska Föreningens i Stockholm Förhändlingar 105, 64-67.

Jakobsson, S.P. (1971) 'Myndun móbergs í Surtsey', Náttúrufræðingurinn 41, 124-128.

Jakobsson, S.P. (1972) 'On the consolidation and palagonitization of the tephra of the Surtsey, volcanic island, Iceland', Surtsey Research Progress Report 6, 121-128.

Jakobsson, S.P. (1978) 'Environmental factors controlling the palagonitization of the Surtsey tephra, Iceland', Bulletin of the Geological Society of Denmark 27, 91-105.

Jones, J.G. (1969) 'Intraglacial volcanoes of the Laugarvatn region southwest Iceland. I', Quarterly Journal of the Geological Society of London 124, 197-211.

Jones, J.G. (1970) 'Intraglacial volcanoes of the Laugarvatn region southwest Iceland. II', Journal of Geology 78, 127-140.

Kjartansson, G. (1943) 'Árnesinga saga', Árnesingafélagið, Reykjavík.

Kjartansson, G. (1955) 'Fróðlegar jökulrákir', Náttúrufræðingurinn 25, 154-171.

Kjartansson, G. (1956) 'Úr sögu bergs og landslags', Náttúrufræðingurinn 26, 113-130.

Kjartansson, G. (1966) 'Stapakenningin og Surtsey', Náttúrufræðingurinn 35, 155-181.

Marcussen, I. (1973) 'Studies on flow till in Denmark', Boreas 2, 213-231.

Marcussen, I. (1975) 'Distinguishing between lodgement till and flow till in Weichselian deposits', Boreas 4, 113-123.

Norðdahl, H. (1983) 'Late Quaternary stratigraphy of Fnjóskadalur, central north Iceland', Lundqua Thesis 12, 78 pp.

Olsson, I.U., Karlsson, M. & Abd-el-Mageed, A. (1972) 'Uppsala natural radiocarbon measurements XI', Radiocarbon 14, 247-271.

Pétursson, H.G. (1979) 'Jarðfræði Núpasveitar', Fjórðaársverkefni, University of Iceland.

Pétursson, H.G. (1986) 'Kvartærgeologiske undersökelser på Vest-Melrakkaslétta, Nord-öst-Island', Cand. real thesis, University of Tromsö.

Pétursson, H.G. (1987) 'Síðjökultími á vesturhluta Melrakkasléttu', abstract in Ísaldarlok á Íslandi, Jarðfræðafélag Íslands, 14-15.

Pétursson, H.G. (1988) 'Eldvirkni á hlýindakafla á síðasta jökulskeiði', abstract in Eldvirkni á Íslandi, Jarðfræðafélag Íslands, 25.

Sæmundsson, K. (1977) 'Geological map of Iceland, sheet 7, Northeast Iceland', Icelandic Museum of Natural History and Iceland Geodetic survey, Reykjavík.

Thoroddsen, Th. (1895) 'Fra det nordöstlige Island, rejseberetning fra sommeren 1985', Geografisk Tidskrift 13, 99-122.

Thoroddsen, Th. (1897) 'Ferð um Norður-Þingeyjarsýslu sumarið 1895', Andvari 22, 17-77.

Thoroddsen, Th. (1905-1906) 'Island, Grundriß der Geographie und Geologie', Petermanns Mitteilungen Ergänzunghefte 152 & 153, 358 pp.

A REVISED MODEL OF WEICHSELIAN DEGLACIATION IN SOUTH AND SOUTH WEST ICELAND

Árni Hjartarson
Orkustofnun
Grensásvegur 9
108 Reykjavik
Iceland

ABSTRACT. New investigations and radiocarbon dating are changing the traditional Quaternary chronology of Iceland. The Búði end moraines of southern Iceland have hitherto been regarded as the type site for the Younger Dryas glaciation. Recent radiocarbon dates now indicate that these moraines are of Preboreal age, around 9,700 B.P. Sediments from the Alleröd Interstadial have not been found in southern Iceland. This is believed to indicate a total glacial coverage of the Southern Lowlands of Iceland during the Younger Dryas stadial. The Reykjavík area is a classic locality in Icelandic geology. The shell-bearing Fossvogur sedimentary layers have long been considered to be of Interglacial age dating from the beginning of the Eemian Interglacial stage, i.e. about 120,000 years old. New radiocarbon age determinations have shown them to date back only to about 11,000 B.P. The former type locality for the Eemian Interglacial, the Fossvogur layers, now seems to be of Late Weichselian age. Consequently the till in the upper part of the Fossvogur sediments must be considered to be of Younger Dryas or Preboreal age. This till is connected to a series of end moraines, one of which is the so-called Álftanes ridge. This ridge has previously been accepted as the type locality of the Older Dryas stadial in Iceland, but must also now be considered to be of Younger Dryas or Preboreal age.

These conclusions indicate a much more extensive glaciation in Late Weichselian and Preboreal times than was previously assumed. The Younger Dryas glacier covered the whole of the Reykjavik area reaching far offshore into Faxaflói bay. Similar results can be seen by plotting radiocarbon dates from Late Weichselian and Preboreal marine shells from Iceland on a time-diagram. The dates show a rather even distribution except for a gap between 10,200-10,700 B.P. which is explained by a period of heavy glaciation during which glaciers mostly prevented deposition of marine sediments inside the present coastlines of the country. The new radiocarbon dates lead to a revision of the traditional Quaternary nomenclature of Iceland. The names Fossvogur Interglacial as a synonym for

J. K. Maizels and C. Caseldine (eds.), Environmental Change in Iceland: Past and Present, 67–77.

the Eemian Interglacial, the Álftanes Stadial for the Older Dryas, and Búði Stadial for the Younger Dryas cannot now be used.

1. THE BÚÐI ADVANCE

The Búði moraines, which can be traced across the Southern Lowlands of Iceland (Fig.1) have long been considered to indicate the last great readvance of the Late Weichselian glacier in Iceland, corresponding to the Ra moraine zone in Norway, the Central Swedish End Moraines and the Sålpausselka in Finland (Kjartansson, 1943, 1958, 1961; Kjartansson et al., 1964; Einarsson, 1968, 1978; Einarsson and Albertsson, 1988).

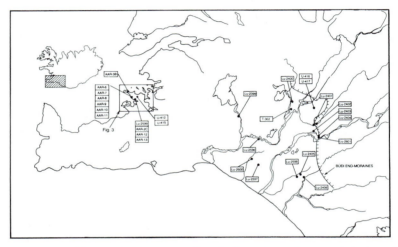

Fig.1 Synoptic map of S. and S.W.Iceland. The Búði end moraines and the locations of the dated samples listed in Table 1 are shown.

Until recently no radiocarbon analyses have been available on material in contact with the Búði moraines, just a few from marine sediments several kilometres outside the Búði limits (T-362, U-416 and U-417 in Table 1). During geological investigations in south Iceland between 1982 and 1984 shell-bearing layers in close contact with the Búði end moraines were discovered in several localities. Samples of shells were taken and analysed at the Radiocarbon Dating Laboratory at Lund University, Sweden to provide new and more precise age determinations for this important formation. These analyses pinpointed the age of the formation even more accurately than expected but they also threw doubt upon the traditional views of Kjartansson and Einarsson that the Búði end moraines in south Iceland are of Younger Dryas age (Hjartarson and Ingólfsson, 1988).

In the following discussion regarding the age of the fossil shells and the shell-bearing strata, the stated age is not the calendar age, but the conventional radiocarbon age based on a ^{14}C half-life of 5,568 years. This radiocarbon age is corrected for ^{13}C and the apparent age of the sea. The latter correction is based upon the most careful and recent measurements by Håkansson (1983) which indicate an apparent age for the sea around Iceland of 365±20 years. The results of all these determinations are presented in Tables 1 and 2.

Among the analysed samples from southern Iceland were shells lying underneath, within and on top of the Búði end moraines (Fig.2). Their dates set narrow limits on the measured age of these moraines:

- shells overlying Búði end moraines: 9,595±160 B.P. (Lu-2402)
- shell fragments from within Búði end moraine: 9,745±140 B.P.
      (Lu-2401)
- shells underlying Búði end moraines: 9,855±90 B.P. (Lu-2402) to
      9,995±90 B.P. (Lu-2403)

The shells over- and underlying the moraines are in situ but the shell fragments from the moraines themselves originate in bottom lyers that theglacier has reworked and deformed; they therefore indicate the maximum age of the moraines. These age measurements differ very little and show that the Búði advance took place in early Preboreal time and that the glacier culminated at some point around 9,700 B.P.

This conclusion indicates that the moraines neither correspond to the zones of Sålpausselka, the Central Swedish nor the Ra Moraines. They are around 400 ^{14}C years younger than the youngest of these zones. In a geological sense 400 years does not appear

Table 1 Radiocarbon dates from south Iceland.

Laboratory No.	Locality	Elevation m a.s.l.	Original ^{14}C date	Corrected ^{14}C date	δ^{13}C
Lu-2399	Sog, Bíldsfell	20	9,420±80	9,055±80	-0.4
Lu-2400	Dynjandi, Brúará	60	10,190±90	9,825±80	+0.9
Lu-2401	Hrepphólar	65-70	10,110±140	9,745±140	-1.2
Lu-2402	Hepphólar	65	9,960±160	9,595±160	+0.1
Lu-2403	Þrándarholt	60	10,360±90	9,995±90	+0.8
Lu-2404	at Minnahof	65	10,220±90	9,855±90	+0.3
Lu-2405	Efri-Rauðalækur	20	10,190±130	9,825±130	+0.8
Lu-2406	Y-Rangá, Bjarg	30-35	10,380±90	10,105±90	+1.0
Lu-2596	Oddgeirshólar	25	10,440±90	10,075±90	+1.0
Lu-2597	Vatnsendi	30	9,840±90	9,475±90	+0.9
Lu-2598	S-Rauðalækur	10	9,870±90	9,505±90	-0.4
Lu-2600	Hróardsholtslækur	35	10,060±70	9,695±70	+1.5
T-362	Spóastaðir, Brúará	55	9,930±140		
U-417	Hellisholtalækur	75	9,800±150	9,435±150	+3

Note: Age calculations are based on Oxalic Acid Factor 0,950NBS, a half-life for ^{14}C of 5,568 years; correction applied for ^{13}C:^{12}C; an assumed age of the sea of 320±20 years (Håkansson, 1983); all Lu-(Lund) dates are from Håkansson (1986, 1987), U-417 from Olsson et al. (1965), T-362 from Kjartansson et al. (1964).

to be a significant difference but in this case a key formation is transferred from one geological epoch to another, i.e. from Pleistocene to Holocene, or more exactly from late Younger Dryas to early Preboreal. If the Búði end moraines were the outermost moraines of the Icelandic Younger Dryas glacier one would rather have expected them to be a little older than the Scandinavian equivalents due to the small size of the Icelandic glacier compared to the Fennoscandian ice shield, thus making it more responsive to climatic changes. It is questionable whether it is necessary to assume that the Búði end moraines are the outermost moraines of the Younger Dryas and Preboreal period, as it is possible that they could represent a short readvance of the retreating Younger Dryas glacier.

The radiocarbon ages of samples near the Búði end moraines fall within a narrow interval of time. This raises the question whether shells from the Late Weichselian exist south of the Búði end moraines. It has been thought that the highest sea level in most parts of Iceland occurred during the Alleröd interstadial (Einarsson, 1968). Ingólfsson (1987) has found strandlines of Alleröd and Older Dryas age in the Borgarfjörður area to be considerably higher than the strandlines of Younger Dryas and Preboreal age. The Southern

Table 2 Radiocarbon dates from Reykjavík.

Locality	Laboratory Number	Year	Uncorrected Age	Corrected Age	$\delta\,^{13}C$
Hótel Loftleiðir	U-412	1964	10,310±260	9,945±260	+0.5
Hótel Loftleiðir	U-415	1964	10,230±190	9,865±190	+7.6
Nauthólsvík	Lu-2599	1987	11,530±100	11,165±100	+3.7
Nauthólsvík	AAR-2C	1988	11,800±150	11,435±150	+1.7
Nauthólsvík	AAR-12	1988	11,330±140	10,965±140	+1.3
Nauthólsvík	AAR-13	1988	11,580±150	11,215±150	+0.9
Skerjafjörður	AAR-6	1988	11,400±160	11,035±160	+1.7
Skerjafjörður	AAR-7	1988	11,380±140	11,015±140	+2.3
Skerjafjörður	AAR-8	1988	11,130±120	10,765±120	+0.7
Skerjafjörður	AAR-9	1988	11,190±270	10,825±270	+2.0
Skerjafjörður	AAR-10	1988	11,170±330	10,805±330	+2.8
Skerjafjörður	AAR-11	1988	11,320±180	10,955±180	+2.3
Austurströnd	AAR-3B	1988	10,180±150	9,815±150	+1.5

Note: All corrections as in Table 1; AAR samples are from Andersen et al. (in press);
Lu- (Lund) samples from Håkansson (1987), U- from Olsson et al. (1965).

Lowlands therefore have been transgressed by the sea if they were deglaciated at all. In a search for late glacial sediments in southern Iceland, analyses of several samples have been made from localities far south of the Búði end moraines. All these determinations (Lu-2405, 2406, 2596, 2597, 2598 and 2600) yielded similar ages to those found in association with the Búði end moraines, i.e. 9,055 - 10,085 B.P. (Table 1).

It is noteworthy that no shells from the Alleröd are found in southern Iceland. Several explanations for this may be suggested:

i) The sea did not transgress southern Iceland during Alleröd and Younger Dryas time.

ii) All Alleröd sediments are buried by younger sediments.

iii) All shells of this period are dissolved in the calcium-poor environments of the sediments of southern Iceland.

iv) The southern Icelandic glacier totally covered the Southern Lowlands during the Younger Dryas Stadial and removed all older sediments.

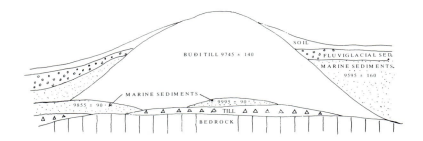

Fig.2 A schematic cross section of the Búði end moraines, the former type locality of the Younger Dryas Stadial in Iceland. Radiocarbon dating has now shown that it represents a Preboreal ice advance between 9,700-9,600 B.P.

Of these the first possibility is unlikely as discussed above. The second is also unlikely as the rivers of southern Iceland have, in many places, cut channels through the surface layers down to bedrock, but in spite of a careful search, no Alleröd sediments have been found. The third explanation also seems out of the question. The Preboreal shells of southern Iceland are mainly very well preserved, as are shells of Alleröd age in western Iceland, so there is no reason for shells in southern Iceland not to have been preserved. The final explanation appears to be the most reasonable, but it involves radical changes in former ideas concerning the late glacial deglaciation of southern Iceland. New aspects of the Quaternary geology of the Reykjavík area do however strongly support this hypothesis.

2. THE FOSSVOGUR LAYERS

The Fossvogur sedimentary layers in Reykjavík are a series of tills and marine sediments that rest on the interglacial Reykjavík Olivine Theoleiite Basalt Group. According to the definition of Einarsson (1968) and Geirsdóttir (1982) they can be found on the shores of the Fossvogur and Skerjafjörður, in the airport area and around the university (Fig.3).

A simplified cross section of the layers is shown in Fig.4 which defines the following broad stratigraphy:

i) The lowest section is a discontinuous layer of till, the lowermost Fossvogur till, resting on striated interglacial lava.

ii) The mid-section is a layer of shell-bearing marine sediment.

iii) The uppermost section consists of coarse, consolidated till or tillite, which was deposited by a glacier that covered the area after the deposition of the marine sediments.

Marine shell-bearing sediments are found in a few places on top of the Fossvogur layers, as at the Hotel Loftleiðir at Reykjavík airport. Shells are also found in consolidated sediments similar to the Fossvogur layers in Kópavogur and Seltjarnarnes. The Fossvogur layers and the overlying marine sediments reflect two glacial periods and two warmer periods with a transgression of the sea.

For decades the Fossvogur layers were thought to consist of formations from the Saalian glaciation (Nauthólsvíkurskeið), Eemian interglacial (Fossvogsskeið) and the Weichselian glaciation (Síðasta jökulskeið) (Einarsson 1968, 1978). It has also been a general view that south west Iceland was ice-free during the Younger Dryas. Einarsson and Albertsson (1988) state that up to one half of the country was ice-free during the Saurbær and Búði stages, i.e. Alleröd and Younger Dryas. Recently radiocarbon dates have been obtained from shell specimens from the Fossvogur layers using both conventional dating (Lu-2599) and accelerator mass spectrometry (AMS) (AAR-2B,2C,6-11 in Table 2). The datings reveal an age 100,000 years younger than expected, and the marine section of the layers was found to be around 11,000 B.P., or from the Late Weichselian, not from the Eemian interglacial.

Consequently the topmost Fossvogur till must be younger than 11,000 B.P., indicating that a Younger Dryas glacier overrode the Reykjavík area, reaching the Faxaflói bay. This glacier is called the SW-glacier in this paper. The radiocarbon dates from Hotel Loftleiðir (U-412 and U-415) and from Seltjarnarnes (AAR-3B) that are around 9,900 B.P. show that the sea transgressed the Reykjavík area immediately after the glaciers retreated. Raised beaches in Reykjavík must therefore date from this time.

3. THE ÁLFTANES ADVANCE AND THE OUTERMOST POSITION OF THE SW-GLACIER

The Álftanes end moraine is the outermost moraine ridge that is undoubtedly visible on dry land in the Reykjavík area. Einarsson (1968) assumed that it was formed by an ice advance that occurred during the Older Dryas stadial, between 12,000 - 11,800 B.P. The Icelandic name for the Older Dryas is Álftanesstig, named after this end moraine. This moraine can be traced across the Álftanes peninsula, past the presidential residency at Bessastaðir and into the sea at Bessastaðanes (Fig.3). According to the age of the Fossvogur layers and their indication of a glacial coverage of the Reykjavík area during the Younger Dryas, it seems clear that the Álftanes end moraine is not from the Older Dryas. Rather, it is connected to a series of end moraines at Kópavogur bay that are younger than the shell-bearing sediments at Kópavogur and Fossvogur.

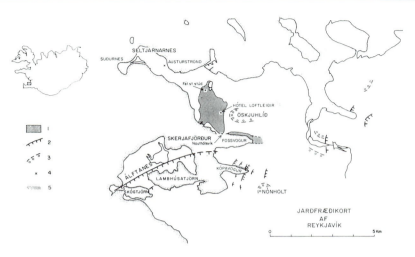

Fig.3 Geological map of the Reykjavík area. 1) Fossvogur layers. 2) End moraine. 3) Raised beach. 4) Dated sample. 5) The Kópavogur shell-bearing layers.

It seems logical to assume that the Álftanes end moraine is from the Younger Dryas or the Preboreal. It does not seem to mark the outermost stage of the SW-glacier during the Younger Dryas and it is situated further inland than the Fossvogur layers. The till in the topmost part of the Fossvogur layers in Skerjafjörður shows that the glacier moved further out after deposition of the Fossvogur marine sediments. The location of the outermost end moraines from the Younger Dryas glacier are still unknown, but it is likely that they lie on the floor of Faxaflói bay further to seaward. The Álftanes and Búði end moraines could be of the same age but direct dating of the Álftanes moraine is still needed.

The Weichselian glaciation ended at 10,000 B.P. and the Holocene began, but it is not known in this area of Iceland when the glaciers of the Younger Dryas retreated back from their outermost position. Their retreat seems to have been rapid, although not continuous. They slowed down and stopped at times, and even advanced again for a while. The Álftanes end moraine, and the group of moraines by Kópavogur are examples of such advances during the late Younger Dryas or early Preboreal.

Fig.4 Schematic cross section of the Fossvogur layers in Reykjavík, the former type locality of the Eemian interglacial in Iceland. The shell-bearing mid-section is about 11,000 B.P. or from the end of the Alleröd. The uppermost part is a Younger Dryas till and indicates a heavy Younger Dryas glaciation in S.W.Iceland. The Fossvogur layers are in some places covered by shell-bearing Preboreal marine sediments.

4. DEGLACIATION AND TRANSGRESSION IN THE ALLERÖD

The extension of glaciers in the Alleröd interstadial in Iceland is more or less unknown. Fig.5 shows the age distribution of the 60-70 available radiocarbon dates for Late Weichselian and Preboreal marine shells from Iceland. The ages range between 12,500 - 9,000 B.P. and may reflect the period of the Late Weichselian and early Holocene marine transgression in Iceland. The dates display a rather even distribution with one striking exception: between 10,700 - 10,200 B.P. there is a total gap in the diagram. Overall the figure represents a transgression during the Alleröd/Bölling periods when marine sediments were deposited inside the present shores of the country, but during the Younger Dryas no such sedimentation is known. This can be explained by a period of sudden ice advance and heavy glaciation during which glaciers filled most fjords and covered the majority of the lowlands of the country. At the end of the Younger Dryas a rapid disintegration of the glaciers took place that was followed by a new transgression of the sea and deposition of marine sediments inside the present coastline.

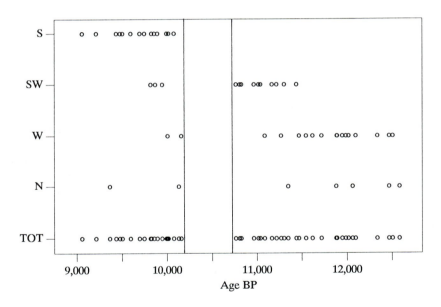

Fig.5 A diagram showing the time distribution of over 60 radiocarbon dates on Icelandic marine shells. In the upper part of the diagram the dates from S, SW, W and N Iceland are shown separately. In the lowermost part all the dates are summarised. The diagram shows rather an even distribution in the period 9,500 - 12,500 B.P. with the exception that in the 500 year interval between 10,200 - 10,700 B.P. no shells have been detected. This gap might reflect a heavy glaciation where the glaciers filled most fjords and prevented deposition of marine sediments inside the present strandlines.

Geological sections reflecting the Alleröd transgression and the Younger Dryas ice advance can be seen in several places in Iceland. In Borgarfjörður and Hvalfjörður Ingólfsson (1987) concluded that the ice extended more or less continuously beyond the present coastline between about 12,000 and 10,300 B.P., with two advances corresponding roughly to the Older and Younger Dryas. In Snæfellsnes, west Iceland, radiocarbon determinations have been performed on shells from Kaldárbru' (U-2227, Olsson, 1972) providing a date of 11,265±160 B.P. Here a glacier has overridden layers of shell-bearing marine sediments, reworked them and deposited a thick till layer containing the shells, hence the age of this advance must postdate 11,265 B.P. Pétursson (1986) studied the type locality for the Kópasker interstadial in northeast iceland, suggested by Einarsson (1968) to correlate with the Bölling interstadial in Scandinavia. Pétursson states that the Younger Dryas ice advance occurred in this area and dates its final deglaciation to 10,000 B.P.

The radiocarbon dates in Table 2 show that in the Reykjavik area the SW-glacier had retreated inside the present shoreline sometime before 11,215±150 B.P. (AAR-13) How far inland it went before its readvance into Faxaflói bay during the Younger Dryas is uncertain. Glacial striae show a NW movement in Reykjavík and an ice divide in the Bláfjöll mountains. The eastern part of this glacier covered Hellisheiði and the district of Ölfus S.Iceland. All these examples from south, south west, west and north Iceland point to an ice advance and the expansion of glaciers throughout Iceland in the Younger Dryas.

5. CONCLUSIONS

The size of the Icelandic Younger Dryas glacier has been greatly underestimated in the past. Several important geological formations have been redated and transposed in the geological time scale. Table 3 shows how the traditional Icelandic order must be rearranged. Consequently traditional Icelandic names for a number of the stadials and stages must be reviewed. The Búði stadial as a synonym for the Younger Dryas stadial, the Álftanes stadial as a synonym for the Older Dryas and the Fossvogur interglacial as a synonym for the Eemian interglacial cannot be used any more.

Table 3 Chronological rearrangement for Icelandic Quaternary.

North West Europe	Traditional Icelandic Order	New Icelandic Order
Preboreal		Búði, Álftanes
Younger Dryas	Búði	
Alleröd	Saurbær	Saurbær, Fossvogur
Older Dryas	Álftanes	
Bölling	Kópasker	Kópasker

ACKNOWLEDGEMENTS

I wish to thank Skúli Víkingsson for a critical reading of the manuscript and Páll Ingólfsson for correcting the English.

REFERENCES

Einarsson, Þ. (1968) Jarðfræði, saga bergs og lands. Heimskringla, Reykjavík.

Einarsson, Þ. (1978) Jarðfræði. Mál og mennig. Reykjavík.

Einarsson, Þ. (1987) Íslensk þjóðmenning I. Myndun og mótun Íslands', Bókaútgáfan Þjóðsaga, Reykjavík, 99-148.

Einarsson, Th. and Albertsson, K. (1988) 'The glacial history of Iceland during the past three million years', Philosophical Transactions of the Royal Society B318, 637-644.

Geirsdóttir, A. (1982) 'Die Fossvogur-Sedimente südlich von Reykajvík, Island', Mathematisch-Naturwissenschaftliche Fakultät der Christian-Albrechts Universität, Kiel.

Hjartarson, Á. (1980) 'Síðkvarteri jarðlagastaflinn í Reykjavík og nágrenni', Náttúrufræðingurinn 50, 108-117.

Hjartarson, Á. (1987) 'Tímatal í ísaldarsögu Reykjavíkur', Abstracts from a conference on the deglaciation of Iceland, Jarðfræðifélag Íslands, Reykjavík, 4-5.

Hjartarson, Á. (1988) Fossvogssyrpa. Deilir um reykvíska jarðfræði. Private edition.

Hjartarson, Á. (1989) 'The ages of the Fossvogur layers and the Álftanes end moraine SW-Iceland', Jökull 59.

Hjartarson, Á. and Ingólfsson, Ó. (1988) 'Preboreal glaciation of southern Iceland', Jökull 38, 1-16.

Håkansson, S. (1983) 'A reservoir age for the coastal waters of Iceland', Geologiska Föreningens i Stockholm Förhandlar105, 64-67.

Håkansson, S. (1986) 'University of Lund Radiocarbon Dates XIX', Radiocarbon 28, 1111-1132.

Håkansson, S. (1987) 'University of Lund Radiocarbon Dates XX', Radiocarbon 29, 353-379.

Ingólfsson, Ó. (1987) 'The Late Weichselin glacial geology of the Melabakkar - Ásbakkar coastal cliffs, Borgarfjörður, W-Iceland', Jökull 37, 57-81.

Kjartansson, G. (1943) Árnesingasaga. Árnesingafélagið, Reykjavík.

Kjartansson, G. (1958) Jarðmyndanir í Holtum og nágrenni. Rit Landbúnaðardeildar Atvinnudeildar Háskólans, 11.

Kjartansson, G. (1961) 'Glefsur úr jarðfræði (Árnessýsla, Grímsnes og Biskupstungur)', Árbók Ferðafélags Íslands, 17-29.

Kjartansson, G., Þórarinsson, S. and Einarsson, Þ. (1964) ^{14}C- aldursákvarðanir á sýnishornum varðandi íslenska kvarterjarðfræði', Náttúrufræðingurinn 34, 97-145.

Olsson, I.U. and Piyanuj, P. (1965) 'Uppsala natural radiocarbon measurements V.', Radiocarbon 7, 315-330.

Olsson, I.U., Klasson, M. and Abd-el-Mageed, A. (1972) 'Uppsala natural radiocarbon measurements XI', Radiocarbon 14, 247-271.

Pétursson, H. (1986) Kvartärgeologiske undersokelser på Vest-Melrakkaslétta, Nordost Island. Unpublished Cand. Real Thesis, University of Tromsö.

SEA-LEVEL CHANGE IN VESTFIRÐIR, NORTH WEST ICELAND

James D. Hansom,
Department of Geography,
University of Canterbury,
Christchurch,
New Zealand*

*Permanent address:
Department of Geography and Topographical Science
University of Glasgow
Glasgow UK

David J. Briggs,
Department of Geographical Sciences,
Polytechnic of Huddersfield,
Huddersfield UK

ABSTRACT. Recent observations in the Vestfirðir area of Iceland have revealed a wealth of raised marine features from ca 70 m a.s.l. to 1m a.s.l. that may reveal a different isostatic uplift pattern from that of the rest of Iceland. At 8.5 m a.s.l. at Hvítahlíð, microplankton-rich marine silts are capped by a peat layer with a radiocarbon age of 6,910 B.P. At Smáhamrar nearby, a suite of raised beaches between ca 70m a.s.l. and present sea level are older than 8,875 B.P. It appears that sea-level dropped rapidly from 70m to 1m some time before ca 10,000 B.P. However, a rise of sea-level to 8.5m occurred at about 9,000 B.P., and peat began to accumulate on beaches at about 8,800 B.P. The ensuing regression was temporarily halted at 6,900 B.P. by a high energy marine event, possibly caused by waves from the 7,000 B.P. Storegga landslide, which deposited a beach ridge full of marine taxa on top of freshwater peats at ca 6m a.s.l. As new regional deglaciation chronologies emerge for Iceland, there is a need to re-evaluate the relative sea-level histories of these regions.

1. INTRODUCTION

Interpretations of the patterns and rates of ice-recession and sea-level change in Iceland have traditionally been based on what might be called a Vatnajökull-centric view. These

79

J. K. Maizels and C. Caseldine (eds.), Environmental Change in Iceland: Past and Present, 79–91.
© 1991 Kluwer Academic Publishers. Printed in the Netherlands.

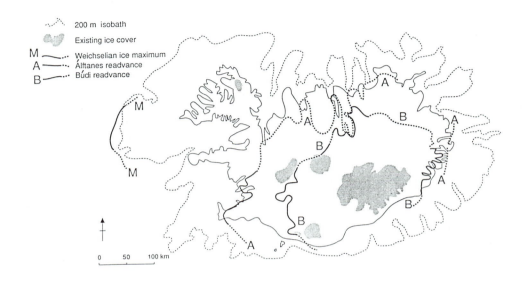

Fig. 1 The Weichselian ice-sheet in Iceland showing the positions of known end moraines (M) in west Iceland: elsewhere the ice-sheet probably extended beyond the 200m isobath. The limits of the Álftanes (A) and Búði (B) readvance stages are also shown (after Einarsson and Albertsson, 1988).

interpretations have assumed a process of isostatic adjustment during Late Weichselian and early Holocene times controlled largely by wastage of the Vatnajökull ice sheet, and resulting in a sequence of raised shorelines inferred to tilt more-or-less uniformly to the north and west (Einarsson and Albertsson, 1988). Such interpretations make little allowance for the effects of secondary ice caps in the northwest of the country in spite of suggestions that, during the Weichselian, Vestfirðir very probably supported an independent ice cap (Einarsson, 1968, 1978; John, 1977; Sigurvinsson, 1983). The most recent synthesis (Fig.1) is that of Einarsson and Albertsson (1988) and shows ice recession from its maximum extent at ca 18,000 B.P. to leave Vestfirðir essentially ice-free by the Álftanes stage at ca 12,000 B.P. with no readvance during the Buði stage (11-10,000 B.P., but recently suggested by Hjartarson and Ingólfsson (1988) to be Preboreal ca 9,500 B.P.).

These interpretations seem to contradict the evidence from the Vestfirðir peninsula itself. Apart from the existence, even today, of the remnant Drangajökull ice-cap (Fig. 1), there is an abundance of well-defined valley moraines indicating previously far more extensive ice limits (John, 1974, 1975; Hjort et al., 1985). In addition, a variety of marine features exist at a wide range of altitudes, indicating previous sea-levels. The alternative hypothesis may thus be advanced that this area experienced an essentially independent history of deglaciation, resulting in a regionally distinctive sequence of features related to sea-level change. This paper presents preliminary results from an ongoing study to test this hypothesis in the Vestfirðir peninsula.

2. THE STUDY AREA

The Vestfirðir peninsula covers an extensive area of land (ca 104 km^2), jutting northwestwards from the Icelandic coast towards Greenland. It is joined to the mainland by a narrow neck of land, ca 10 km wide, and with a maximum altitude of ca 300 m a.s.l. (Fig.1). Northwards, it rises to 957 m a.s.l., and is dominated by the Glama and Dranga plateaux, both of which occupy substantial areas at about 600-900 m a.s.l.

Raised shoreline features occur widely in the area. John (1974), for example, reported a marine terrace at 135m a.s.l. on the west coast, whilst other shorelines have been mapped at up to 26m a.s.l. in Hornstrandir (Hjort et al., 1985). On the other hand, a range of marine sediments have been identified closer to modern sea-level, from 8.5m a.s.l. at Reykhólar in Breiðafjörður to -2m at Bær in Hrútafjörður (John, 1974) (Fig.2). The age and stratigraphic relationships of these features are, however, far from clear. Few dates have been obtained from the deposits, although ^{14}C analysis of shells from the 8.5m sediments at Reykhólar gave a date of 10,460 B.P., and may thus indicate the existence of Búði (Younger Dryas) ice close to sea-level in southern Vestfirðir at this time (John, 1974). Similarly, at Asmundarnes in Bjarnafjörður, shells in glacio-marine clay at 1m a.s.l. yielded a date of 9,930 B.P. The most widely recognised feature would nevertheless appear to be a shell-rich beach comprising abundant remains of *Nucella* sp. at ca 4 m a.s.l. dated to ca 4,000 BP (John, 1974).

Given the hypothesis that Vestfirðir represented an independent centre of deglaciation during the Lateglacial and Holocene, attention in this study was focused on evidence for sea-level change in the southern area of the peninsula. This, it can be postulated, might represent a hinge-line between the regional isostatic system in Vestfirðir, and the main Vatnajökull-related system to the south. To this end, a reconnaissance survey was undertaken in 1985 to assess the distribution of marine shoreline features in the area, to select key sites for preliminary analysis and identify sites for further investigation.

3. RESULTS

The reconnaissance survey confirmed the presence of a wide range of raised shoreline features in the Vestfirðir peninsula. Examples of the *Nucella* beach were found at a number of sites, including Bær, Hvítahlíð, Broddanes, Smáhamrar, Asgardsgrund and Ennisdalur (Fig.2). Higher beaches and marine deposits were also identified, at (i) Smáhamrar, (ii)

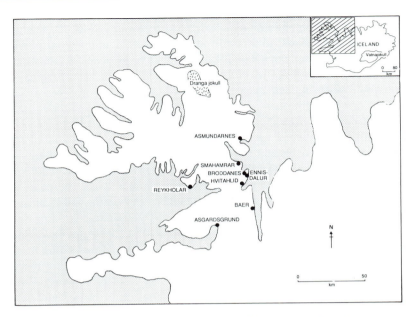

Fig. 2 Location of the sites in Vestfirðir mentioned in the text.

Asgardsgrund, and (iii) Hvítahlíð. Results of analyses from these latter sites are summarised below:

(i) Smáhamrar. Smáhamrar, in Steingrimsfjörður, provides an exceptionally fine example of raised shoreline features. Here, an unbroken series of ca 30 beach ridges rises from sea-level to 70 m a.s.l. (Fig. 3). Each ridge consists of 1-2 m of moderate-coarse beach gravels, with the intervening swales being filled with peat (Fig.4). Samples were taken for analysis from a number of these swales: that from a swale at 40 m a.s.l. was submitted for ^{14}C analysis but yielded a relatively young date of 8,875±50 B.P. (GrN-15846). This may indicate a considerable interval between ridge formation, beach abandonment and peat development, but at least fixes a 'no younger than' date for the 40 m sea-level.

(ii) Asgardsgrund. A thick deposit (more than 7.5 m) of lagoonal silts and sands occurs as a wide beach and sand-dune capped marine terrace at Asgardsgrund in Hvamms-fjörður (Fig. 3). The surface lies at about 6 m a.s.l. and cores were collected to a depth of -1.4 m a.s.l. and used for sedimentological and palynological analyses (Burns, 1990). A shell-bearing horizon containing the marine species *Nucella* occurs at ca 4.1 m a.s.l. within the silts (Fig.3). Results of the sedimentological analyses showed a broadly cyclical change in depositional conditions, possibly reflecting minor changes in the energy environment and sediment supply. Recovery of pollen was disappointingly limited, but in general indicated an arctic-alpine environment dominated by Gramineae and Cyperaceae. Single

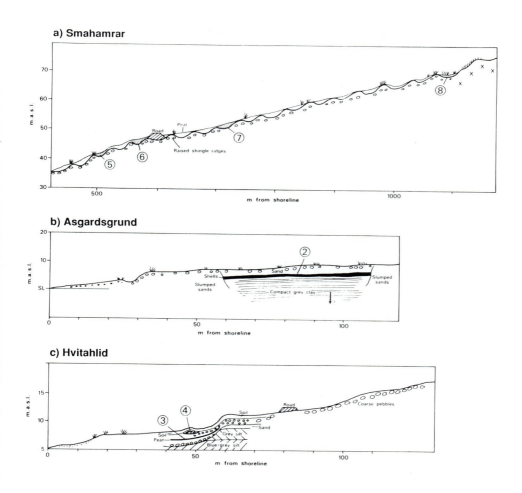

Fig.3 Section diagrams from Smáhamrar, Asgardsgrund and Hvítahlíð. Horizontal and vertical scales vary.

grains of *Pinus* (together with *Salix*), however, occur at two levels in the core. More significantly, *Plantago lanceolata* was identified in three samples from the cores, at -1.2, -1.1 and 0.0 m a.s.l. As will be mentioned below, this species has previously been regarded in Færoe and Iceland as evidence of human occupation (Johansen, 1986; Hallsdóttir, 1987). Its occurrence at this site may thus imply a very late (post-settlement) date for the deposits

Fig.4 The raised beach ridges and intervening peat-filled swales at 40m a.s.l. in Smáhamrar are part of an unbroken series of ridges from 70m a.s.l. to ca 1m a.s.l.

at Asgardsgrund. It may alternatively demonstrate the much earlier occurrence of *Plantago lanceolata* in Iceland than has been believed until now, and this is supported by the stratigraphic position of the pollen finds below the shells of the *Nucella* beach. These shells are as yet undated at Asgardsgrund but are assumed, on the basis of radiocarbon dating elsewhere, to have been deposited about 4,000 B.P.

(iii) Hvítahlíð. On the north shore of Bitrufjörður, at Hvítahlíð, a 3 m deep stream section cut into silts of at least 8 m thickness (they are also exposed downstream close to sea-level) is truncated at its landward and upper surface by an erosional scarp which is itself draped by a sand and gravel deposit (Fig.3). The deposit contains a 10 cm thick layer of peat which slopes seawards from 8.5 m a.s.l. to about 6.5 m a.s.l. Stratigraphically above the peat layer is a distinct beach ridge composed of coarse marine sands, gravels and cobbles. Contained within this ridge is a 10 cm thick layer of peat whose upper surface gives way to gravels and cobbles capped by soil (Fig.5). The ridge constitutes the small upper section and is shown as a gap on the pollen stratigraphy (Fig.6). Incremental samples were taken through both parts of the entire section for sediment analysis, pollen and microfauna. The pollen diagram from these samples suggests a sedge-grass tundra, with some pine, dwarf willow and birch, juniper, heaths and heathers (Fig 6a). The diagram can be divided into two: a basal assemblage with grasses around 20% and sedges thereby reduced, and an upper assemblage heavily dominated by sedges and a considerable incidence of spores. As with Asgardsgrund, the pollen from Hvítahlíð suggests an essentially arctic-alpine environment throughout the period represented by the samples. It also shows the early occurence, at 3.5 m depth, within the silts, of *Plantago lanceolata* pollen, as distinct from *Plantago maritima* which also occurred in the samples.

Fig. 5 Marine silts at ca. 8.5 m a.s.l. at Hvítahlíð are capped by two phases of freshwater peat deposition indicating abandonment by the sea prior to ca 8,800 B.P. Stratigraphically above these peats is a storm ridge composed of cobbles and sand.

Examination of the section for microfauna revealed a rich, decidedly boreal, faunal assemblage and since diatoms were largely absent from the sediments, attention was focused upon the dinoflagellates or 'microplankton'. These could be divided into three assemblages (Fig.6a). The basal section, corresponding to the silt horizon, yielded predominantly marine taxa derived almost exclusively from a shallow lagoonal or estuarine environment, with an absence of 'open marine' taxa and only limited amounts of non-marine species from freshwater nearby. This was replaced at about the stratigraphic level of the lower peat by smaller amounts of an almost 100% freshwater assemblage. The uppermost assemblage above the upper peat shows a return to marine deposition with very high values of the marine boreal species *Peridinium faroense*. Thus the microplankton reveals a sequence of environmental changes not shown by the pollen. The microplankton indicate a shallow lagoonal marine environment up to 8 m a.s.l. followed by two periods of freshwater peat development that were truncated by a return to marine conditions when the lower beach ridge was deposited. Fortunately dating of the two peat layers allows these events to be bracketed. The regression which led to the erosional cliffing (Fig.5) of the pre-

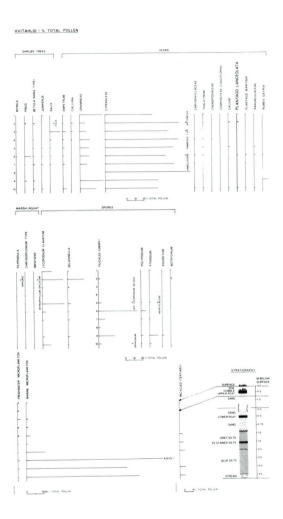

Fig. 6a Pollen and microplankton diagram from Hvítahlíð, pollen and summary micro-plankton. Note that the upper section of the stratigraphy refers to the storm ridge sediments and the upper peat.

existing marine silts occurred prior to the development of the lower peats, the lower 2 cm of which are dated to 8,830±60 B.P. (GrN-15844). However, the regression was temporarily halted by a high energy marine event which deposited a beach ridge full of marine taxa

HVITAHLID - MICROPLANKTON

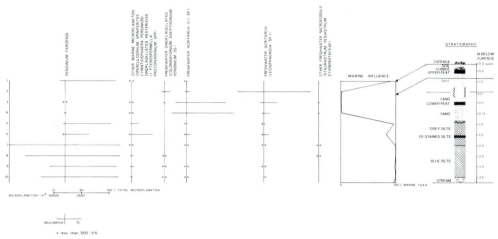

Fig. 6b Microplankton diagram from Hvítahlíð.

at ca 6 m a.s.l. The upper 2 cm of the peat buried by the beach ridge yielded a radiocarbon
age of 6,910±100 B.P. (GrN-15843).

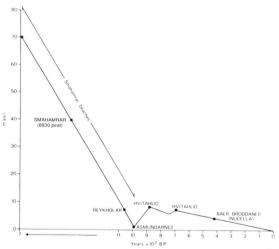

Fig. 7 Relative sea-level changes in southeastern Vestfirðir. How far sea-level fell
below ca 1 m a.s.l. in the period ca 10 - 9,000 B.P. is not known.   Note the anomal-
ously 'young' date from 40 m a.s.l. at Smáhamrar. Shell dates are unadjusted for the
reservoir age of the coastal waters of Iceland (Håkansson, 1983).

4. DISCUSSION

The limited information presented above is used only to provide a first approximation of sea-level change in Vestfirðir since the data are, as yet, very incomplete and need verification at many more sites and altitudes than reported here.

4.1 18,000 - 9,000 B.P.

At the Weichselian maximum, the whole of Iceland was glacier-covered to well beyond the present coastline (Fig. 1) and sea-level was low (Ólafsdóttir, 1975; Andersen, 1981). At this time, and for some time afterwards, the Vestfirðir peninsula was probably covered by an independent ice cap (Einarsson, 1968, 1978; John, 1977; Sigurvinsson, 1983) centred on the Dranga and Glama plateaux. However, by about 12,000 B.P., ice retreat and sea-level rise led to the formation of marine limits outwith the ice-filled troughs at 135 m a.s.l. in western Vestfirðir (John, 1975) and at least 70 m a.s.l. in southeastern Vestfirðir at Smáhamrar. The unbroken nature of the beach sequence at Smáhamrar suggests that the subsequent sea-level fall from 70 m a.s.l. to at least 1 m a.s.l. was rapid and continuous; probably in association with ice recession from the outer reaches of the fjords (Fig.7). This may partly explain why Hjort et al. (1985) found the pre-Younger Dryas marine limit in Hornstrandir, north Vestfirðir, to be no higher than 26 m a.s.l. followed by a fall to below 1 m a.s.l. At Reykhólar, Breidafjörður, the falling sea-level lay at 8.5 m a.s.l. at 10,460 B.P. as indicated by the radiocarbon age of shells found in marine sediments (John, 1975). The age of the shells found in similar sediments at Asmundarnes, Bjarnarfjörður show that sea level stood at 1 m a.s.l. by 9,930 B.P. (John, 1975). At Bær, Hrútafjörður, undated freshwater peats at -2 m found beneath a well-formed *Nucella* beach supports this possibility as do the results of research activity outside Vestfirðir. For example, Thors (pers. comm.) shows that rapid isostatic recovery in Eyjafjörður, North Iceland, between 12,000 B.P. and ca 10,500 B.P. led to a sea-level fall from 20 m a.s.l. to -30 m a.s.l. as shown by submarine terraces; similar submerged terraces occur at -10 m to -35 m in Hvalfjörður, southwest Iceland. In northeast Iceland at Melrakkaslétta, Pétursson suggests a fall in sea-level from 50-60 m a.s.l. at ca 13,000 B.P. to 10 m a.s.l. at 11,200 B.P. and 0 m by 9,000 B.P. In southeast Iceland, Jónsson (1957) demonstrated an early Holocene sea-level of -7 m at Dryholæy whilst Boulton (pers. comm.) uses erosional planing of the continental shelf to imply an early Holocene sea-level of -60 to -65 m. Whereas there appears to be general agreement for a rapid sea-level fall in Iceland between 12,000 B.P. and the early Holocene, there is little agreement on the regional amount of that sea-level fall. In spite of the fact that there is an obvious lack of dated altitudinal sea-level information, there is a clear suggestion above of very different sea-level and isostatic recovery histories between the regions of Iceland. This is at variance with the 'Vatnajökull-centric' view implying an isostatic gradient sloping southeastwards with the highest shorelines occurring about 100 m a.s.l. in south Iceland and 30-60 m elsewhere (Einarsson and Albertsson, 1988).

4.2 9,000 - 4,000 B.P.

Immediately prior to 9,000 B.P., sea-level was about -2 m a.s.l. or possibly more in Vestfirðir. However, the occurrence of freshwater peats dated to 8,830 B.P. on top of shallow-water marine silts at Hvítahlíð suggests that sea-level rose rapidly from below sea-level to about 8.5 m a.s.l. Unfortunately the Hvítahlíð silts are separated from the overlying peat by an erosional unconformity and thus could be either part of a transgression just prior to 8,830 B.P. or the eroded remnants of earlier deposition as sea-level fell from 70 m a.s.l. to -2 m a.s.l. Two lines of evidence support the first interpretation. The rapidity of sea-level fall from 70 m a.s.l. makes it difficult to envisage the existence of conditions conducive to continued shallow water lagoonal deposition within Bitrufjörður at that time. In addition the resultant microplankton asemblage would then contain a substantial deepwater component, yet none exists. Secondly, the same sequence of lagoonal silts occurs from below sea-level to ca 4.1 m a.s.l. at Asgardsgrund incorporating, as its uppermost horizon, a layer of shells high in *Nucella* spp. These are as yet undated but are probably of *Nucella* beach (ca 4,000 B.P.) age. Since no unconformity exists between the shell horizon and the underlying silts, they may form part of an incomplete shallow-water depositional unit which formed during the 8.5 m transgression. The erosional cliffing of the Hvítahlíð silts contrasts with the unbroken Asgradsgrund silts but may be explained in terms of greater relative exposure to waves at Hvítahlíð. The marine abandonment of the Hvítahlíð silts and the subsequent freshwater peat development is graphically shown by the fluctuation in the microplankton assemblage from almost 100% marine taxa within the silts to almost 100% freshwater taxa immediately above the silt (Fig.6b).Extensive peat development at this time may well have extended onto higher unvegetated surfaces; the 'young' date for peat held within the ridge swales at Smáhamrar (8,875 B.P.) suggests either a time-lag between beach abandonment and peat development or a dating error. During this time period, the pollen of *Plantago lanceolata* is present, though not abundant, in silts from both Hvítahlíð and Asgardsgrund. In both of these cores, the ocurrence of *Plantago lanceolata* clearly substantially pre-dates the Landnám settlement period and casts doubt upon the continued use of this species as an anthropogenic indicator heralding the arrival of man (Johansen, 1986; Hansom and Briggs, 1990).

Regression from the 8.5 m sea-level at Hvítahlíð has continued until the present time and has been interrupted by two events (Fig. 7). The first event appears to be a high energy marine transgression at 6,900 B.P., suggested by freshwater peats buried beneath a distinct storm ridge full of cobbles, sand and containing 100% marine taxa. The event was probably very short-lived and evidence for its occurrence is unknown elsewhere in Iceland. However, recent work has documented the occurrence at ca 7,000 B.P. of coarse marine sand full of deepwater fauna contained within peat and lagoonal clays at ca 200 coastal sites in eastern and northern Scotland (Dawson et al., 1988). The deposit is thought to represent a single high-energy marine event, and in all probability the result of a tsunami generated by the failure off the Norwegian coast at ca 7,000 B.P. of the vast Storegga submarine landslide. Just as the eastern and northern coast of Scotland lies in the path of any wave generated by this event, the eastern coasts of Iceland and of Vestfirðir might be expected to be affected. Whilst it is tempting to interpret the Hvítahlíð storm ridge as the Icelandic representation of this event, such an assertion must await the discovery of more widespread evidence. The second event which temporarily halted continued regression was a small, though widespread, transgression which deposited the widely-found *Nucella* beach at ca 4 m a.s.l. about 4,000 B.P. (John, 1974) (Fig.7).

5. CONCLUSION

The evidence for sea-level change presented above and summarised in Fig. 7 indicates that Vestfirðir, probably on account of a different deglaciation history, has experienced a different sea-level history from other areas of Iceland. It seems reasonably certain that sea-level fall from ca 136 m a.s.l. in western Vestfirðir and from at least 70 m a.s.l. in southeastern Vestfirðir to at least -2 m a.sl. was very rapid indeed between 12,000 B.P. and 9,000 B.P. However, the characteristics and amount of sea-level drop in Vestfirðir differs from that experienced in north and southwest Iceland, northeast Iceland, southeast Iceland and from that suggested by Einarsson and Albertsson (1988) for the whole of Iceland.

This sea-level fall was followed shortly before 8,800 B.P. by a transgression to ca 8.5 m a.s.l. resulting in deposition of lagoonal silts at two locations in Vestfirðir. Regression has dominated in Vestfirðir since then and has been temporarily reversed only twice. The first reversal occurred shortly before ca 6,900 B.P., when a high energy marine event (possibly related to the waves created by the Storegga submarine landslide of ca 7,000 B.P.), deposited a distinct cobble ridge at ca 6 m a.sl. at Hvítahlíð. The second transgression to ca 4 m a.s.l. is widely represented in Vestfirðir as the *Nucella* beach of ca 4,000 B.P.

As distinctive regional deglaciation chronologies begin to emerge (Hjartarson and Ingólfsson, 1988), there seems a clear need to establish the accompanying regional sea-level signature as distinct from a view based on one dominant uplift centre. This need is most pressing in Vestfirðir where separation from the mainland ice mass was probably effected early in the deglaciation sequence and thus created conditions conducive to an independent isostatic recovery history.

ACKNOWLEDGMENTS

This research was supported by a grant from the University of Sheffield when both authors were members of staff at the Department of Geography, University of Sheffield. Radiometric dating was carried out by Professor W.G. Mook of C.I.O. in Gröningen and the pollen analysis was expertly performed by Dr. C.O.Hunt. Anna Moloney of the Department of Geography, University of Canterbury, efficiently word-processed the final draft which was produced whilst Dr. J.D.Hansom held the post of visiting Lecturer at the Department of Geography, University of Canterbury, New Zealand.

REFERENCES

Andersen, B.G. (1981) 'Late Weichselian Ice Sheets in Eurasia and Greenland,' in G.H. Denton and T.J. Hughes (eds), The Last Great Ice Sheets, Wiley, 3-65.

Burns, K.M. (1990) Analysis of environmental change in Vestfirðir, Iceland, Unpublished B.Sc. dissertation. Department of Human Ecology, Polytechnic of Huddersfield.

Dawson, A.G., Long, D. and Smith, D.E. (1988) 'The Storeggan slide: evidence from eastern Scotland for a possible tsunami', Marine Geology 82, 271-276.

Einarsson, Þ. 1968 Jarðfræði. Saga bergs og lands, Mál og Menning, Reykjavík.

Einarsson, Þ. 1978 Jarðfræði. Mál og Menning, Reykjavík.

Einarsson, Th. and Albertsson, K.J. (1988) 'The glacial history of Iceland during the past three million years', Philosophical Transactions of the Royal Society of London 318, 637-644.

Håkansson, S. (1983) 'A reservoir age for the coastal waters of Iceland', Föreningens i Stockholm Förhandlingar 105, 65-68.

Hallsdóttir, M. (1987) Pollen analytical studies of human influence on vegetation in relation to the Landnám Tephra Layer in Southwest Iceland, Lundqua thesis 18.

Hansom, J.D. and Briggs, D.J. (1990) 'Pre-Landnám Plantago lanceolata in Iceland', Fróðskaparrit, in press.

Hjartarson, Á. and Ingólfsson, Ó. (1988) 'Preboreal Glaciation of Southern Iceland', Jökull 38, 1-13.

Hjort, C., Ingólfsson, Ó. and Norðdahl, H. (1985) 'Late Quaternary geology and glacial history of Hornstrandir, Northwest Iceland: A reconnaissance study', Jökull 35, 9-28.

Johanssen, J. (1986) 'Plantago lanceolata in the Faroe Islands and its significance as an indicator of prehistoric settlement', Fróðskaparrit 34-35, 68-75.

John, B.S. (1974) 'Northwest Iceland reconnaissance 1973 (Durham University Vestfirðir Project)', Department of Geography Durham University Special Publication, 54pp.

John, B.S. (1975) 'Durham University Vestfirðir Project, 1975', Department of Geography Durham University Special Publication.

John, B.S. (1977) 'The extent of Weichselian Ice in Iceland', QRA symposium abstract - The last glaciation in the northern hemisphere, Durham.

Jónsson, J. (1957) 'Notes on changes of sea-level of Iceland', Geografiska Annaler 34, 143-212.

Ólafsdóttir, Þ. (1975) 'Jökulgarður á sjávarbotni út af Breiðafirði', Náttúrufræðingurinn 45, 31-36.

Sigurvinsson, J.R. (1983) 'Weichselian glacial lake deposits in the highlands of North-western Iceland', Jökull 33, 99-109.

EVIDENCE FROM SOUTH WEST ICELAND OF LOW SEA LEVEL IN EARLY FLANDRIAN TIMES

Kjartan Thors and Guðrún Helgadóttir
Marine Research Institute
Skúlagata 4
121 Reykjavík
Iceland

ABSTRACT. An erosional unconformity and drowned coastal features observed in seismic profiles from south west Iceland indicate a former sea level 30-35 metres below present. Radiocarbon dates of marine shells and freshwater peat combined with field evidence on land indicate that the low sea level occurred after a very rapid regression in early Flandrian times. We think that an unusually rapid postglacial isostatic rebound due to low vicosity of the Icelandic crust and upper mantle was responsible for this anomalous sea level history.

1. INTRODUCTION

A sequence of submerged littoral features identified in seismic profiles from Eyjafjörður, North Iceland, led Thors and Boulton (in press) to define a sea level of -40 m postdating Late Weichselian glaciation of the area, and a subsequent transgression to present sea level. The inferred timing of these events was based i.a. on imperfect knowledge of the history of glacier changes at the end of the Weichselian glaciation in Eyjafjörður.

The following account describes an erosional unconformity and inferred coastal deposits 30-35 m below sea level in Faxaflói, Kollafjörður, and Hvalfjörður, South West Iceland. These features are thought to represent a regression of sea level analogous to that in Eyjafjörður. In this area Late Weichselian events on land are better documented than in Eyjafjörður and the time window for the drop in sea level is therefore more easily definable. Furthermore, the evidence of low Late Weichselian sea level in South West Iceland suggests that the processes which led to the regression were not limited to central northern Iceland.

93

J. K. Maizels and C. Caseldine (eds.), Environmental Change in Iceland: Past and Present, 93–104.
© 1991 *Kluwer Academic Publishers. Printed in the Netherlands.*

1.1. The Study Area

This study is based on seismic profiles from Faxaflói, Kollafjörður, and Hvalfjörður which were obtained in three separate surveys by the use of an EG&G Uniboom system and Racal Micro-Fix radiolocation. The surveys were made in 1983 (Thors, 1983), 1985 (Thors and Helgadóttir, 1986), and 1988 (Thors and Helgadóttir, 1990).

Faxaflói is a large, shallow bay open to west (Fig. 1). The Syðra-Hraun shoal occupies a large portion of the southern part of the bay. A channel defined by the 40 m isobath separates this shallow ground from the adjacent land to the east and north and thus from Hvalfjörður, a 35 km long fjord entering the bay from the east. Hvalfjörður is mostly shallow but in Galtarvíkurdjúp, a depression half way along the fjord, a depth of 84 m has been recorded. Kollafjörður is a smaller re-entrant in the Faxaflói shoreline. It is shallow with depths ranging down to 35 metres.

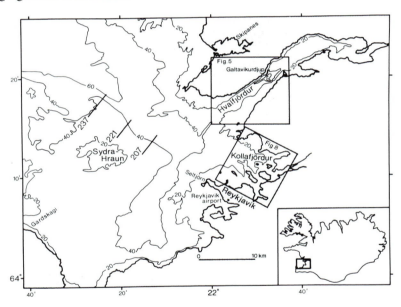

Fig. 1 Index map of the southern part of Faxaflói showing Hvalfjörður and Kollafjörður and the location of the three profiles across the terrace north of Syðra-Hraun. Isobaths at 20, 40, and 60 metres shown.

2. THE EVIDENCE

2.1. The Hvalfjörður unconformity

The seismic survey of Hvalfjörður extended from the mouth of the fjord to the overdeepened

Galtavíkurdjúp. The morphology of this stretch is such that extensive shallows border the southern coastline and the main channel of the fjord is situated in the northern part. The channel is relatively flat-bottomed with depths of 30-35 metres. A rock threshold is exposed near the mouth of the fjord at about 30 metres below sea level. From there, bedrock falls gently to reach -110 metres in Galtavíkurdjúp. Sediments take up the space between bedrock and sea bottom. A seismic refraction survey (Guðmundsson, 1989) revealed a wide threshold of consolidated sediments to the west of the Galtavíkurdjúp depression, probably of morainic origin.

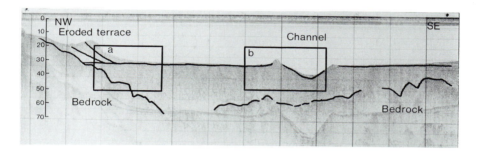

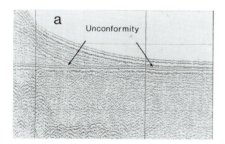

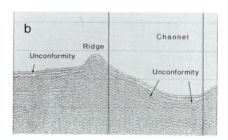

Fig 2 Profile across Hvalfjörður showing an erosion surface at approximately 33 metres depth. The surface is almost perfectly horizontal but for a channel reaching below 40 metres and a low ridge running parallel to the channel. Under this surface an approximately 30 m thick succession of sediments without apparent internal layering lies on top of bedrock. A gravel terrace merging into a thin sediment blanket overlies the unconformity. Length of section 2700 m. For location see Fig.5.

Three seismic reflection profiles across the main channel in Hvalfjörður are illustrated in Figs. 2, 3, and 4. A prominent erosional unconformity is observed in the sedimentary succession. A thin blanket of sediment overlies this surface, but grades into thick and in places extensive gravel terraces on both sides of the fjord. The unconformity may be traced in all profiles from the fjord and in Fig. 5 depth contours have been drawn on this surface. The contours outline a relatively flat surface at 30-35 metres depth into which is carved a narrow channel reaching below 40 metres. The 35 metre contour traces a bifurcation of the channel with a northern, subsidiary arm dying out to the west.

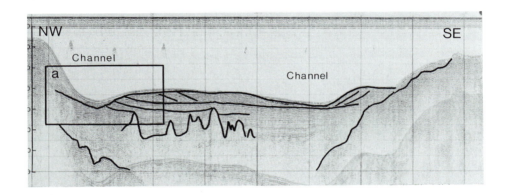

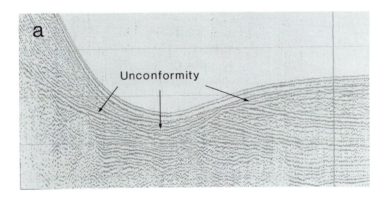

Fig. 3 The erosional unconformity in Hvalfjörður is well brought out in this section where the underlying strata are seen to be truncated (detail a). Two channels appear in this profile, slightly shallower than in the previous Figure. Length of section 2000 m. For location see Fig.5.

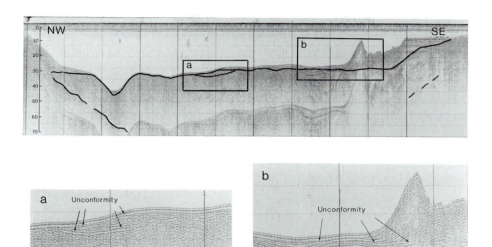

Fig. 4 The channel is on the northern side of the fjord in this section. In the southern half the erosion surface is undulating and rises higher than elsewhere in the fjord. A lens of sediment (detail a) is thought to be a drowned beach. The large gravel terrace on the SE side has been hollowed out by dredging. Length of section 2600 m. For location see Fig.5.

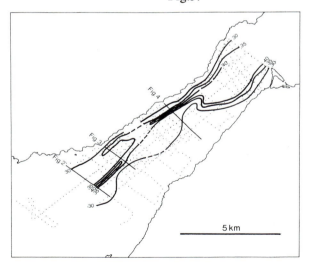

Fig. 5 Depth of unconformity in Hvalfjörður. Position of profiles in Figs. 2, 3, and 4 indicated.

We believe that the unconformity and overlying sediments imply erosion at lower sea level and a later rise to present level. The channel helps formulate a guess as to how low sea level stood during the formation of the erosion surface. The size and shape of the channel suggests that water currents were responsible for its sculpturing and maintenance. Although wind-induced currents were no doubt important modifiers of sediments at low sea level, it is suggested that tidal currents and freshwater runoff combined to form currents strong enough to carve the channel in Hvalfjörður and prevent sedimentation in it. The bifurcation of the channel allows the speculation that the flood current entered the fjord along the southern arm of the channel, whereas part of the ebb current retreated by the northern arm. This scenario would place mean sea level between -30 and -35 metres with high tide near -30 and low tide near -35 metres. In a few profiles from Hvalfjörður a lenticular body of sediment was observed associated with the unconformity (Fig. 4). This feature is thought to be a drowned beach and it occurs at 30-35 metres depth.

2.2. Kollafjörður

Kollafjörður is situated between Reykjavík in the south and the Kjalarnes headland in the north. Several islands rise above sea level in the southern part of the area. The landscape here is heavily influenced by glacier erosion and the islands sit on ridges between glacially carved troughs (Thors, 1983). The troughs carry thick accumulations of sediments.

The unconformity observed in adjacent Hvalfjörður also appears here (Figs. 6,7). It is quite flat but in contrast to Hvalfjörður, slopes gently seawards from land (Fig. 8). In the western part of the area the unconformity descends below -30 m and the -35 m contour describes the end of a channel in the westernmost profiles.

If, as seems likely, sea level fell as low here as suggested for Hvalfjörður, then most of the Kollafjörður area was dry land during the period of low sea level. In some of the profiles there are indications of shallow channels in the unconformity suggesting that rivers flowed down this surface. The profiles are not of sufficiently high quality to trace the path of the channels but it is possible that the channel at -35m represents the point of entry of surface runoff to the sea.

2.3. Faxaflói

The Syðra-Hraun shoal in southern Faxaflói is made up of consolidated sedimentary rock, possibly of glacial origin (Thors, 1978). The highest point of the shoal is at 10 m depth. The rough surface of the shoal is replaced at 25-30 m depth by a flat, fringing platform, which occurs to the north, east and south of Syðra-Hraun.

The three profiles shown in Fig. 9 are taken across the platform to the north of Syðra-Hraun and show the platform here to be a wide terrace of cross-bedded sediment. This terrace is large; its width is of the order of 3.5 km and it occurs in all profiles along the 20 km stretch to the north of the shoal covered by the survey. The lens of sediments making up the terrace is over 30 m thick. Internal stratification suggests truncation of foresets in the inner (older) part of the lens. A side-scan sonar survey (Thors, 1978) revealed sand patches

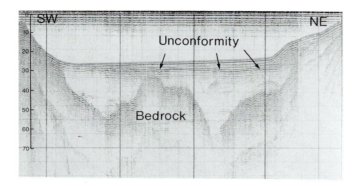

Fig. 6 Profile across Kollafjörður from Lundey to Brimnes showing unconformity overlain by modern muds and gravel terrace. For location see Fig. 8.

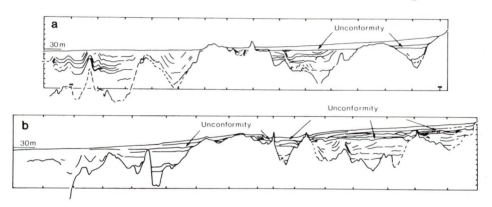

Fig. 7 Interpreted profiles along two glacial troughs in the Kollafjörður area. Unconformity is seen to slope gently away from land. Overlying younger sediments wedge out to the west. For location see Fig. 8.

on the surface of this terrace, and their orientation and shape suggested transport to the east. The sand patches are visible in the seismic profiles (Fig. 9).

The volume of the terrace is of the order of 2 km³, so its origin requires a large source of sediment. It is suggested here that the terrace started forming at a time of falling sea level. Lower sea level allowed waves entering Faxaflói bay to break down the relatively soft rocks of Syðra-Hraun and deposit the debris on the lee side of the shoal. As sea level fell, the first-deposited sediments were reworked and added to the prograding front of the terrace. As sea level rose again less sediment became available from the Syðra-Hraun shoal and progradation of the terrace all but ceased. The sand presently in transit across the terrace is pre-

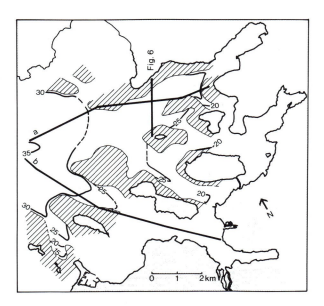

Fig. 8. Map of Kollafjörður. The profiles illustrated in Figs.6 and 7 a and b are indicated. Depth contours to the Kollafjörður unconformity are shown. The unconformity
abuts against bedrock highs (shaded).

dominantly biogenic in composition and most probably owes its origin to the local shell
fauna. The sand is utilised as a source of lime.

The elevation of the surface of the terrace should give an indication of the magnitude
of the fall in sea level. Fig. 9 shows that the westernmost part of the terrace lies below -35
metres whereas in the two eastern profiles the terrace rises above -30 m near the shoal. This
could stem from differential erosion as the eastern profiles are more sheltered by Syðra-
Hraun proper. A minimum sea level of -30 to -35 metres as in Hvalfjörður is feasible, but
would mean that the part of the terrace closest to Syðra-Hraun was above sea level for some
time.

3. DISCUSSION

The evidence presented here for low sea level in South West Iceland is circumstantial. It
remains to be established whether the unconformities in Hvalfjörður and Kollafjörður and
the terrace in Faxaflói are coeval and when they were formed. Existing knowledge of the
area and published radiocarbon dates do, however, allow us to speculate on the conditions
that led to lowered sea level, and its timing.

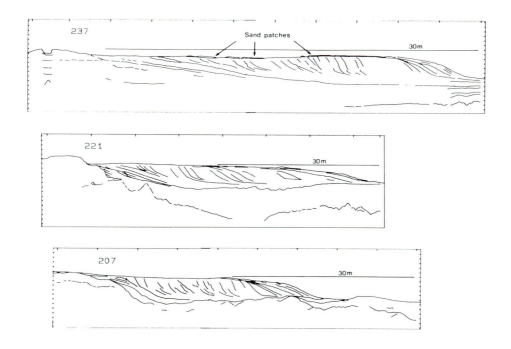

Fig. 9. Profiles across terrace to the north of Syðra-Hraun in Faxaflói which is to the left. The truncation of the foresets is particularly clear in the inner part of the terrace. Sand patches on the terrace are visible in places, but not always resolved due to thinness (for location see Fig.1).

According to Ingólfsson (1988) glacial ice was still present in Hvalfjörður during the local Skípanes, Latrar and Skorholtsmelar events correlated with the Older Dryas stadial, the Alleröd interstadial, and the Younger Dryas stadial in northwest Europe. Sea level was high during all this period. The Skorholtsmelar glacier started to retreat about 10,300 B.P., and Ingólfsson associated a sea level of 60-70 m with the retreat. Similarly, Hjartarson (1989) suggests that the Reykjavík (Kollafjörður) area was covered by ice during the major part of the Younger Dryas chronozone and that a raised beach at 43 m a. s.l. in Reykjavík is of early Preboreal age.

It is not likely that the flat and smooth unconformities in Hvalfjörður and Kollafjörður would have survived the advances and retreats of late Weichselian glaciers and consequently they are thought to postdate glaciation. There are several indicators to help us follow sea level down to the -30 to -35 metres suggested by our data. These are plotted on Fig. 10

and numbered as follows:

1. Ingólfsson (op.cit.) identified a regional marine limit at 60-70 m on the northern side of Hvalfjörður, and associated it with the final deglaciation of the lowlands which he suggests started about 10,300 B.P.

2. A marine mollusc (Lu-2197) from 11-13 m a.s.l. at Skípanes was dated to 10,005±190 B.P. (Ingólfsson op. cit.).

3. Samples of *Pecten islandicus* collected from Reykjavík airport (13 m a.s.l.) by Guðmundur Kjartansson in 1963 were dated in Uppsala (Einarsson, 1964, Olsson and Piyanuj, 1965). When corrected for reservoir age, the dates of 9945±1260 and 9865±1190 B.P. were obtained (Hjartarson, 1989). Here, 9865 B.P. is used.

4. A suction dredger, ms. Sandey, working northeast of Syðra-Hraun in Faxaflói in 1968 brought up some peat from underneath shelly sand. Capt. L. Einarsson brought a sample to Þorleifur Einarsson (pers.comm.) who sent it to Uppsala for dating. The peat gave ages of 9,120±1180 and 9,460±1100 B.P. (Olsson et al., 1972). In the area where the peat was found, (Recent) carbonate sand lies on top of the Faxaflói terrace which here rises to about -30 m. The depth of 40 metres quoted by Olsson et al. is likely to be too great, as the dredger was not equipped to work at that depth (Hreinn Hreinsson, pers. comm.). The provenance of the peat is by no means certain. Twenty-odd years after its discovery a former officer of the ms. Sandey remembers that there was a considerable amount of it about, causing problems with dredging operations. This suggests that the peat was in situ. If not, the peat need not have travelled far; with the Syðra-Hraun shoal above sea level peat could well have formed a few hundred metres away from the place of discovery. Once in place, with rising sea level the sand transport along the terrace would then have buried the peat. On Fig. 10 the peat is represented by a dot at -25 metres. This is a compromise, as most of the Syðra-Hraun occurs at 17-30 metres depth and the likely depth of retrieval of the peat sample is near 30 m.

5. The bottom layer of peat, now accessible only at Low Water spring tide in Seltjörn near Reykjavík, gave ages of 9,030±1280 and 8,780±1150 B.P. (Thórarinsson, 1964). On Fig. 10 the greater age is used.

Using these five points, a sea level retreat from 65 metres to ca 30 metres has been sketched on Fig. 10. Curves for global eustatic sea level have also been drawn on this diagram. The outcome warrants a few comments:

a. The maximum slope of the suggested sea level curve represents isostatic rise at the rate of over 20 cm per year. This is extremely fast rebound and requires that the substrate is very inviscid. Little is known about crustal viscosity in Iceland, but Einarsson (1966) and Tryggvason (1973) argued for low crustal and sub-crustal viscosities. Recent modelling work (Sigmundsson,in prep.) indicates that the high thermal gradient encountered in the mid-ocean ridge/hot spot environment of Iceland is likely to give rise to exceptionally low viscosities of the crust and upper mantle.

b. The fact that the sea level curve cuts the curves of Fairbridge (1961) and Shepard (1963) suggests that these are not directly applicable to South West Iceland.

We conclude that there are strong indications that sea level in Faxaflói, Hvalfjörður, and Kollafjörður fell from ca 65 m a.s.l. to ca -30 metres after the retreat of Late Weichselian glaciers from the area, and it looks as if this regression took place within a time span of 700 years. More work needs to be done in the area for this sequence of events to be confirmed. We would, for example, like to see a programme of coring in strategic locations, and a closer study of the Faxaflói peat.

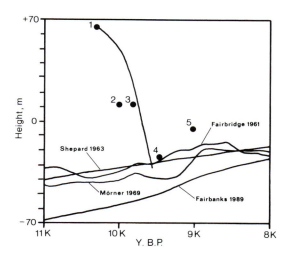

Fig. 10 Suggested path of sea level change in southern Faxaflói after the disappearance of glaciers from the area. Horizontal scale thousands of years B.P., vertical scale height of sea level. Four curves of eustatic sea level shown. For numbered points see text.

ACKNOWLEDGEMENTS

The survey in Faxaflói was commissioned by the Iceland State Cement Works and the Hvalfjörður survey by the Public Roads Administration. The dredging company Björgun h.f. and the Reykjavík City Engineer financed the Kollafjörður survey. Prof. Þórleifur Einarsson generously allowed us to use the ¹⁴C dates for the Faxaflói peat. We thank Freysteinn Sigmundsson for his advice on viscosity. Halldór G.Pétursson made valuable comments on the manuscript.

REFERENCES

Einarsson, Th. (1964) 'Radiocarbon dating of subfossil shells', (In Icelandic, English summary), in G.Kjartansson, S.Thórarinsson, and Th.Einarsson, '^{14}C datings of Quaternary deposits in Iceland', Nattúrufræðingurinn 34, 97-145.

Einarsson, Tr. (1966) 'Late- and post-glacial rise in Iceland and sub-crustal viscosity', Jökull 16, 157-166.

Fairbanks, R.G. (1989) 'A 17,000-year glacio-eustatic sea level record: influence of glacial melting rates on the Younger Dryas event and deep-ocean circulation', Nature 342, 637-642.

Fairbridge, R.W. (1961) 'Eustatic changes of sea level', Physical Chemistry of the Earth 5, 99-185.

Guðmundsson, M.T. (1989) 'The Hvalfjörður Tunnel. A seismic refraction survey', (In Icelandic), Mimeographed report OS-89047VOD-09 B. National Energy Authority, Reykjavík.

Hjartarson, Á. (1989) 'The age of the Fossvogur layers and the Álftanes end-moraine, SW-Iceland', Jökull 39, 21-31.

Ingólfsson, Ó. (1988) 'Glacial history of the lower Borgarfjörður area, western Iceland', Geologiska Föreningens i Stockholm Förhandlingar 110, 293-309.

Mörner, N.A. (1969) 'The Late Quaternary history of the Kattegat Sea and the Swedish west coast', Sveriges Geologiska Undersøken C-640, 1-487.

Olsson, I.U. and Piyanuj, P. (1965) 'Uppsala natural Radiocarbon measurements V', Radiocarbon 7, 315-330.

Olsson, I.U., Klasson, M. and Abd-el-Mageed, A. (1972) 'Uppsala natural Radiocarbon measurements XI', Radiocarbon 14, 247-271.

Shepard, F.P. (1963) 'Thirty-five thousand years of sea level', in T. Clements, (ed.) Essays in marine geology in honour of K.O. Emery, University of Southern California Press, 1-10.

Thórarinsson, S. (1964) '^{14}C datings connected with tephrochronology' (In Icelandic, English summary), in G.Kjartansson, S.Thórarinsson and Th.Einarsson (eds.) '^{14}C datings of Quaternary deposits in Iceland', Náttúrufræðingurinn 34, 97-145.

Thors, K. (1978) 'The sea-bed of the southern part of Faxaflói, Iceland', Jökull 28, 42-52.

Thors, K. (1983) 'Seismic stratigraphy of the Kollafjörður area', (In Icelandic), Mimeographed Report Marine Research Institute, Reykjavík.

Thors, K. and Boulton, G.S. (in press) 'Deltaic and sublittoral sedimentation associated with rising sea level - late Quaternary examples from northern Iceland'.

Thors, K. and Helgadóttir, G. (1986) 'Seismic reflection profiling in Faxaflói 1985', (In Icelandic), Mimeographed Report Marine Research Institute, Reykjavík.

Thors, K. and Helgadóttir, G. (1990) 'Stratigraphy of Hvalfjörður', (In Icelandic), Mimeographed Report Marine Research Institute, Reykjavík.

Tryggvason, E. (1973) 'Surface deformation and crustal structure in the Mýrdalsjökull area of south Iceland', Journal of Geophysical Research 78, 2488-2497.

PART 2

ENVIRONMENTAL CHANGES DURING THE POSTGLACIAL

2.1 Biotic Changes, Climatic Change and Human Settlement

'IF THIS IS A REFUGIUM, WHY ARE MY FEET SO BLOODY COLD?'
THE ORIGINS OF THE ICELANDIC BIOTA IN THE LIGHT OF RECENT RESEARCH

Paul Buckland
Department of Archaeology and Prehistory
University of Sheffield
Sheffield S10 2TN
UK

Andrew Dugmore
Department of Geography
University of Edinburgh
Drummond Street
Edinburgh EH8 9KP
UK

ABSTRACT. Two main models have been advanced as explanation for the origins of the Icelandic biota, either post-glacial colonization or survival in situ through at least the last glaciation. Iceland has been an island for at least fifteen million years but it lacks endemic species; given the frequency and severity of glacial conditions, and the cool temperate character of the biota, survival in situ appears most unlikely. The low powers of dispersal and the European affinities of most of the biota, coupled with the westerly flow of current atmospheric and oceanic circulation patterns argue against recent arrival as aerial plankton. It therefore seems most probable that colonization occurred during the earliest Holocene on ice rafts and in flood debris from a rapidly decaying Fennoscandinavian ice sheet. Whilst not excluding other mechanisms of dispersal, this model appears to offer the most complete explanation for the origins of the biota, and is supported by the fossil evidence.

'Most biogeographers try to check their conclusions against the background of geological facts to the best of their ability. It seems to me that geologists, in their turn, should spend some time on attempts to understand and evaluate the biological way of thinking which, actually is not very complicated'.

Lindroth (1972)

J. K. Maizels and C. Caseldine (eds.), Environmental Change in Iceland: Past and Present, 107–125.
© 1991 Kluwer Academic Publishers. Printed in the Netherlands.

1. INTRODUCTION

The hypothesis that elements of both flora and fauna were able to survive periods of glaciation in ice-free areas developed in Scandinavia during the latter part of the last century. In 1860, Lovén had published a paper concerning the survival of glacial relict species of Crustacea through the Holocene in some Scandinavian lakes (Segerstråle, 1962) and this led to a consideration of problems of the origin of the Postglacial biota. Both Blytt (1876, 1893) and Sernander (1896, 1908) approached the problems of the origins and survival of the Scandinavian flora through the Pleistocene and Warming (1888) had proposed that the Greenlandic flora had survived glaciation in ice-free areas. The theory initially largely concerned arctic-alpine elements in the flora surviving upon nunataks but it was later expanded to include discussion of the fauna, survival in areas marginal to the ice sheets, and the origins of the biota of North Atlantic Islands. Whilst there remained, particularly amongst geologists, some support for the hypothesis of total elimination of terrestrial biota, a *Tabula rasa* or clean slate, and post-glacial re-immigration, the prevailing model rapidly became that of survival in Refuges. In 1963, Áskell Löve felt able to write, in the conclusion to a Reykjavík symposium on North Atlantic island biota (Löve & Löve, 1963), 'Out of this emerged the theory of the survival of plants within the glaciated areas of Scandinavia replacing the now merely historical *tabula rasa* idea'; only Hoppe (1963), and to some extent Faegri (1963), in the same symposium, raised contrary notes.

Ives (1974) in a full review of the problem emphasised the dichotomy of views between biologists and geologists, with the former tending to argue in favour of in situ survival and the latter pointing out the overwhelming extent of Weichselian glaciation in northern Europe. The Refugia hypothesis has been strongly restated by E. Dahl (1987), and he has attempted to synthesise the geological, geomorphological and botanical data for Scandinavia. The model of survival through glaciation, both in northern Europe and on North Atlantic islands, has come under increasing pressure as the result of examination of the palaeontological record. In 1969, Coope had suggested that the refuges for the Scandinavian insect fauna lay in southern England, beyond the margins of the ice and invoked ice-rafting as part of the mechanism for dispersal, a method which Lindroth (1963a) had also favoured as a means of dispersal for some ground beetles to Newfoundland; ten years later, Coope (1979) further amplified the model to include North Atlantic islands. Examination of the invertebrate fossil record, principally that of the insects, has modified and extended this model (Coope, 1986; Buckland et al., 1986; Buckland, 1988) and a mechanism of initial ice-rafted dispersal of biota at the Lateglacial /Holocene boundary (Fig. 1) has received a measure of acceptance amongst neontologists (e.g. Böcher, 1988; Downes, 1988). Arguing from a purely botanical standpoint, Danielsen (1971) had also doubted the need for refugia and Nordal (1987) has recently similarly challenged the accepted view in Scandinavia.

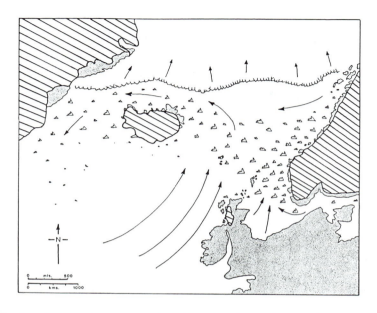

Fig.1 The North Atlantic at ca 10,000 B.P., with the mechanism for dispersal of biota to the islands. Modified from Buckland et al. (1986).

2. REFUGIA VERSUS *TABULA RASA* IN ICELAND

In Iceland, the work of the late Carl Lindroth upon the terrestrial and freshwater invertebrate fauna, published in 1931 (see also Lindroth, 1957; 1965), firmly established the Refugia Hypothesis as the prevalent model for the origins of the biota, despite the earlier support by both the geologist Thorvaldur Thoroddsen (1914) and botanist Stefán Stefánsson (1913) for a *Tabula Rasa* after glaciation. Botanists, like Steindórsson (1937, 1962), in common with their Scandinavian colleagues (e.g. Nannfeldt, 1935; Nordhagen, 1936; Hultén, 1937; E. Dahl, 1946; Gjærevöll, 1963), also developed the model, isolating areas of potential survival, and defining new endemic species (Óskarsson, 1979) or sub-species, often based upon only minor cytological characters (Löve & Löve, 1947, 1956). In 1978, Friðriksson, much involved in monitoring the immigration of plants and animals to the new island of Surtsey (Friðriksson, 1975; Lindroth et al., 1973) and once a supporter of Holocene origins for much of the biota (cf. Fridriksson, 1962), commented that survival in Refugia was the most widely held view.

Whilst ideas of permanence of land masses and a relatively short, simple Pleistocene chronology prevailed, the Refugium hypothesis appeared unassailable (e.g. D. Löve, 1963). Biotic origins were pushed back along land-bridges in either undefined interglacials or the late Tertiary (e.g. Lindroth, 1963b; see also Matthews, 1980) and biologists tended to regard it as a geological problem to support their apparently irrefutable evidence for survival. Ives (1974) suggested that there was general agreement that the last interglacial (Eemian/Ipswichian/ Sangamon) was 'of sufficient duration to permit gradual migration of plants into Pre-Wisconsin-Würm positions to account for their present-day distribution, assuming last Ice Age survival in refuges'. Ives' comment might have found some support in North America, but the North Atlantic islands are much less easily reached and the presence of biota of virtually wholly European affinities in the face of climatic and oceanic systems, which move essentially from west to east, had been stressed by Lindroth (1931, 1960a). To him (Lindroth, 1963b), a land bridge from the European continent appeared the only option available. This, in itself, left the additional problem as to why the fossil record of the Tertiary in Iceland, consisting largely of plants, but including an aphid (Freidrich et al., 1972), had clear affinities with the Eastern Deciduous Forests of North America (Simonarson, 1979). These problems, however, always remained what might be best defined as an S.E.P., someone else's problem (Buckland, 1988).

On geomorphological grounds, Hoppe (1968, 1982) was not slow to dismiss most potential refuges by showing clear evidence that they had been over-ridden by ice but, in Scandinavia, R. Dahl's (1963) suggestion that the position of Refugia varied through time, as the centres of ice accumulation shifted, essentially cut the rug from beneath the feet of those adopting this approach. In Iceland, the Refugia hypothesis has gained support from geologists, as well as botanists. Thórleifur Einarsson (1977, 1985) consistently favoured the existence of ice-free areas, which were able to support much of the pre-Landnám biota. In his latest summary of Icelandic geology (Einarsson & Albertsson, 1988), he maintains a certain ambivalence, pointing out that Sigurvinsson (1983) obtained secure evidence for ice-free areas on a mountainous peninsula in north-west Iceland and suggesting that similar areas may have existed elsewhere; he also refers to the Postglacial immigration model discussed by Buckland et al. (1986).

Increasing knowledge, of both the ocean floor and the complexity of Pleistocene climatic events has inevitably led to problems for the Refugia hypothesis. The replacement of a static land mass model with one involving plate tectonics and an expanding North Atlantic, beginning in the Labrador Sea region during the Late Cretaceous, some 80 million years ago (Laughton, 1971), has pushed the latest putative land connections, even by island hopping along the line of the Wyville-Thompson Ridge via the Faroes, back into the mid-Miocene (Nilson, 1978), at least 15 million years ago. E. Dahl (1987), however, has continued to dispute this, interpreting the geological evidence in favour of a land bridge as late as the Early Pleistocene. That element of the fauna and flora which may disperse in what Crowson (1981) terms the 'aerial plankton' might be explained in terms of sweepstake (sensu Simpson, 1940) arrival during the Holocene. This process is probably still occurring (cf. Gíslason, 1981), although it is increasingly difficult to distinguish from anthropochorous elements. The ground living fauna, including several relatively heavy, flightless ground beetles and weevils, and much of the flora are ill-adapted to such methods of dispersal and survival in situ appeared the only alternative.

Such a hypothesis leads to some disquiet amongst biologists with an interest in rates of evolution, for the North Atlantic islands, despite their long isolation and in contrast with other oceanic islands, appear to lack endemics in all groups other than those, like members of the Compositae genera *Hieraceum* and *Taraxacum* (Steindórsson, 1962; Óskarsson, 1979), in which species definition is frequently a contentious issue between those intent upon the description of new taxa and the taxonomically conservative. In his recent extensive review of the insect faunas of the North Atlantic islands, Downes (1988), like Lindroth (1957) previously, discounts any evidence for endemic species and suggests that a number which have been claimed, particularly amongst the Diptera (cf. Nielsen, Ringdahl & Tuxen, 1954), may be the result of insufficient knowledge of the adjacent continental faunas, a point recently emphasised by the synonomy of the previously claimed Icelandic endemic fly, *Crumomyia tuxeni* Coll., with the widespread Palaearctic species *C. nigra* (Meig.) (Norrbom & Kim, 1985). Although this position contrasts with that of many botanists, it is largely at the subspecific level that endemics have been claimed (e.g. Löve, 1957, 1970), and one of the two apparent endemic plants of Iceland, *Alchemilla faroensis* Bus., is also known from the Faroes (Löve & Löve, 1979). The implication, from the refugial viewpoint, has to be that a wide range of plants and invertebrates have shown no significant change, either in habitat or morphology, since their populations were last separated from the European mainland over 15 million years ago. Such stability might not appear remarkable in the more conservative groups, such as the insects (cf. Buckland & Coope, 1990), and could be taken as supporting evidence for a punctuated equilibrium rather than gradualist view of evolution, yet the intense selective pressures which isolated populations on remote islands face has led to rapid speciation elswhere (cf. Stanley, 1979, 40-47). Crowson (1981), in a discussion of the geographic distribution of beetles, draws out the contrast between the Antarctic islands of Kerguelen and Campbell, with their largely endemic faunas, and Iceland; he opts for an anthropochorous origin for those species lacking effective dispersal mechanisms, an hypothesis which work upon the fossil fauna has since dispelled (Buckland et al., 1986). It was no doubt this same problem which made Lindroth, who also had worked on the Azores with their large number of endemic taxa (Lindroth, 1960b), insist on a Late Pleistocene land connection (Lindroth, 1963b). The same arguments apply to the East Greenland Refugia evidence (Funder, 1979).

Löve and Löve (1979), in an attempt to update the Refugium model to take account of plate tectonics, accept long term species stability on the premise of reduced rates of evolution in cold climates, and push down the latest date of a land connection between Iceland and Europe into the Pliocene and with the Faroes to the late Pleistocene; both are untenable on geological grounds. The idea that evolutionary rates are lower in cold climates, very much a reflection of human attitudes to lower temperatures and analogy with chemical reactions, is not supported by the evidence from a wide range of groups. Cool temperate and arctic environments were restricted to isolated high altitude localities until the Late Tertiary in the Northern Hemisphere and the tundra biome evolved as a result of cooling temperatures into the Quaternary (Matthews, 1979; Wolfe, 1985). Whilst much of the change may result from the biogeographic mixing of formerly discrete groups to make new assemblages, many taxa evolved in the new environment. This is particularly evident for mammals (cf. Kurtén, 1968), but, even in a group where the prevailing strategy was change in geographic range rather than evolutionary change, namely the insects, there is good evidence for evolution in the new environment (Matthews, 1977). There is no reason to assume that Icelands biota would have been immune from these selection pressures, indeed, the fossil record of the Late Tertiary indicates the progressive extinction of an American temperate forest environment (cf. Símonarson, 1979). The presence of a cool temperate flora and fauna including North American forest elements, estimated at two million years old, at Kap København on the north eastern tip of Greenland (Bennike, 1985; Funder et al., 1985) might be taken as evidence for a northern Refugium but, as Downes (1988) has pointed out, with the development of the ice cap, continuously present, as the ice core data shows, since before the last interglacial (Dansgaard et al., 1971), north Greenland effectively becomes part of the Nearctic and is thereafter marginally relevant to the history of the remaining islands of the North Atlantic.

The progressive cooling into the Quaternary led to an increasingly amplified series of cold to warm oscillations leading to a sequence of glacial /interglacial cycles (Berger et al., 1984). In Iceland, in the Tjörnes sequence on the north coast (Eiriksson, 1980), there is evidence for the development of ice sheets by two million years ago and the earliest tillites, probably of glacial origin, may be slightly older than three million years (McDougall & Wensink, 1966). Einarsson and Albertsson (1988) suggest that there have been between fifteen and twenty three glaciations in Iceland during the past three million years. The extent of the earlier glaciations is difficult to assess on the terrestrial record, but evidence from deep sea cores indicates that large amounts of ice rafted debris first appeared in the North Atlantic about 2.4 million years ago (Ruddiman & Raymo, 1988). In such a rapidly evolving landscape as that of Iceland, it is unlikely that evidence will survive to allow direct estimation of the extent of ice cover during glaciations prior to the last one. By extrapolation from the deep sea evidence, however, four previous glaciations during the last 450,000 years were at least of equal magnitude to the most recent and were equally likely to have resulted in the extermination of North Atlantic island biota. The existence of a number of interglacial deposits, with similar floras to the present interglacial (e.g. Thórarinsson, 1963) cannot be used to either support or deny the Refugia hypothesis since the evidence can easily be accommodated by both models. In the case of the Elliðaárvogur site near Reykjavík, this evidence also extends to an insect fauna of apparent Middle Pleistocene age (Henriksen, in Thorkellsson, 1935; Hjartarson, 1980).

It has been suggested that the nature of the polar yearly cycle of increased summer day length might provide sufficient heat input to allow survival in refugia, which would not be viable further south and it is relevant to examine the biota of existing high Arctic islands as possible full glacial analogues. The biota of some of the islands of the Canadian Arctic Archipelago have been the subject of detailed examination. Savile (1961), in his study of Ellef Ringnes Island (79°N), found only 49 species of vascular plant and concluded that all could have immigrated post-glacially, despite the fact that the island showed no evidence of having been glaciated, although extensive perennial snowfields leave little geomorphological trace.   The insect faunas of islands in similar high arctic situations are also depauperate. McAlpine (1964) found no Coleoptera on the whole group of islands of which Ellef Ringnes is one, and only 55 species of insect in total; he similarly favoured a postglacial origin for the fauna. At the same latitude on Ellesmere Island, Oliver (1963) recorded over three hundred species yet the cool temperate elements which so characterise the Icelandic fauna are lacking.

The maxima of periods of glaciation episodically brought the Polar Front down to the south west coast of Portugal (Ruddiman & McIntyre, 1976), locking the North Atlantic in ice. The intensity of glacial maxima were such that exposed areas of landscape, potential refuges, must have resembled the ice-free valleys of southern Victoria Land, Antarctica (Calkin, 1973), where, away from the coast, life is reduced to a few algae living in the surface of the rock (Llano, 1962). It is also relevant that there appears to be a significant gap in the biological record in lowland Britain between ca 18,000 and 15,000 B.P., in such apparently suitable deposits as the varves of Proglacial Lake Humber.   In the nearby cave deposits at Creswell, Coles (1987) has noted a similar gap in the pollen record.   Local amelioration of climate, provided by Iceland's hot springs, might be seen as a means of creating Refugia but, as Tuxen (1944) in his study of the springs' biota realised, these would be unable to curtail the expansion of large ice sheets since several geothermal areas presently lie beneath existing glaciers (e.g. Björnsson & Kristmannsdóttir, 1984).

On balance, the frequency of climatic oscillation, the probability that most of the Pleistocene climate was typified by cold, arctic tundra, when not by actual glaciation, and the cool temperate, rather than arctic nature of much of the biota (cf. Buckland 1988; Downes, 1988) would seem to militate against survival in Refugia without substantial evolutionary changes in both flora and fauna during the Quaternary. If this is accepted, the presence or absence of ice-free areas in Iceland becomes merely of academic interest in terms of the history of the biota, but it still leaves a number of pieces of purely biological data for interpretation.

3. THE ICELANDIC BIOLOGICAL EVIDENCE - AN ALTERNATIVE VIEW

If the *Tabula rasa* hypothesis is to be reinstated effectively, then the biological evidence adduced in favour of survival in Refugia has to be re-examined, in the light of the fossil record. Terrestrial organic sediments of Lateglacial age are unknown in Iceland. Moraines of putative Lateglacial age in the south-west of the country have recently been re-interpreted as early Holocene (Preboreal) in age (Hjartarson & Ingólfsson, 1988). Despite sparse research upon the postglacial plant macrofossil record before the arrival of Man, the broad outline of the history of the Icelandic flora has been obtained by palynology (Th. Einarsson, 1961, 1963; Hallsdóttir, 1987; Vasari & Vasari, 1990). The pollen record indicates an expansion of birch forest from the north-east during the early Holocene. The detailed biogeography of other elements in the biota are less easily obtained but insects are preserved extensively in suitable anaerobic deposits (Buckland et al., 1986) and Árni Einarsson (1982, 1985) has examined the mid-and later Holocene freshwater record of several other invertebrate groups from Mývatn.

        In the modern biota, one of the apparently convincing pieces of information in
support of the Refugium hypothesis is Lindroth's (1957, 1963b) examination of Carabid
(ground beetle) distribution and morphology. He used the ratio of macropterous to
brachypterous individuals in several ground beetle species to identify putative Refugia.
Arguing from experiments with another ground beetle, he reasoned that the inevitably
flightless short-winged forms were genetically dominant and that long-winged individuals
would be more frequent towards the perimeter of a species' distribution as a result of their
greater dispersal ability; dominance of short-winged individuals therefore indicated old
populations. Using the varying wing length of a large sample of the small ground beetle
Bembidion grapei Gyll., he was able to localise two refuges along the south coast, around
Eyjafjallajökull, and along the coastal fringe south-east of Vatnajökull (Fig. 2), where other
lines of evidence also appeared to support his interpretation (Lindroth, 1931, 1968). He
similarly was able to support the idea that hot springs provided refuges; Greenlandic and
Norwegian specimens added further support to the model.

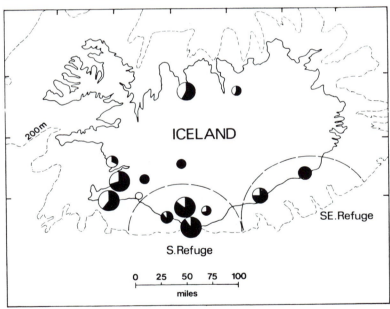

Fig.2   The frequency of long- and short-winged forms of *Bembidion grapei* in Iceland
        (white = long-winged; black = short-winged). Redrawn from Lindroth (1957).

        Research with another Carabid, *Notiophilus biguttatus* F., also by Lindroth (1957,
1963a), provides an alternative, equally plausible explanation of such distributional
patterns. This species is also wing-dimorphic and has been introduced to Newfoundland
by Man. Its distribution does not include the North-West Peninsula, where Norse settlement

has been proven at the site of L'Anse aux Meadows (Ingstad, 1977), and the general picture (Fig. 3) is one of expansion from coastal localities on the Avalon Peninsula, where English fishermen established land stations from the late sixteenth century onwards (Grant Head, 1976). The pattern of macropterous individuals dominating at localities remote from the primary stations is evident. This has been established in a maximum of a little over three hundred years and the Icelandic data can be usefully examined in the light of this. Although *B. grapei* lacks a pre-Landnám fossil record (Buckland, 1988), this need not indicate that it is anthropochorous, since its usual habitat, on dry ground poor in vegetation (Larsson & Gígja, 1958; Lindroth, 1965), is unlikely to be adjacent to suitable localities to provide fossil assemblages. Whilst Lindroth (1957) suggests that the Icelandic distribution of *B. grapei* is relatively mature, it remains uncertain as to how long the present pattern has taken to emerge. In the light of the Newfoundland evidence, admittedly of another species, the development of the present pattern of distribution is likely to have been a rather more rapid process than the ten thousand years since deglaciation favoured by Lindroth.

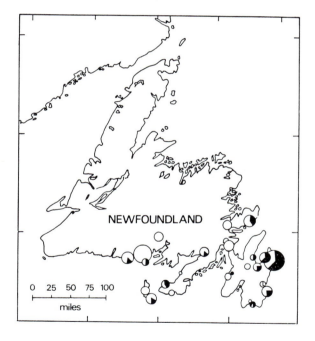

Fig.3 The frequency of long- and short-winged *Notiophilus biguttatus* in Newfoundland (white = long-winged; black = short-winged). Redrawn from Lindroth (1957).

It is not the aim of this paper to dismiss the Refugium hypothesis out of hand, since there are many contexts outside the North Atlantic region where the evidence is more than convincing (e.g. Hopkins et al., 1982). The problems lie in identifying what had constrained the species to the so-called refugial areas in the first place. Lindroth (1957), like many of his contemporaries, tended to underestimate the impact of the Norse settlers upon the natural distribution of the fauna. In the late twelfth century Ári the Wise, compiler of Íslendingabók, wrote,

' Í þann tíð var Ísland viði vaxið á milli fjalls og fjöru'.
(Valtysdottir, 1980)
In that time (Landnám) Iceland was forested from mountain to sea shore.

The present, denuded landscape of much of Iceland is best described as an ovigenic one[1], created by the destruction of vegetation and soils since sheep were first introduced by the Norse settlers during the ninth century (Dugmore & Buckland, this volume). After immigration with most of the rest of the biota at the Lateglacial /Holocene transition (Fig. 1) (Buckland et al., 1986), *B. grapei* must have spread across the island but then been curtailed and communities isolated by the development of woodland and the loss of open ground habitat. Populations would have been restricted to refugia from the forest, in places adjacent to the ice sheets and above the tree-line; hot springs would similarly have had reduced vegetation cover around them. The present distribution reflects a stage in the expansion of habitat made available to the species as Man and his domestic animals progressively stripped the soil from the landscape (Þórarinsson, 1961). A similar hypothesis was advanced in Britain by Piggott and Walters (1954) to account for disjunct patterns in the flora.

In contrast with the increased habitat availability for bare ground species, modern patterns of species rarity must also reflect human interference in the environment. The fossil record provides evidence in support of this. The small rove beetle *Lathrobium brunnipes* (F.) was known to Lindroth (1965; Lindroth et al., 1988) only from Hornafjörður in the extreme south-east of the country and Ólafsson (in Buckland et al., 1986) has recently taken an individual at Landbrot in Vestur-Skaftafellssýsla; it is a common fossil in sediments through to at least the sixteenth century across the whole of southern Iceland (Buckland et al., 1985). Widespread drainage of its main habitat, lowland *Sphagnum* bog, is likely to have led to its apparently drastic decrease in frequency, a process which may have contributed to the extinction of one species, *Hydraena britteni* Joy, from the same area (Buckland, Perry & Sveinbjarnardóttir, 1983). Although some of the contrasts between fossil and modern assemblages may reflect taphonomic problems (Buckland, Dugmore & Sadler, this volume), it is evident that Man has had a major impact, both directly and indirectly upon the distribution and abundance of insects and great care must be exercised when consideration is restricted to modern collecting data, which at a maximum only allows consideration of changes over the last two centuries (Larsen & Gígja, 1958; Gíslason & Ólafsson, 1989).

As Downes (1988) has pointed out, most of the problems concerning endemic taxa have disappeared as the taxonomic status of the fauna has become better known. Most of these concern the Diptera, of which only recently has it become possible to examine the fossil record (Skidmore, in prep.). One species of Ephydrid fly, *Scatella thermarum* Coll., is restricted in its distribution to hot spring areas in Iceland and Greenland (Nielsen, Ringdahl & Tuxen, 1954), where it is able to exploit algae growing in water up to 47° Centigrade. If the taxonomy is correct and this fly is endemic, it presents one of the few convincing pieces of evidence for long term survival in Refugia, yet its apparent ability to have dispersed to all the hot spring areas in the country up to 900 m.a.s.l. suggests that it might equally form part of the 'aerial plankton' and that a Nearctic origin should be sought.

Attempts to resolve the botanical data have tended to revolve around the idea that unicentric, bicentric and even polycentric distribution patterns, particularly of arctic-alpine taxa, are reflections of the location of refuges from the ice sheets (e.g. Steindórsson, 1962), rather than the dynamic end result of changing patterns of distribution through the Holocene. Like the case of the brachypterous form of *B. grapei*, some species must be isolated as a result of the expansion of birch and willow forest during the Postglacial, whilst others reflect isotherms and microclimatic factors. The sweepstake element in dispersal, survival and extinction also requires that too much weight should not be placed upon individual occurrences without the evidence from the fossil record. The species concept varies amongst botanists, often in relation to their viewpoint with regard to survival in Refugia. Nordal (1986), in her examination of the Scandinavian evidence, used Haldane's (1949) concept of an evolutionary unit, the Darwin. This was defined as the quantitative change in a character to the base *e* per million years. Nordal pointed out that a very wide range of values for rates of evolution had been substantiated in the more easily documented groups, particularly among some mammals, and that the degree of taxonomic differentiation apparent within Refugia species of plants fell well within the maximum rate found in such groups if a postglacial immigration was to be assumed.

This view is equally applicable to the Icelandic evidence but, on isolated islands, a further factor requires consideration. If the processes of colonisation involve a significant sweepstake element, it is probable that the full range of genetic diversity present in the Continental population need not be introduced to the island. Much of Iceland's biota is not adapted to anemochorous dispersal and the episode of immigration is likely to have been restricted to a very short period at the Lateglacial to Holocene transition, when ice-rafted debris from the decaying Scandinavian ice sheet was widespread in the North Atlantic, a result of rapid warming and a turning back of the North Atlantic Drift along the retreating edge of the Polar pack (Fig. 1) (Buckland et al., 1986; Buckland, 1988). Such a dispersal mechanism provides explanation for the apparently disharmonic nature of the North Atlantic island faunas, which are markedly deficient in phytophages. If the transport mechanism operated for only a few years, insufficient time may have been available for the host plant to have become sufficiently established to provide habitat for its feeders, which need not necessarily have made the same landfall; the inevitable result is that polyphages dominate markedly over mono- and oligophages.

One implication of the ice-rafting hypothesis for the flora is that populations of some species may be descended from very few, if not single individuals. Such 'bottle-necking' may lead to the phenotypic form in the isolated population appearing noticeably different from the norm of the parent population. In part, this may be a result of differing natural selective pressures on the island but, equally, the norm in that group, which might be rare on the Continent, may have been largely suppressed by intraspecific competition in the parent population.

4. CONCLUSION

Much of the dispute over the Refugia hypothesis on the North Atlantic islands has stemmed from the failure of individual specialists to adopt what entomologists might call a holoptic approach. Any explanation of the distribution pattern of a taxon has to take account of that of all associated elements in the biota. What appears to be overwhelmingly the case for the insects must also have a direct bearing upon the flora and any other group. The climatic, geological, geomorphological and palaeontological evidence would seem to militate against significant survival of biota in refugia on the North Atlantic islands. The biological evidence is also amenable to interpretation in terms of Postglacial immigration. The bulk of the pre-Landnám biota arrived as a result of a short-lived episode of ice-rafted dispersal at the Lateglacial /early Holocene transition around 10,000 BP, perhaps making landfall in the north-east part of Iceland before the final decay of the islands main ice sheet. Thereafter, chance dispersal in the aerial plankton may have added a number of species, but these form an insignificant part of the biota when compared with the scale of introduction provided by Man from the ninth century onwards (cf. Davídsson, 1967; Buckland, 1988; Buckland, Dugmore & Sadler, this volume).

The rapid transition from high arctic conditions as far south as southern England (Coope, 1981) to a climate probably warmer than present day (Osborne, 1980) at the Holocene boundary led to the elimination of most arctic elements in the fauna, leaving a largely cool temperate pool of potential emigrants, as passive dispersal was initiated from along the south-western edge of the Scandinavian ice sheet. A similar pattern has been noted in the development of the North American fauna, where many cold stenotherms found no refuges south of the Laurentide ice sheet when intense warming began around 15,000 B.P. and species became restricted to the Beringian refugium (Schwert & Ashworth, 1988).

The palaeoclimatic evidence, in particular, for several episodes of intense cold, with Iceland locked in a frozen North Atlantic, should be conclusive against the survival of its cool temperate fauna in refugia. Further research on the fossil record, as well as the geomorphology, of north-east Iceland should lead to further refinements in the model. Raised beaches should contain the erratics rafted from the European mainland and allow some further definition of source areas for the biota, perhaps resolving the problems of whether there were Scottish (Coope, 1986) as well as southern Scandinavian components in the ice-rafting process. A dynamic approach to the botanical data, unconstrained by the refugia model and extended to include studies of DNA, is also an essential aspect of further studies on the origins of North Atlantic island life. Local extinction and restriction of available gene pool are also likely to have occurred as a result of deposition of tephra from major eruptions and patterns of plant distribution could be profitably examined in relation to the tephro-chronological record.

Future research should combine a multidisciplinary approach with an open mind, ready to abandon one model in the face of overwhelming evidence from another. As Lindroth (1972) noted, there has been a marked reluctance for the various proponents to indulge in a meaningful dialogue, leading inevitably to polarisation of views. It is time for a repeat of the 1962 Reykjavík symposium on the North Atlantic Biota (Löve & Löve, 1963), perhaps leading to a rather different, if less categoric conclusion.

ACKNOWLEDGEMENTS

The research upon the history of the biota of the North Atlantic islands began as a subsidiary element in a study of man, climate and environmental change, funded initially by the Leverhulme Trust and carried out under Icelandic research permits 1/80, 48/80, 22/81, 16/82, 28/83 and 29/84; subsequent work has been carried out as part of archaeological projects with the National Museum of Iceland, under the auspices of Guðmundur Ólafsson and Guðrún Sveinbjarnardóttir, to whom special thanks are due. The comments of David Gilbertson, Gísli Már Gíslason, Jim Hansom, Tom McGovern, Erlingur Ólafsson, Jon Sadler, Peter Skidmore and Pat Wagner on various aspects of the research are gratefully acknowledged.

[1]Ovigenic, from Latin ovis, sheep and Greek γενεσιζ, 'made by sheep' (note the ablative case) may not be acceptable to the linguistic purists, but is more easily understood and universal than, for example, Icelandic beitt á af fé.

REFERENCES

Bennike, O. (1985) 'Skov-tundra i Nordgronland i Plio-Pleistocaen -plantegeografiske implikationer', Dansk geologisk Foreningen Årsskrift for 1984, 111-112.

Berger, A.L., Imbrie, J., Hays, J., Kukla, G. & Salzman, B. (eds.) (1984) Milankovitch and Climate. D.Reidel, Boston.

Blytt, A. (1876) Essay on the immigration of the Norwegian flora during alternating rainy and dry periods. Cammermeyer, Oslo.

Blytt, A. (1893) 'Zur Geschichte der nordeuropaischen, besonders der norwegischen Flora', Englers Botanische Jahrbuch 17, no.41.

Björnsson, H. & Kristmannsdóttir, H. (1984) 'The Grimsvötn Geothermal Area, Vatna-jökull, Iceland', Jökull 34, 25-50.

Böcher, J. (1988) 'The Coleoptera of Greenland', Meddelelser om Grønland Bioscience 26.

Buckland, P.C. (1988) 'North Atlantic faunal connections - introduction or endemics?', Entomologica Scandinavica Supplement 32, 7-29.

Buckland, P.C. & Coope, G.R. (1990) A Bibliography and Review of Quaternary Entomology. University of Sheffield.

Buckland, P.C., Gerrard, A.J., Larsen, G., Perry, D.W., Savory, D.R. & Sveinbjarnardóttir, G. (1985) 'Late Holocene palaeoecology at Ketilsstaðir in Myrdalur, South Iceland', Jökull 36, 41-55.

Buckland, P.C., Perry, D.W., Gíslason, G.M. & Dugmore, A.J. (1986) 'The pre-Landnám fauna of Iceland; a palaeontological contribution', Boreas 13, 173-184.

Buckland, P.C., Perry, D.W. & Sveinbjarnardóttir, G. (1983) 'Hydraena britteni Joy (Coleoptera, Hydraenidae) fundin á Íslandi í setlögum frá því seint á Nútíma', Náttúrufræðingurinn 52, 37-44.

Calkin, P.E. (1973) 'Glacial processes in the ice-free valleys of Southern Victoria Land, Antarctica', in B.D.Fahey & R.D.Thompson (eds.), Research in Polar and Alpine Geomorphology, 167-186. Geoabstracts, Norwich.

Coles, G. (1987) Aspects of the Application of Palynology to Cave Deposits in the Magnesian Limestone Region of North Nottinghamshire. Unpublished Ph.D. Thesis, University of Sheffield.

Coope, G.R. (1969) 'The contribution that the Coleoptera of glacial Britain could have made to the subsequent colonisation of Scandinavia', Opuscula entomologica 34, 95-108.

Coope, G.R. (1979) 'The Carabidae of the glacial refuge in the British Isles and their contribution of the Post Glacial colonization of Scandinavia and the North Atlantic Islands', in T.L.Erwin, G.E.Ball & D.R.Whitehead (eds.), Carabid Beetles, Their Evolution, Natural History and Classification, 407-424. Junk, The Hague.

Coope, G.R. (1981) 'Episodes of local extinction of insect species during the Quaternary as indicators of climatic changes', in J.Neale & J.Flenley (eds.) The Quaternary in Britain, 216-221. Pergamon, Oxford.

Coope, G.R. (1986) 'The invasion and colonisation of the North Atlantic Islands; a palaeoecological solution to a biogeographic problem', Philosophical Transactions of the Royal Society of London B314, 619-635.

Crowson, R.A. (1981) The Biology of the Coleoptera. Academic Press, London.

Dahl, E. (1946) 'On different types of unglaciated areas during ice ages and their significance to phytogeography', New Phytologist 45, 225-242.

Dahl, E. (1987) 'The nunatak theory reconsidered', Ecological Bulletins 38, 77-94.

Dahl, R. (1963) 'Shifting ice culmination, alternating ice covering and ambulant refuge organisms?', Geografiska Annaler 45, 122-138.

Danielsen, A. (1971) 'Skandinaviens fjellflora i lys av senkvartær vegetasjonshistorie', Blyttia 29, 183-209.

Dansgaard, W., Johnsen, S.J., Clausen, H.B. & Langway Jr., C.C. (1971) 'Climatic record revealed by the Camp Century Ice Core', in K.K.Turekian (ed.) Late Cenozoic Glacial Ages, 37-56. Yale University Press.

Davídsson, I. (1967) 'The immigration and naturalization of flowering plants in Iceland since 1900', Greinar IV 3, 1-36.

Downes, J.A. (1988) 'The post-glacial colonization of the North Atlantic Islands', Memoirs of the Canadian Entomological Society 144, 55-92.

Einarsson, Á. (1982) 'The palaeolimnology of Lake Mývatn, northern Iceland: plant and animal microfossils in the sediment', Freshwater Biology 12, 63-82.

Einarsson, Á. (1985) 'Botn Mývatns: Fortíð, nutíð, framtíð', Náttúrufræðingurinn 55, 153-173.

Einarsson, Th. (1961) 'Pollenanalytische Untersuchungen zur spät- und postglacialen Klimageschichte Islands', Sonderveröffentlichungen des Geologisches Institutes der Universität Köln, 6, 1-52.

Einarsson, Th. (1963) 'Pollen-analytical studies on the vegetation and climate history of Iceland in Late and Post-glacial times', in A.Löve & D.Löve (eds.), North Atlantic Biota and their History, 337-354, Reykjavík.

Einarsson, Þ. (1977) 'Um gróður á isöld á Íslandi. Skogarmál, 56-72. Reykjavík.

Einarsson, Þ. (1985) Jarðfræði. Mál og menning, Reykjavík.

Einarsson, Th. & Albertsson, K.J. (1988) 'The glacial history of Iceland during the past three million years', Philosophical Transactions of the Royal Society of London B318, 411-430.

Eiríksson, J. (1980) 'Tjörnes, North Iceland: A Bibliographical Review of the Geological Research History', Jökull 30, 1-20.

Faegri, K. (1963) 'Problems of immigration and dispersal of the Scandinavian flora', in Löve & Löve, 221-232.

Freidrich, W.L., Símonarsson, L.A. & Heie, O.E. (1972) 'Steingervingar í millilögum í Mókolsdal', Náttúrufræðingurinn 42, 4-17.

Friðriksson, S. (1962) 'Um adflutning íslenzku florunnar', Náttúrufræðingurinn 32, 175-189.

Friðriksson, S. (1975) Surtsey: Evolution of life on a Volcanic Island. Butterworth, London.

Friðriksson, S. (1978) 'The degradation of Icelandic ecosystems', in M.W.Holdgate & M.J.Woodman (eds.) The Breakdown and Restoration of Ecosystems, 145-156. Plenum Press, New York.

Funder, S. (1979) 'Ice-Age Refugia in East Greenland', Palaeogeography, Palaeoclimatology, Palaeoecology 28, 279-295.

Funder, S., Abrahamsen, N., Bennike, O. & Feyling-Hanssen, R.W. (1985) 'Forested Arctic: evidence from North Greenland', Geology 13, 542-546.

Gíslason, G.M. (1981) 'Predatory exclusion of Apatania zonella (Zett.) by Potamophylax congulatus (Steph.) (Trichoptera: Limnephilidae) in Iceland', in G.P.Moretti (ed.) Proceedings of the 3rd Symposium on Trichoptera, 93-98. Junk, The Hague.

Gíslason, G. M. & Ólafsson, E. (1989) 'Entomology in Iceland', Fauna norvegica series B 36, 11-16.

Gjærevöll, O. (1963) 'Survival of plants on nunataks in Norway during the Pleistocene glaciation', in A.Löve & D.Löve (eds.) North Atlantic Biota and their History, 261-283, Reykjavík.

Grant Head, C. (1976) Eighteenth Century Newfoundland. Carleton University Library.

Haldane, J.B.S. (1949) 'Suggestions as to the quantitative measurement of rates of evolution', Evolution 3, 51-56.

Hallsdóttir, M. (1987) 'Pollen Analytical Studies of Human Influence on Vegetation in Relation to the Landnám Tephra Layer in Southwest Iceland', Lundqua Thesis 18.

Hjartarson, Á. & Ingólfsson, Ó. (1988) 'Preboreal Glaciation of Southern Iceland', Jökull 38, 1-16.

Hopkins, D.M., Matthews Jr., J.V., Schweger, C.E. & Young, S.B. (1982) Paleoecology of Beringia. Academic Press, London.

Hoppe, G. (1963) 'Some comments on the ice-free refugia of northern Scandinavia', in Löve & Löve, North Atlantic Biota and their History, 321-336, Reykjavík.

Hoppe, G. (1968) 'Grimsey and the maximum extent of the last glaciation of Iceland', Geografiska Annaler 50A, 16-24.

Hoppe, G. (1982) 'The extent of the last inland ice sheet of Iceland', Jökull 32, 3-11.

Hultén, E. (1937) Outline of the history of Arctic and Boreal biota during the Quaternary period. Stockholm.

Ingstad, A.S. (1977) The Discovery of a Norse Settlement in America: Excavations at L'Anse aux Meadows, Newfoundland, 1961-68, I. Universitetsforlaget.

Ives, J.D. (1974) 'Biological refugia and the nunatak hypothesis', in J.D.Ives & R.G.Barry (eds.) Arctic and Alpine Environments, 605-636. Methuen, London.

Kurtén, B. (1968) Pleistocene Mammals of Europe. Weidenfeld & Nicholson, London.

Larsson, S.G. & Gígja, G. (1958) Coleoptera 1. Synopsis. Zoology of Iceland, III, 46a. Ejnar Munksgaard, Copenhagen.

Laughton, A.S. (1971) 'South Labrador Sea and the evolution of the North Atlantic', Nature 232, 612-617.

Lindroth, C.H. (1931) 'Die Insektenfauna Islands und ihre Probleme', Zoologiska Bidrag Uppsala 13, 105-600.

Lindroth, C.H. (1957) The Faunal Connections between Europe and North America. Wiley, New York.

Lindroth, C.H. (1960a) 'Is Davis Strait - between Greenland and Baffin Island - a floristic barrier?', Botaniske Notiser 113, 129-140.

Lindroth, C.H. (1960b) 'The ground-beetles of the Azores (Coleoptera Carabidae) with some reflections on overseas dispersal', Boletim do Museu Municipal Funchal 13, 5-48.

Lindroth, C.H. (1963a) 'The fauna history of Newfoundland (illustrated by Carabid Beetles)', Opuscula entomologica Supplement, 23.

Lindroth, C.H. (1963b) 'The problem of late land connections in the North Atlantic area', in Löve & Löve, North Atlantic Biota and their History, 73-85, Reykjavík.

Lindroth, C.H. (1965) 'Skaftfell, Iceland a living glacial refugium', Oikos Supplement, 6.

Lindroth, C.H. (1968) 'The Icelandic form of Carabus problematicus Hbst. (Col. Carabidae). A statistic treatment of subspecies', Opuscula entomologica 33, 157-182.

Lindroth, C.H. (1972) 'Reflections on Glacial Refugia', Ambio Special Report 2, 51-54.

Lindroth, C.H., Andersson, H., Bodvarsson, H. & Richter, S.H. (1973) 'Surtsey, Iceland - the development of a new fauna. 1963-1970. Terrestrial Invertebrates', Entomologica Scandinavica Supplement, 5.

Lindroth, C.H., Bengtson, S.-A. & Enckell, P. (1988) 'Terrestrial faunas of four isolated areas: A study in tracing old faunal centres', Entomologica Scandinavica Supplement, 32.

Llano, G.A. (1962) 'The terrestrial life of the Antarctic', Scientific American 207, 212-230.

Lovén, S. (1860) 'Om nagra i Vettern och Venern funna Crustaceer. Stockholm.

Löve, A. (1963) 'Conclusion', in A.Löve, & D.Löve (eds.), North Atlantic Biota and their History, 391-398, Reykjavík.

Löve, A. (1970) Íslenzk Ferðaflora. Almenna Bókafélag, Reykjavík.

Löve, A. & Löve, D. (1947) 'Studies on the origin of the Icelandic flora. 1.Cyto-ecological investigations on Cakile', Rit Landbunaðardeildar, atvinnudeilð Háskólans, B2.

Löve, A. & Löve, D. (1956) 'Cytotaxonomical conspectus of the Icelandic flora', Acta Horticulturalis Gotobergensis 20, 65-291.

Löve, A. & Löve, D. (1963) North Atlantic Biota and their History. Pergamon Press, Oxford.

Löve, D. (1963) 'Dispersal and survival of plants', in A.Löve, & D.Löve, (eds.), North Atlantic Biota and their History, 189-206, Reykjavík.

Löve, A. & Löve, D. (1979) 'The history and geobotanical position of the Icelandic flora', Phytocoenologia 6, 94-105.

Matthews Jr., J.V. (1977) 'Tertiary Coleoptera from the North American Arctic', Coleopterists' Bulletin 31, 297-308.

Matthews Jr., J.V. (1979) 'Fossil beetles and the Late Cenozoic history of the tundra environment', in J.Gray & A.J.Boucot (eds.), Historical Biogeography, Plate Tectonics and the Changing Environment, 371-378. Oregon State University Press.

Matthews Jr., J.V. (1980) 'Tertiary land bridges and their climate: Backdrop for development of the present Canadian insect fauna', Canadian entomologist 112, 1089-1103.

McAlpine, J.F. (1964) 'Arthropods of the bleakest barren lands: composition and distribution of the Arthropod Fauna of the Northwestern Queen Elizabeth Islands', Canadian entomologist 96, 127-129.

McDougall, I. & Wensink, H. (1966) 'Paleomagnetism and geochronology of Pliocene-Pleistocene lavas in Iceland', Earth and Planetary Science Letters 1, 232-236.

Nannfeldt, J.A. (1935) 'Taxonomical and plant-geographical studies in the Poa laxa group', Symbolae Botanicae Upsaliensas 1, 5.

Nielsen, P., Ringdahl, O. & Tuxen, S.L. (1954) Diptera 1. (exclusive of Ceratopogonidae and Chironomidae). Zoology of Iceland, III, 48a. Ejnar Munksgaard, Copenhagen.

Nilson, T.H. (1978) 'Lower Tertiary laterite on the Iceland-Faroe ridge and the Thulean land bridge', Nature 274, 786-788.

Nordal, I. (1987) 'Tabula rasa after all? Botanical evidence for ice-free refugia in Scandinavia reviewed', Journal of Biogeography 14, 377-388.

Nordhagen, R. (1936) 'Skandinavias Fjellflora og dens Relasjoner til den Siste Istid', Nordiska (19. skand.) naturforskening i Helsingfors, 93-124. Helsinki.

Norrbom, A.L. & Kim, K.C. (1985) 'Systematics of *Crumomyia* Macquart and *Alloborborus* Duda (Diptera: Sphaeroceridae)', Systematic Entomology 10, 167-225.

Oliver, D.R. (1963) 'Entomological studies in the Lake Hazen Area, Ellesmere Island, including lists of species of Arachnida, Collembola and Insecta', Arctic 16, 175-180.

Óskarsson, I. (1979) 'A new *Hieracium* species found in South Iceland (Eu-Hieracia - Sect. Tridentata)', Acta Botanica Islandica 5, 71-72.

Osborne, P.J. (1980) 'The late Devensian-Flandrian transition depicted by serial insect faunas from West Bromwich, Staffordshire, England', Boreas 9, 134-147.

Piggott, C.D. & Walters, S.M. (1954) 'On the interpretation of the discontinuous distribution of certain British species of open habitats', Journal of Ecology 42, 95-116.

Ruddiman, W.F. & McIntyre, A. (1976) 'Northeast Atlantic palaeoclimatic changes over the past 600,000 years', Geological Society of America. Memoir 145, 111-146.

Ruddiman, W.F. & Raymo, M.E. (1986) 'Northern Hemisphere climate regimes during the past 3 Ma: possible tectonic connections', Philosophical Transactions of the Royal Society of London B318, 411-430.

Savile, D.B.O. (1961) 'The botany of the northwest Queen Elizabeth Islands', Canadian Journal of Botany, 39, 909-942.

Schwert, D.P. & Ashworth, A.C. (1988) 'Late Quaternary history of the northern beetle fauna of North America: a synthesis of fossil and distributional data', Memoirs of the Entomological Society of Canada 144, 93-107.

Segerstråle, S.G. (1962) 'The immigration and prehistory of the glacial relicts of Eurasia and North America. A survey and discussion of modern views', Internationale Revue der Gesamten der Hydrobiologie und Hydrographie 47, 1-25.

Sernander, R. (1908) 'On the evidence of postglacial changes of climate furnished by the peat-mosses of northern Europe', Geologiska Föreningens i Stockholm Förhandling 30, 465-478.

Sigurvinsson, J.R. (1983) 'Weichselian glacial lake deposits in the highlands of Northwestern Iceland', Jökull 33, 99-109.

Símonarson, L.A. (1979) 'On climatic changes in Iceland', Jökull 29, 44-46.

Simpson, G.G. (1940) 'Mammals and landbridges', Journal of the Washington Academy of Science 30, 137-163.

Stanley, S.M. (1979) Macroevolution Pattern and Process. Freeman, San Francisco.

Stefansson, S. (1913) Plönturnar. Kennslubók í grasafræði. Copenhagen.

Steindórsson, S. (1937) 'Jurtagroðurinn og jökulstíminn', Náttúrufræðingurinn 7.

Steindórsson, S. (1962) 'On the age and immigration of the Icelandic flora', Vísindafélag Íslendinga, 35.

Thórarinsson, S. (1963) 'The Svínafell layers. Plant-bearing Interglacial sediments in Öræfi, southeast Iceland', in Löve, A. & Löve, D. (eds.) North Atlantic Biota and their History, 377-390, Reykjavik.

Thorkellsson, T. (1935) 'A fossiliferous interglacial layer at Elliðaárvogur Reykjavík', Greinar 1, 78-91.

Thoroddsen, Th. (1914) An Account of the Physical Geography of Iceland. Botany of Iceland, I, pt.I,2. Copenhagen.

Tuxen, S.L. (1944) 'The Hot Springs, their animal communities and their zoogeographical significance', Zoology of Iceland, I, 11. Ejnar Munksgaard, Copenhagen.

Valtyrsdóttir, H. (1980) '˜ Í þann tíð var Ísland viði vaxið.˜. ', Náttúrufræðingurinn, 50, 1-12.

Vasari, Y. & Vasari, A. (1990) 'L'Histoire Holocène des Lacs Islandais', in S.Devers (ed.) Pour Jean Malaurie, 102 temoignages en hommage á quarante ans d'etudes arctiques, 277-293. Plon, Paris.

Wolfe, J.A. (1985) 'Distribution of major vegetation types during the Tertiary', in E.T.Sundquist & W.S.Broecker (eds.) The carbon cycle and atmospheric CO_2 natural variations Archaean to present, 357-376. American Geophysical Union Monograph 32, Washington.

Warming, E. (1888) 'Om Grønlands Vegetation', Meddelelser om Grønland 12.

Þórarinsson, S. (1961) 'Uppblastur á Íslandi í ljósi öskulagarannsókna', Árskrit Skógræktarfélag Íslands (1961), 17-54.

FAUNAL CHANGE OR TAPHONOMIC PROBLEM?
A COMPARISON OF MODERN AND FOSSIL INSECT FAUNAS FROM SOUTH EAST ICELAND

Paul Buckland
Department of Archaeology and Prehistory
University of Sheffield
Sheffield S10 2TN
UK

Andrew Dugmore
Department of Geography
University of Edinburgh
Drummond Street
Edinburgh EH8 9XP
UK

Jon Sadler
Department of Archaeology and Prehistory
University of Sheffield
Sheffield S10 2TN
UK

ABSTRACT Study of the insect fauna of Iceland includes casual records as far back as the late seventeeth century. More recently, the study of fossil assemblages has extended the evidence back over much of the Holocene and it is now possible to assess changes in both distribution and species content over a more extended timescale. Human impact is seen as the major factor forcing change, ranging from the destruction of the forests and soils to improvements in the nature of exploitation of farmlands. Although there is more evidence for introductions than for extinctions, most of the former consist of species which are only able to maintain themselves in the artificially cushioned environments created by Man.

1. INTRODUCTION

Despite the increasing amount of research carried out on fossil insect faunas (cf. Buckland & Coope, 1990), there remain serious problems in using the fossil record to assess changes

127

J. K. Maizels and C. Caseldine (eds.), Environmental Change in Iceland: Past and Present, 127–146.
© 1991 *Kluwer Academic Publishers. Printed in the Netherlands.*

in the recent past. Many of these are the result of differing collecting methods and the paucity of quantified data upon the present biota. Rarely are entomological records quantified beyond the vague comments of common, rare, or very rare and a list of localities (e.g. Larsson & Gígja, 1959), and the fossil record has an equally enigmatic set of problems concerning the taphonomy of assemblages (cf. Kenward, 1976). Preservational problems and limits upon the levels of identification presently achieved further impede the comparison between fossils and modern assemblages and largely restrict study to the better known groups, particularly the Coleoptera (beetles). Recent work in the use of modern Coleopterous assemblages in the assessment of localities for conservation purposes (e.g. Garland, 1983; Hutcheson, 1990) further emphasises the need to bridge the gap between neontologist and palaeontologist by examining assemblages of the 'recently dead' and using these both as a means of quantifying the modern record and sorting out the taphonomic problems of the fossils. In the absence of a full Icelandic checklist (Ólafsson, in prep.), taxonomy follows Kloet and Hincks (1977).

2. THE ICELANDIC BEETLE FAUNA

Records of individual species of insect in Iceland extend back to the seventeenth century (Gíslason & Ólafsson, 1989), but it is only with the detailed studies of the late Carl Lindroth, as part of his investigation into the origins of the fauna (Lindroth, 1931), that a general overview of the Coleopterous fauna of the island, with some indications of frequency, could be obtained. Lindroth's (1931) compilation, with later records, largely by Gígja, Tuxen and Björnsson, formed the basis for the Coleoptera volume in the *Zoology of Iceland* series (Larsson & Gígja, 1959), which provided distribution and habitat data for a total of 160 species, with additional comments on a number of casual records. A further volume in the same series by Larsson (1959) was concerned largely with the zoogeography of the beetle fauna. Subsequent work (Lindroth, 1965; Lindroth et al.,1973, 1988; Ingólfsson, 1976; Ólafsson, 1979; Dugmore & Buckland, 1984) has added a number of species to both the indigenous and anthropochorous lists, but a critical assessment of the records by Lindroth (1965) reduces the total to 141 species. The examination of fossil insect assemblages has not only contributed to the list of casual introductions (cf. Buckland, 1988), but also provided evidence for possible extinction (Buckland et al., 1983) and severe contractions in frequency; several of the species recorded from single localities by Lindroth (1965; Lindroth et al., 1973; Lindroth et al., 1988) are remarkably common fossils in Late Holocene sediments. The frequency with which further species are added to the list, the difficulties of establishing whether a recent introduction has become established and taxonomic problems in certain groups, such as the Aleocharinae, makes any estimate of species number a transient, if not contentious affair. At least fourteen species, one of which, *Hydraena britteni* Joy, is only known as a fossil (Buckland et al., 1983), can be added to Larsson and Gígja's (1959) list, as revised by Lindroth (1965), giving a total of 155 species. Of these about one half, 77, may have been present pre-Landnám, of which, allowing for problems of identification to the species level of some groups, again largely the Aleocharinae, one third are known from pre-settlement deposits (Buckland, 1988). A number, like the recent occurrence of the ground beetle *Bembidion tetracolum* Say around hothouses in the Reykjavík area (Ólafsson, pers. comm.) and at Reykir in Skagafjörður, may be casual introductions in the ballast and dunnage of shipping (Lindroth, 1957; Sadler, in press), which have found suitable habitats in the 'natural' landscape. With the exception of a few species restricted to the south east and north of the country, most of the 77 species regarded tentatively as indigenous occur in the region between Mýrdalssandur and the Reykjanes peninsula, bounded to the north by Mýrdalsjökull, Eyjafjallajökull and the upland plateau, and on the off-shore islands of Vestmannæyjar. To the list may be added a further fifty species which occur either in wholly synanthropic situations, the interiors of barns, byres and foodstores, or in 'natural' habitats, which clearly owe their origin to human activity, for example, the introduction of large herbivores.

Although the species list might be regarded as fairly comprehensive, knowledge of relative and absolute frequency of individual taxa remains rudimentary. Most works lack anything but the broadest indications of relative numbers, and single captives of rare species, like *Lathrobium brunnipes* (F.), from Hornafjörður (Lindroth, 1965) and *Stenus umbratilis* Casey from northern Iceland (Ólafsson, pers. comm.), are better documented than many apparently more ubiquitous animals; both species are relatively frequent fossils in the late Holocene of south east Iceland. As part of the Surtsey research project, Lindroth and his co-workers (1973) compiled their records from the region between Mýrdalssandur and Markarfljót and from Heimæy into ranking orders for the top twenty outdoor species. In addition, frequency on the other islands in the Vestmannæyjar group was expressed in terms of captives per unit time of search. As the study was aimed at potential dispersal to Surtsey, little regard was taken either of habitat type or frequency and it is not possible to measure directly any elements of bias in the collecting, which largely employed search and capture techniques. Similarly, the influence of weather upon activity and emergence over the five years, 1965-70, collecting period on the mainland (and in 1966 on Heimæy) cannot be assessed. Lying outside the objectives of the project, synanthropic faunas were not quantified in the same way, although habitat and distributional data on the fauna of Eyjafjallasveit and Heimæy were included in the final monograph. These problems were partly addressed in 1980 by one of us (Dugmore, 1981) when a pitfall trapping exercise was conducted across a range of habitats in Eyjafjallasveit and in the birch woodland of Þórsmörk. Whilst it is recognised that the results are not directly comparable with those employed earlier, the differing rank order provides not only a cautionary tale but a bridge to the fossil record. In 1981, Dugmore carried out further research on the faunas of barns and byres both in the same region and in Skagafjörður in the north. This provides a further set of data for comparison with materials recovered from archaeological sediments (cf. Sveinbjarnardóttir et al., 1980; Perry et al., 1985).

Research upon fossil insect assemblages in Iceland was begun in 1979 with a preliminary examination of two samples from post-medieval deposits from the farm mound at Stóraborg, on the coast, south of Eyjafjallajökull (Sveinbjarnardóttir et al., 1980) and this work has been expanded to include archaeological deposits at Holt in Eyjafjallasveit (Buckland et al., in press a), Bessastaðir (Sadler & Ólafsson, in prep.) and Reykholt (Sveinbjarnardóttir et al., in prep.). In parallel, a series of natural successions, both pre- and post-Landnám, were examined across the region to provide information upon changes in the 'natural' landscape (Buckland, 1988; Buckland et al., 1986a).

3. THE LANDSCAPE

It is perhaps not a little ironic that tourism in Iceland focuses on what is regarded as the natural landscape and yet the bulk of that visited is the result of human management over the last 1100 years. The destruction of the birch forests (Hallsdóttir, 1987) was followed by the progressive loss of soils (Dugmore & Buckland, this volume). A large proportion of the bare rock was once clothed in soils and vegetated over. Much of the desert is man, or rather sheep, made, and the unconstrained nature of sandur appears also to be the result of forest clearance and overgrazing (Buckland et al., in press a). Bjarnason (1978) perceptively comments that the Icelanders owe their country over a quarter of its soils. Such radical alterations in habitat have inevitably led to changes in the distribution and frequency of species in the indigenous fauna, as well as creating and expanding niches for introductions. Changes in the flora have been the subject of extensive palynological research, initially by Thórarinsson (1944) and, more comprehensively, by Einarsson (1961) and recently by Hallsdóttir (1987). It is evident that the vegetation has suffered major change, to the extent that woodland has largely disappeared, surviving in a few small refuges, one of which, Þórsmörk, lies within the area of this study. Large areas of the grassland, created and modified by the colonists has, within the last century, suffered further improvement by drainage and reseeding and several of the dominant grasses are recent introductions (Davídsson, 1967); large areas have also been reduced to bare rock by overgrazing. It is against this background that changes in the beetle fauna may be considered. Fig. 1 provides a visual summary of the evolution of the Icelandic landscape.

4. THE PRE-LANDNÁM FAUNA

Fossil insect assemblages from deposits which clearly predate the arrival of Norse settlers in the mid-ninth century are surprisingly uniform in composition, dominated by species of *Stenus* (Table 1). Limitations upon the identification of the disarticulated fragments, which are recovered by the paraffin (kerosene) flotation technique (Buckland & Coope, 1990), frequently restricts assignment to the generic level, although forty individuals belong to the species most commonly recorded in Iceland at the present day, *S. carbonarius* Gyll. Two further species, *S. impressus* Germ. and *S. umbratilis* Casey (Fig. 2), are recent additions to the Icelandic list, the former from a single specimen from Seljalandsfoss in Eyjafjallasveit (Lindroth et al., 1973), supplemented by more recent material from Vík

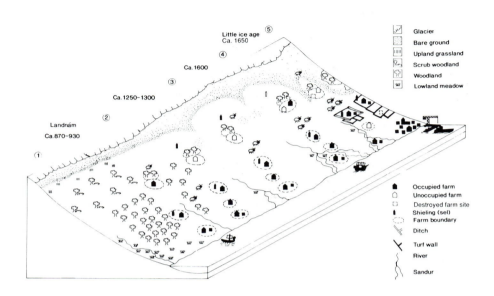

Fig.1 Flow diagram for the development of the Icelandic landscape.

in Mýrdalur (Lindroth et al., 1988) and from Þórsmörk (AJD), and the latter presently unrecorded in southern Iceland. This apparent rarity is shared by other species which are among the most frequent fossil taxa. The dominance of *Hydraena britteni* Joy would be more marked if only sites east of Markarfljót were considered since this water beetle, now apparently extinct in Iceland, does not occur in assemblages further west (Buckland et al., 1983). Although the rove beetle *Quedius umbrinus* Er. is widely distributed in Iceland, it tends to be very local (Lindroth et al., 1973) and its dominance in the fossil assemblages may be an artifact, the result of its habitat, wet meadowland, being most likely to form the basis of peat accumulation. *Lathobium brunnipes* (F.) is currently only recorded from two localities, one in the extreme south east of the country, in Horna-fjörður (Lindroth et al., 1988) and the other at Landbrot in Vestur-Skaftafellsýsla (Ólafsson, pers. comm.); it occurs in virtually all pre-Landnám samples examined and is present in many later successions, for example at Ketilsstaðir in Mýrdalur (Buckland et al., 1986b) which have accumulated away from human interference in the habitat.

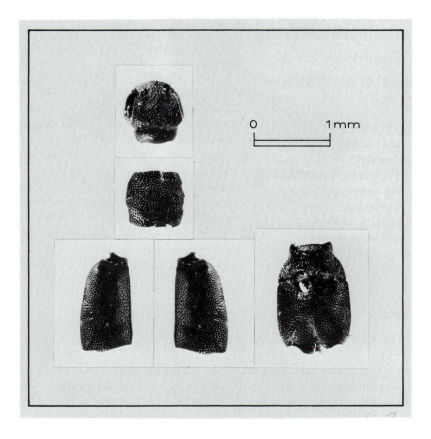

Fig.2 *Stenus impressus* Casey, a fossil specimen from pre-Landnám sediments at Holt in Eyjafjallasveit, S.W.Iceland.

The point of its final decline in frequency has yet to be fixed, since material which covers the period from the seventeenth to the twentieth century has not been studied, but this radical difference between the pre-human interference natural assemblages and the modern ones requires consideration. At Holt in Eyjafjallasveit, where the stratigraphy and tephrochronology allow some correlation between two farm mounds some 500m apart (Buckland et al., in press a), this assemblage disappears after Landnám, probably as a result of eutrophication, induced by waste from Man and his animals. At Ketilsstaðir, in a bog grazed and cut for peat (Buckland et al., 1986b), it survives throughout the sequence examined, through at least to the late sixteenth century.   The differences cannot be

Table 1. Ranking list for the 10 most abundant species of Coleoptera from
Pre-Landnám assemblages in South East Iceland (No. of sites 8, no. of
samples 53).

1. *Hydraena brittreni* Joy   (79 examples)
2. *Stenus* spp.   (78)
3. *Quedius umbrinus* Er.   (74)
4. *Lathrobium brunnipes* (F.)   (66)
5. Aleocharinae indet.   (54)
6. *Hydroporus nigrita* (F.)   (44)
7. *Stenus carbonarius* Gyll.   (40)
8. *Otiorhynchus nodosus* (Mull.)   (22)
9. *Agabus bipustulatus* (L.)   (13)
10. *Pterostichus diligens* (St.)   (10)

dismissed as taphonomic, and those who might seek to exonerate Man might seek to
implicate the cooler period of the 'Little Ice Age' in the faunal changes. *H. britteni*, on
its European mainland distribution, appears close to its northern limit in southern Iceland,
a point further emphasised by its apparent absence west of Markarfljót, a pattern shared
with a number of other species, including *Trechus obtusus* (Er.), also present in the
Ketilsstaðir pre-Landnám assemblages. Extinction as a result of climatic change in a group
of animals known for their sensitivity to climate (cf. Coope, 1987) might appear the better
option (Buckland et al., 1983) but such fails to account for the other faunal changes. The
association is one of a habitat particularly difficult to collect from, tussocks and pools on
the active bog surface, and modern distribution patterns are therefore likely to be under-
estimates. This tussock environment may also include two other species which appear
to be more common as fossils than in the modern landscape, the Scydmaenid *Stenichnus
collaris* (Müll. & Kunz.) and the Pselaphid *Bryaxis puncticollis* Denn. Examination of the
available habitat data does little to extend the general image of a wetland assemblage,
although individual species are not all restricted to such localities. Drainage and improvement of grassland has certainly contributed to the decline of these species, perhaps leading
one, *H. britteni* , to extinction from the Icelandic fauna and leaving others close to the
edge.
    Not all of Iceland's wetlands, however, have been drained and reseeded, and even in
Eyjafjallasveit and Mýrdalur, which include some of the most extensively utilised land for
pastoral agriculture, there remain areas where evident human impact is restricted to the
intermittent passing grazing of sheep. At least three of the pre-Landnám beetles,
however, have a connection with that rarest of surviving Icelandic habitats, the birch
forest. The absence of *Urwaldreliktiere*, the fauna associated with primary, undisturbed
woodland, from the island is a biogeographic accident consequent upon the immigration

of the biota during early Holocene deglaciation (Buckland et al., 1986a; Buckland & Dugmore, this volume), but several animals were able to exploit the ameliorated climate that forest cover would have provided.   Larsson and Gigja (1959) record *Bryaxis puncticollis* from 'shaded ground, particularly in withered leaves at the base of old tree stumps' in Central Europe and, in Scandinavia, it shows the same apparent liking for forest habitats (Lindroth et al., 1973). In Britain, in contrast, Pierce (1957) notes it from similar environments to the Icelandic captures, 'in moss and grass tussocks'. A comparable pattern is noted for *Hydraena britteni*. Lohse (1971) describes it from Central Europe in 'laubgefullten Tumpeln' and, in Scandinavia, Hansen   (1987) further stresses the woodland connection, 'in or at the edge of predominantly stagnant waters, mainly in more or less eutrophic shallow, somewhat shaded pools among moss, leaf litter and the like, often in woodland'. The British occurrences of the species (Balfour-Browne, 1958; Friday, 1988) indicate a preference for eutrophic localities, fens and grassy pools, something of a contrast with the apparently oligotrophic habitat suggested by the Icelandic fossil records. Ecological information upon individual species of the large rove beetle genus *Stenus* is less easily obtained but Lindroth (Lindroth et al., 1973) records his one specimen of *S. impressus* Germ. on 'somewhat moist grassland in a shaded position'; collecting and pit-fall trapping by one of us (AJD) in Þórsmörk in 1980 found that this species was not infrequent amongst the forest floor vegetation and leaves at the base of birches throughout the park. The tenuous link with woodland may be inappropriate, conditioned by the known losses from the British fauna as a result of forest clearance (Buckland, 1979). For many species, it is the relative permanence of their core habitat area which is of overriding significance. Peripheral localities, contrained by a cushioned thermal regime, may maintain populations in the face of declining temperatures until disturbed by Man or his stock. The low dispersal potential of species at the edge of their range, coupled with an apparent increased habitat specificity in such locations, severely restricts possibilities of recolonisation, until both core area densities and climatic thresholds are passed (cf. den Boer, 1977). In an interesting and innovative experiment, Perry (1986; see also Coope, 1986) reversed the climatic change modelling package devised by Atkinson and his co-workers (1986, 1987), and predicted what species of ground, water, rove and dung beetle might, given unlimited access, establish themselves in Iceland. He then proceeded to theoretically cool the climate, and noted what species thereby became extinct. A one degree centigrade depression, less than that which occurred during the 'Little Ice Age' (Grove, 1988) was sufficient to force *H. britteni* off the south coast of Iceland. As this beetle formed part of the natural fauna, the implication, even allowing for the protection provided by a former forest cover, is that the coldest part of the Holocene in Iceland was during the last four hundred years. Based upon a single species, which may yet be found living in the country, this is a fragile argument and many more pre-Landnám samples require analysis to substantiate any hypothesis, but there is some support for a cold late Holocene for other, geomorphological lines of evidence (Dugmore, 1987).

The taphonomic bias in any fossil assemblage is more easily understood when a multidisciplinary approach, utilising several lines of evidence is applied to the sample (cf. Buckland et al., 1986b), but quantification remains a problem. The rank order of ten most frequent fossil beetle taxa (Table 1) is not directly comparable with similar lists for post- Landnám localities. Despite Ári the Wise's comment in the thirteenth century that at Landnám the country was forested from mountain top to sea shore (Valtysdóttir, 1980), the associated woody peats have been rather elusive in south west Iceland. The three principal localities, Ketilsstaðir in Mýrdalur, Einhyrningur in Fljótshlíð, and Kópavogur near Reykjavík, are in raised bog situations, with little evidence of the invasion of trees during the late Holocene. Holt in Eyjafjallasveit lies on coastal sandur of Markarfljót and is associated with silt deposition until after Settlement (Buckland et al., in press a). Only at Ósnabakki, below Vörðufell in Arnessýsla, do large birches appear in the pre-Landnám sequence and the associated fauna is sparse. The nature of the forest cannot therefore be regarded as adequately assessed from the entomological perspective. The primal nature of the interior plateau, where sheep have been progressively extending the desert coastwards for over a thousand years is also uncertain. A sampling strategy designed to include these broad habitat groupings, together with those of dry to bare ground and the marine littoral, may modify the rank order of species, and add those which, like *Bembidion grapei* Gyll. (cf. Coope, 1986), presently lack a fossil record in Iceland.

5. THE POST-LANDNÁM FAUNA

The problems of assessing the nature of the pre-Landnám fauna are minor compared with those which attend assemblages from the mid-ninth century onwards. As well as direct introductions to habitats wholly introduced by Man, like the hay and dung faunas (cf. Lindroth, 1957), human interference created a mosaic of environments, which allowed species of restricted distribution, such as those of the eutrophic grasslands of bird cliffs, to become widespread. Species diversity in samples from localities away from the archaeological sites increases immediately after Landnám (cf. Buckland et al., 1986b, fig.7), and samples from the sites themselves are massively weighted by the introduced fauna of stored hay and its residue, present, on the Holt evidence, by before A.D.1000 (Buckland et al., in press a). Several of the indigenous species, including several Aleocharines, with other decaying vegetation species, like *Tachinus corticinus* Grav., are able to join undoubted introductions, like the dung beetle, *Aphodius lapponum* Gyll., in the new habitats, and several incidental stowaways on the colonists' ships must have moved into 'natural' environments, finding niches unoccupied since their origin in the early Holocene.

The Leiodid *Catops fuliginosus* Er. is by far the most frequent beetle in archaeological assemblages and is usually regarded as wholly synanthropic in Iceland (Larsson & Gígja, 1959). Lindroth, however, (Lindroth et al., 1973) found the species in the lush vegetation below the cliffs at Víkurhamrar, one kilometre west of Vík in Mýrdalur, where the association seems to be with puffin burrows, a habitat from which it is also recorded in the Faroe Islands. This locality, one of the warmest places in Iceland, includes a number of otherwise largely synanthropic taxa, including *Ocalea picata* Steph. and

Cryptophagus pilosus Gyll. Both *C. fuliginosus* and *C. pilosus* were again found at the site in 1984. As all are macropterous (Lindroth et al., 1973), their presence may reflect active dispersal and successful colonisation from nearby farms.

Table 2. Ranking list for the 10 most abundant species of Coleoptera from 'natural' post-Landnám assemblages in South East Iceland (no. of sites 2, no. of samples 34).

1.	*Quedius umbrinus* Er.	(57 examples)
2.	Aleocharinae indet.	(56)
3.	*Hydroporus nigrita* (F.)	(40)
4.	*Hydraena britteni* Joy	(39)
5.	*Stenus* spp.	(27)
6.	*Lesteva longoelytrata* (Goez.)	(22)
7.	*Hypnoidus riparius* (F.)	(14)
8.	*Otiorhynchus nodosus* (Mull.)	(13)
-.	*Lathrobium brunnipes* (F.)	(13)
10.	*Othius angustus* Steph.	(12)
-.	*Gabrius* cf. *trossulus* (von Nord.)	(12)

Any assessment of species frequency from 'natural' habitats post-Landnám has to rely upon those samples which are uninfluenced by archaeological deposits (Table 2). The latter not only include taxa which may belong to either synanthropic or outdoor communities, but also have the additional problem of the widespread use of peat, both as litter and as fuel, leading to the inclusion of individuals which may be several thousand years older than the enclosing deposit (Buckland et al., in press b). Some of the differences between Tables 2 and 3 are reflections of these factors, particularly the large number of Aleocharines in Table 3 and high frequencies of *Hydraena britteni* and the other species common in pre-Landnám assemblages, but other differences are a result of human modification of the environment. Those species associated with drier grassland and hayfield appear in the top ten most frequent taxa which includes archaeological samples, a factor only partly reflected in the 'natural' assemblages. The expansion of grassland, consequent upon forest clearance (Hallsdóttir, 1987), and its maintenance by the grazing of introduced animals, produced low-altitude meadowland, as well as nutrient enriched areas. The group of species, *Lesteva longoelytrata* (Goez.), *Calathus melanocephalus* (L.), *Othius angustus* Steph. (? *melanocephalus* (Grav.)), *Tachinus corticinus* Grav., and the click beetle, *Hypnoidus riparius* (F.), provide an index of this change. The pre-Landnám wetland assemblage, however, maintains its frequency in all situations away from farm sites and its decline is evidently a feature of the modern period.

6. THE SYNANTHROPIC FAUNA

Norse colonisation of the North Atlantic islands relied upon the export of a self-contained, almost exclusively pastoral agricultural system from Scandinavia and the Scottish islands, and it is hardly surprising that much of the synanthropic insect fauna

Table 3. Ranking list for the 10 most abundant species of Coleoptera from post-Landnám fossil assemblages in South East Iceland (no. of sites 5, no. of samples 137.

1. Aleocharinae indet.	(866 examples)
2. *Lesteva longoelytrata* (Goez.)	(173)
3. *Tachinus corticinus* Grav.	(161)
4. *Hydraena britteni* Joy	(95)
-. *Othius angustus* Steph.	(95)
6. *Stenus* spp.	(94)
7. *Quedius umbrinus* Er.	(92)
8. *Hydroporus nigrita* (F.)	(89)
9. *Calathus melanocephalus* (L.)	(68)
10. *Hypnoidus riparius* (F.)	(10)

was inadvertently carried with the primary settlers (Sadler, in press). Samples include a number of species previously (Larsson & Gígja, 1959) regarded as recent introductions to Iceland. *Phyllodrepa floralis* Payk. noted (op.cit.) as appearing 'to have immigrated in rather recent time and to have not yet attained its final distribution', occurs in medieval deposits at Bessastaðir. The same site has produced *Oryzaephilus surinamensis* L., regarded by Gígja (1944) as of American origin but common in Roman deposits in England (Buckland, 1981), and *Sitophilus granarius* (L.), not listed by Larsson and Gígja (1959). Both species occur in most samples from the site, implying the regular import of cereals (Sadler & Ólafsson, in prep.); elsewhere they only appear sporadically. Archaeological samples also include a number of potential colonists which failed to establish themselves. Two additional species of dung beetle appear at Stóraborg (Buckland, in Coope, 1986) and another Scarabaeid associated with rotting vegetation, *Oxyomus silvestris* (Scop.), occurs at Bessastaðir. Two species of ant, from post-medieval deposits at Reykholt in Borgarfjörður, were presumably adventitious in packing. One of these, *Hypoponera punctatissima* (Rog.), has recently established itself in heated buildings in Reykjavík (Ólafsson & Richter, 1985). Ólafsson (1976) has also discussed the casual occurrence of the ladybird *Coccinella septempunctata* L. which has failed to establish itself.

Synanthropic insect assemblages have been extensively examined from archaeological deposits at Stóraborg, Holt and Bessastaðir. The samples, covering the period from shortly after Landnám to the seventeenth century, provide much detailed evidence to improve archaeological interpretation (cf. Buckland & Perry, 1989), and also allow a ranking of species frequency (Table 4) which may be compared with modern data collected by Dugmore (Table 5). Although the comparative samples from actively used byres and barns are small, the general pattern is confirmed by other, unquantified collecting.

Table 4. Ranking list of the 10 most abundant synanthropous species of Coleoptera from archaeological assemblages in South East Iceland.

1.	*Catops fuliginosus* Er.	(2741 examples)
2.	*Typhaea stercorea* (L.)	(1057)
3.	*Lathridius minutus* (group)	(678)
4.	*Corticaria elongata* (Gyll.)	(397)
5.	*Atomaria* spp.	(275)
6.	*Cryptophagus* spp.	(240)
7.	*Omalium rivulare* (Payk.)	(187)
8.	*Quedius mesomelinus* (Marsh.)	(176)
9.	*Omalium excavatum* Steph.	(99)
10.	*Xylodromus concinnus* (Marsh.)	(92)

The dominance in the fossil assemblages of *Catops fuliginosus* remains even when one atypical sample, containing a minimum number of 1,046 individuals from Stóraborg, is excluded; the beetle was not taken during the summers of 1980 and 1981 and Lindroth and others (1973) do not record it as common. Perry and others (1985) have discussed the possible reasons for this apparent major decline in frequency. The nature of hay storage and byre interiors, on the archaeological and ethnographic evidence, has not changed substantially, and the analogues for its natural habitats provided by Man's farms do not seem obviously to have declined. Changes in rank order of other elements in the hay fauna are also relevant, in that the bulk of the species consist of feeders upon the fungal microflora and their predators. Species of *Cryptophagus*, largely *C. scanicus* (L.), are now much more frequent than *Typhaea stercorea* (L.). *Lathridius minutus* (L.), with *L. pseudominutus* Strand noted in small numbers amongst the material from south east Iceland (Tozer, pers. comm.), dominates the assemblage. This shift in frequency must indicate a very recent change in the nature of the hay microflora. Mechanisation of farming has certainly led to a much drier crop entering storage but, whilst recovery from the fields may be more efficient, the same buildings have continued in use and damp, mouldy residues fail to produce assemblages of beetles similar to the fossil ones. Since the synanthropic fauna exploits an artificially warmed environment, that created by the fungal decay of hay, climatic change is unlikely to be relevant. The major change that has taken place lies in the drainage and reseeding of hayfields. Where once fields were rich in a range of species, both native and anthropochorous, the modern hayfield tends towards a monoculture, with lush, fast-growing grasses replacing the former small, scythed fields of semi-natural grassland, in which the sedges, Cyperaceae, were at least as frequent as the grasses. Change in the synanthropic insect fauna, therefore, probably reflects changes in the fungal microflora of decay in the now grass dominated barns. Stefan Adelsteinsson (pers. comm.) has pointed out that grass hay in storage tends to generate much higher temperatures during decay than sedge-dominated material and this must inevitably have some impact on the faunal assemblage. Examination of barns used for silage, a technique introduced to Iceland about a century ago, has produced relatively few Coleoptera, suggesting that temperatures therein are beyond the range of tolerance of many synanthropic species.

Table 5. Ranking list for the 8 most abundant species of Coleoptera found by Dugmore in 1980-81 in synanthropic situations in South East Iceland (modified by comparison with fossil assemblages.

1.	*Lathridus minutus* (group)	(154 examples)
2.	*Atomaria apicalis* Er.	(28)
3.	*Philonthus sordidus* Grav.	(6)
4.	Aleocharinae indet.	(5)
5.	*Cryptophagus* spp.	(4)
6.	*Typhaea stercorea* (L.)	(3)
7.	*Quedius mesomelinus* (Marsh.)	(2)
8.	*Corticaria elongata* (Gyll.)	(2)

A number of other changes are a direct reflection of increased levels of human hygiene, although these are more evident in the Dipterous faunas than the Coleopterous ones (Skidmore, in prep.). The spider beetle, *Tipnus unicolor* (Pill & Mitt.), on evidence from Reykholt (Sveinbjarnardóttir et al., in prep.), appears characteristic of, but not exclusive to (Lindroth et al., 1973), the marginally drier habitats provided by human abodes, rather than those of his stock. It is perhaps less frequent now than in the past and is supplemented, if not replaced in drier habitats by the Australasian import *Ptinus tectus* Boied. Whether this is a case of competitive exclusion is doubtful, since the much drier habitat, created by adequate heating, favoured by *P. tectus* would have been unavailable in the past and it appears merely to have moved successfully into an unoccupied niche.

7. THE RECENT FAUNA OF THE LANDSCAPE

The basis for the interpretation of fossil assemblages of insects has to be a thorough understanding of the living biota. It is therefore instructive to compare sets of data obtained by different collecting techniques and at different times in south west Iceland and on Heimæy (Tables 6-8). Although varying positions within the ten most frequent taxa may be the result of stochastic factors for much of the assemblage, there do appear to be significant differences between Heimæy and the mainland. The absence of *Patrobus atrorufus* (Str.) from the 1980-1 collections must be significant. Dugmore (1981) relied almost entirely upon pitfall trapping in his research and concentrated upon changes across ecotones, ranging from bare ground, over erosion scars onto grazing lands and hayfields. His samples should have included sufficient of the rich, more or less weedy vegetation preferred by this species (Lindroth et al., 1973) to obtain a representative sample; only two specimens were taken.

Table 6. Ranking list for the 10 most abundant species of Coleoptera in South East Iceland (modified from Lindroth et al. (1973) for comparison with fossil assemblages).

1.	*Hypnoidus riparius* (F.)	(416 examples)
2.	*Trechus obtusus* Er.	(303)
3.	*Calathus melanocephalus* (L.)	(267)
-.	*Nebria gyllenhali* (Sch.)	(267)
5.	Aleocharinae indet.	(162)
6.	*Patrobus atrorufus* (Str.)	(138)
7.	*Gabrius* sp.	(117)
8.	*Geostiba circellaris* (Grav.)	(111)
9.	*Othius angustus* Steph.	(98)
10.	*Otiorhynchus arcticus* (F.)	(97)

Table 7. Ranking list for the 10 most abundant species of Coleoptera on Heimæy, Iceland (modified from Lindroth et al. (1973) for comparison with fossil assemblages).

1.	Aleocharinae indet.	(460 examples)
2.	*Amara quenseli* (Sch.)	(309)
3.	*Calathus melanocephalus* (L.)	(199)
4.	*Nebria gyllenhali* (Sch.)	(174)
5.	*Otiorhynchus arcticus* (F.)	(149)
6.	*Hypnoidus riparius* (F.)	(118)
7.	*Trechus obtusus* Er.	(105)
8.	*Patrobus atrorufus* (Str.)	(101)
9.	*Othius angustus* Steph.	(99)
10.	*Barynotus squamosus* Germ.	(83)

This ground beetle is also an infrequent fossil, occurring only in the post-medieval deposits at Stóraborg. Although the latter may be a purely taphonomic problem, in that its habitat rarely contributed to the deposits which made up the farm mounds, its virtual absence from Dugmore's collections is more likely to be an artifact of the time of year that fieldwork was carried out. Thiele (1977) notes that imagines are active in the early summer and then aestivate until reproduction takes place in late summer. Lindroth and his co-workers (1973) had extended field seasons, not only over five years but also over the greater part of some summers. Whilst the latter set of results must therefore be more representative of the Coleopterous fauna overall, it is doubtful whether they reflect a true indication of species frequency, in that their project was not intended to assess the frequency of habitat type as well as individual species.

Table 8. Ranking list for the 10 most abundant species of Coleoptera in South East Iceland, collected by Dugmore (1980-81) (modified for comparison with fossil assemblages).

1.	*Nebria gyllenhali* Sch.	(192 examples)
2.	*Patrobus septentrionis* (Dej.)	(163)
3.	Aleocharinae indet.	(111)
4.	*Calathus melanocephalus* (L.)	(106)
5.	*Amara quenseli* (Sch.)	(49)
6.	*Trechus obtusus* Er.	(43)
7.	*Tachinus corticinus* (Grav.)	(40)
8.	*Hypnoidus riparius* (F.)	(33)
9.	*Pterostichus adstrictus* Esch.	(21)
10.	*Quedius fulvicollis* (Steph.)/ *boops* (Grav.)	(16)

8. CONCLUSION

The fossil record indicates major changes in the beetle fauna of Iceland, on a scale at least equivalent to that noted in the flora (Hallsdóttir, 1987). Similar changes may be expected in other groups, as human impact has been effective across the spectrum of animal life. The freshwater breeding flies are the subject of detailed study at Mývatn and periodic changes in frequency have been noted (cf. Einarsson, 1982; Gíslason & Jóhannsson, 1985). Changes in animal husbandry are also likely to have modified the introduced synanthropic Dipterous fauna (Skidmore, in prep.). There are also indications of change in the caddis fly (Trichoptera) fauna (Gíslason, 1981). Both groups, in the form of larval exuviae are common fossils in peats and lacustrine sediments and are important in the assessment of water quality and human impact (cf. Williams, 1988; Brodin, 1986). Other groups are better documented. The impact of one mammalian introduction into a previously unoccupied niche, the American mink, extends across the food chain from the Crustacea to the fish and birds (Skírnisson, 1980). Similar studies have been the cause of worldwide concern (e.g. Elton, 1958; Crosby, 1986) and were the subject of a recent symposium in London (Kornberg & Williamson, 1987). Much of the change, however, is insidious and can only be assessed by recourse to the fossil record.

The interpretation of the evident changes and their effective quantification remains problematic, beyond the wider conclusions and this paper is an attempt to provide hypotheses for the testing, as well as suggesting directions for further research. The ranked lists of species (Tables 1-8) undoubtedly include data upon faunal change which are obscured by a lack of knowledge of taphonomic processes, which convert a living assemblage of animals to a fossil one. There remains the gap in the record between the early post-medieval period and the beginnings of modern collecting during the nineteenth century and the latter are poorly quantified, whilst the former are time averaged over a variable span. Sample resolution can be improved by the careful use of the frequent tephra horizons, which provide isochrones over large parts of the country and beyond (Buckland et al., 1981), although these are rarely available in the more disturbed deposits of archaeological sites. Modern collecting requires a concerted effort to classify habitat type and frequency, best achieved by recourse to remote sensing techniques, before quantified species lists for each are achieved. Thereafter, monitoring of change and assessment of conservation aspects might be more easily obtained. The examination of the recently dead, that is, the fossil assemblages in the making, will bridge the gap between the living and the dead.

ACKNOWLEDGEMENTS

This research was carried out under the Icelandic Research Council permits 1/80 and 22/81. Later work, upon insect faunas from archaeological sites was carried out as part of a joint ongoing project with the National Museum of Iceland, under the auspices of Mjóll Snæsdóttir at Stóraborg, Guðmundur Ólafsson at Bessastaðir and Guðrún Sveinbjarnardóttir at Reykholt; their assistance and interest in the research is gratefully acknowledged. Our knowledge of the natural history of Iceland owes much to discussions with Gísli Már Gíslason and Erling Ólafsson, although the errors, and some deliberate windmills for others to tilt at, remain our own. The research was primarily funded by a research grant from the Leverhulme Trust to the University of Birmingham in 1979. Later work has included funding from the Universities of Birmingham, Edinburgh and Sheffield, NATO Scientific Affairs Division and the National Museum of Iceland. J.P.Sadler is in receipt of an SERC research award and A.J.Dugmore an SERC research fellowship. Finally, the farmers of Eyjaflallasveit, who tolerated our intrusion onto their land and into their barns are gratefully acknowledged.

REFERENCES

Atkinson, T. C., Briffa, K. R., Coope, G. R., Joachim, J. M. & Perry, D. W. (1986) 'Climatic calibration of coleopteran data', In B. E. Berglund (ed.) Handbook of Holocene Palaeoecology and Palaeohydrology, 851-858. J. Wiley & Son, Chichester.
Atkinson, T. C., Briffa, K. R. & Coope, G. R. (1987) 'Seasonal temperatures in Britain during the past 22,000 years, reconstructed using beetle remains', Nature (London) 325, 587-592.
Balfour-Browne, F. (1958) British Water Beetles. III. Ray Society, London.

Bjarnason, A.H. (1978) 'Erosion, tree growth and land regeneration in Iceland', in M.W.Holdgate & M.J.Woodman (eds.) The Breakdown and Restoration of Ecosystems, 241- 248. Plenum Press, New York & London.

Boer, P.J. den (1977) 'Dispersal power and survival. Carabids in a cultivated countryside', Miscellaneous papers 14, Landbouwhogeschool Wageningen.

Brodin, Y. (1982) 'Palaeoecological studies of the recent development of the Lake Växjösjön. 4. Interpretation of the eutrophication process through the analysis of subfossil chironomids', Archiv für Hydrobiologie 93, 313-326.

Buckland, P. C. (1979) 'Thorne Moors : a palaeoecological study of a Bronze Age site (a contribution to the history of the British insect fauna)', University of Birmingham, Department of Geography Occasional Publication 8, Birmingham.

Buckland, P. C. (1981) 'The early dispersal of insect pests of stored products as indicated by archaeological records', Journal of Stored Product Research 17, 1-12.

Buckland, P. C. (1988) 'North Atlantic faunal connections - introduction or endemics?', Entomologica Scandinavica Supplement 32, 7-29.

Buckland, P.C. & Coope, G.R. (1990) A Review and Bibliography of Quaternary Entomology. University of Sheffield.

Buckland, P. C., Foster, P., Perry, D. W. & Savory, D. (1981) 'Tephrochronology and Palaeoecology: the value of isochrones', in S. Self & R. S. J. Sparks (eds.) Tephra studies, 381-389. N.A.T.O. Advanced Study Institutes Series. Reidel, Dordrecht.

Buckland, P.C., Gerrard, A.J., Larsen, G., Perry, D. W., Savory, D.R. & Sveinbjarnardóttir, G. (1986) 'Late Holocene Palaeoecology at Ketilsstaðir in Mýrdalur, South Iceland', Jökull 36, 41-55.

Buckland, P. C. & Perry, D. W. (1989) 'Ectoparasites of Sheep from Stóraborg, Iceland and their interpretation. Piss, parasites and people, a palaeoecological perspective', Hikuin 15, 37-46.

Buckland, P. C., Perry, D. W., Gíslason, G. M. & Dugmore, A. J. (1986) 'The pre-Landnám Fauna of Iceland: a palaeontological contribution', Boreas 15, 173-184.

Buckland, P. C., Perry, D. & Sveinbjarnardóttir, G. (1983) '*Hydraena britteni* Joy (Coleoptera, Hydraenidae) fundin á Íslandi í setlögum frá því seint á nútíma', Náttúrufræðingurinn 52, 37-44.

Buckland, P.C., Dugmore, A.J., Perry, D., Sveinbjarnardóttir G. & Savory, D. (in press a) 'Holt in Eyjafjallasveit, Iceland: a palaeoecological study of the impact of Landnám', Acta Archaeologica Lundensis.

Buckland, P.C., Sadler, J. & Smith, D. (in press b) 'An insect's eye-view of the Norse farm', in C.Batey (ed.) Proceedings of the Eleventh Viking Congress, Kirkwall.

Coope, G. R. (1986) 'The invasion and colonization of the North Atlantic islands: a palaeoecological solution to a biogeographic problem', Philosophical Transactions of the Royal Society of London B314, 619-635.

Coope, G.R. (1987) 'The response of Late Quaternary insect communities to sudden climatic changes', in J.H.R.Gee & P.S.Giller (eds.), Organisation of Communities - Past and Present, 421-438. Blackwell Scientific Publications, Oxford.

Crosby, A.W. (1986) Ecological Imperialism.The Biological Expansion of Europe, 900-1900, Cambridge University Press, Cambridge.

Davídsson, I. (1967) 'The immigration and naturalization of flowering plants in Iceland since 1900', Greinar 4(3).

Dugmore, A.J. (1981) Icelandic Coleoptera: a study of the Coleopteran Fauna of Southern Iceland. Unpublished B.Sc. dissertation, University of Birmingham.

Dugmore, A.J. (1987) Holocene glacier fluctuations around Eyjafjallajökull , south Iceland: a tephrochronological study. Unpublished Ph.D. Thesis, University of Aberdeen.

Dugmore, A.J. & Buckland, P.C. (1984) 'Vatnabjallan *Oreodytes sanmarki* (Sahl.) (Col.,Dytiscidae) fundin á Íslandi', Náttúrufræðingurinn 53, 7-12.

Einarsson, A. (1982) 'The palaeolimnology of Lake Mývatn, northern Iceland: plant and animal microfossils in the sediment', Freshwater Biology 12, 63-82.

Einarsson, Th. (1961) 'Pollenanalytische Untersuchungen zur spät- und postglacialen Klimageschichte Islands', Sonderveröffentlichungen der Geologisches Instituten Universitäts Köln 6, 1-52.

Elton, C. S. (1958) The Ecology of Invasions by Animals and Plants. Methuen, London.

Friday, L.E. (1988) 'A Key to the Adults of British Water Beetles', Field Studies 7, 1-151.

Garland, S.P. (1983) 'Beetles as primary woodland indicators', The Sorby Record 21, 3-38.

Gígja, G. (1944) Meindýr í Husum og Gródri og varnir gegn ðeim. J.Guðbjörnsson, Reykjavík.

Gíslason, G.M. (1981) 'Predatory exclusion of *Apatania zonella* (Zett.) by *Potamophylax congulatus* (Steph.) (Trichoptera: Limnephilidae) in Iceland', in G.P.Moretti (ed.) Proceedings of the 3rd. Symposium on Trichoptera, 93-98. Junk, The Hague.

Gíslason, G.M. & Jóhannsson, V. (1985) 'Bitmyið í Laxá í S-Þingeyjarsýslu', Náttúrufræðingurinn 55, 175-194.

Gíslason, G.M. & Ólafsson, E. (1989) 'Entomology in Iceland', Fauna norvegica Series B36.

Grove, J. (1988) The Little Ice Age. Methuen, London.

Hallsdóttir, M. (1987) 'Pollen Analytical Studies of Human Influence on Vegetation in Relation to the Landnám Tephra Layer in Southwest Iceland', Lundqua Thesis 18. Lund.

Hansen, M. (1987) 'The Hydrophiloidea (Coleoptera) of Fennoscandia and Denmark', Fauna Entomologica Scandinavica 17.

Hutcheson, J. (1990) 'Characterization of terrestrial insect communities using quantified Malaise trapped Coleoptera', Ecological Entomology 15, 143-151.

Ingólfsson, A. (1976) 'Smáðyrlif og gróður á sjávarfitjum við Gálgahraun', Náttúrufræðingurinn 46, 223-237.

Kenward, H. K. (1976) 'Reconstructing ancient ecological conditions from insects remains: some problems and an experimental approach', Ecological Entomology 1, 7-17.

Kloet, G.S. & Hincks, W.D. (1977) 'A Checklist of British Insects, part 3. Coleoptera and Strepsiptera. (rev. R.D.Pope)', Handbooks for the Identification of British Insects, XI 3. Royal Entomological Society of London.

Kornberg, H. & Williamson, M.H. (1987) Quantitative Aspects of the Ecology of Biological Invasions, Royal Society, London.

Larsson, S.G. (1959) 'Coleoptera 2. General Remarks ', Zoology of Iceland III, 46b. Ejnar Munksgaard, Copenhagen.

Larsson, S.G. & Gígja, G. (1958) 'Coleoptera 1. Synopsis', Zoology of Iceland III, 46a. Ejnar Munksgaard, Copenhagen.

Lindroth, C.H. (1931) 'Die Insektenfauna Islands und ihre Probleme', Zool Bidr. Uppsala 13, 105-600.

Lindroth, C. H. (1957) The Faunal Connections between Europe and North America, Wiley & Sons, New York.

Lindroth, C.H. (1965) 'Skaftfell, Iceland a living glacial refugium', Oikos, Supplement 6.

Lindroth, C.H., Andersson, H., Böðvarsson, H. & Richter, S.H. (1973) 'Surtsey, Iceland - the development of a new fauna. 1963-1970. Terrestrial Invertebrates', Entomologica Scandinavica Supplement 5.

Lindroth, C.H., Bengtson, S.-A. & Enckell, P. (1988) 'Terrestrial faunas of four isolated areas : A study in tracing old faunal centres', Entomologica Scandinavica Supplement 32.

Lohse, G.A. (1971) 'Fam. Hydraenidae', in H. Freude, K.W. Harde & G.A. Lohse (eds.), Die Käfer der Mitteleuropas, 3, 95-125. Goecke & Evers, Krefeld.

Ólafsson, E. (1976) 'Mariudeplugengdar í NV-Evrópu verður vart á Íslandi', Náttúrufræðingurinn 46, 134-138.

Ólafsson, E. (1979) 'Hambjalla, *Reesa vespulae* (Mill.) (Coleoptera, Dermestidae), nytt meinðyr á Íslandi', Náttúrufræðingurinn 49, 155-162.

Ólafsson, E. and Richter, S.H. (1985) 'Húsamaurinn *Hypoponera punctatissima*', Náttúrufræðingurinn 55, 139-146.

Perry, D.W. (1986) The Analysis of sub-fossil insect assemblages: a Numerical Approach. Unpublished Ph.D. Thesis, University of Birmingham.

Perry, D. W., Buckland, P. C. & Snæsdóttir, M. (1985) 'The application of numerical techniques to insect assemblages from the site of Stóraborg, Iceland', Journal of Archaeological Science 12, 335-345.

Pierce, E.J. (1957) 'Coleoptera (Pselaphidae)', Handbooks for the Identification of British Insects IV, 9. Royal Entomological Society of London.

Sadler, J.P. (in press) 'Beetles, ballast and biogeography: insect invaders of the North Atlantic', Acta Archaeologica Lundensis.

Skirnisson, K. (1980) 'Fæðuval minks í Grindavík', Náttúrufræðingurinn 49, 194-203.

Sveinbjarnardóttir, G., Buckland, P. C., Gerrard, A. J., Greig, J. R. A., Perry, D., Savory, D. & Snæsdóttir, M. (1981) 'Excavations at Stóraborg : a palaeoecological approach', Arbók hins Íslenzka Fornleifafélags (1980), 113-129.

Thiele, H.U. (1977) Carabid Beetles in their Environment. Springer, Berlin.

Thórarinsson, S. (1944) Tefrokronologiska studier på Island. Thorsardulur och dess foroderse. Ejnar Munksgaard, Copenhagen.

Valtysdóttir, H. (1980) ' ˜Í þann tíð var Ísland viði vaxið ', Náttúrufræðingurinn 50, 1-12.

Williams, N. E. (1988) 'The use of caddisflies (Trichoptera) in paleoecology', Palaeogeography, Palaeoclimatology, Palaeoecology 62, 493-500.

TEPHROCHRONOLOGY AND LATE HOLOCENE SOIL EROSION IN SOUTH ICELAND

Andrew Dugmore
Department of Geography
University of Edinburgh
Drummond Street
Edinburgh EH8 9XP
UK

Paul Buckland
Department of Archaeology and Prehistory
University of Sheffield
Sheffield S10 2TN
UK

ABSTRACT. The scale of soil erosion in Iceland since the Norse Settlement is generally well-known, but the detailed patterns of change are less clear. Through the application and refinement of tephrochronological frameworks, established by the late Professor Sigurður Þórarinsson and Guðrún Larsen, it has been possible to study in some detail the pattern of aeolian sediment accumulation around Eyjafjallajökull, and infer a pattern of soil erosion on a local scale. Later pre-historic rates of sediment accumulation seem to have been comparatively uniform in the zone between ca 500m above sea level (a.s.l.) and the sandur edge, but historic rates vary with altitude. In the early historic period, dramatic increases in upland sediment accumulation rates imply the early onset of acute local soil erosion in natural grasslands. These high rates decline through time, presumably as slopes stabilise at bedrock or gravel surfaces. At lower elevations inferred onsets and peaks of severe local soil erosion occur progressively later, indicating that a zone of severe instability moved slowly downhill, reaching low lying areas during the last two centuries. The timing and location of these changes reinforces the view that they are primarily anthropogenic. Settlement in the area has always tended to be restricted to the coastal strip and the lower slopes rising from major floodplains. Extensive early denudation in the uplands may have been an important contributory factor in the early abandonment of Landnám farm sites in the Þórsmörk area. Although later abandonments appear to be ultimately the result of social and cultural factors, some may have also been encouraged by the effects of anthropogenic soil erosion.

147

J. K. Maizels and C. Caseldine (eds.), Environmental Change in Iceland: Past and Present, 147–159.
© 1991 *Kluwer Academic Publishers. Printed in the Netherlands.*

1. INTRODUCTION

In this paper a regional scale model of soil erosion in South Iceland is proposed, based on the interpretation of aeolian sediment accumulation rates at sites up to 400m a.s.l. Since the Norse colonisation of Iceland, about half of the pre-existing soil cover has been stripped away (Runolfsson, 1978) and a provisional map of the denuded areas of the south-eastern half of the country has recently been published by Árnalds (1988). In both relative and absolute terms soil erosion has been extensive and acute, turning some 20,000km² of formerly vegetated land into semi-desert and creating patchworks of erosion in much of the surviving grassland. In Iceland, winds are particularly important, if not dominant agents of soil erosion (Bjarnarson, 1942, 1976, 1978; Thórarinsson, 1961) and in areas of accumulation aeolian sediments can form deposits several metres deep. The presence of numerous tephra layers within these profiles offers a means of determining rates of sediment accumulation, and therefore of inferring rates of erosion. Pioneering tephro-chronological studies by Thórarinsson (1944, 1961, 1981) highlighted the dramatic accelera-tion of aeolian sediment accumulation rates following Norse settlement, clearly indicated the

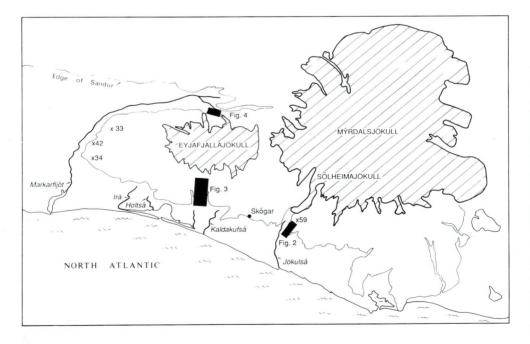

Fig.1 The study area showing the locations of outlying soil sections and the detailed maps in Figs. 2-4.

primary role of anthropogenic impact in post-Landnám soil erosion and suggested that spatial variations in erosion relate directly to intensity of settlement. The impact of Landnám on the vegetation of Iceland, first extensively studied by Einarsson (1961), has recently been the subject of more detailed research by Hallsdóttir (1987). The correlation between destruction of the vegetation cover and ensuing soil erosion is immediately apparent in any comparison of these two lines of study. Spatial models of erosion have been developed further by Ashwell (1966, 1972) who has proposed that belts of acute erosion are associated with local wind patterns determined by ice caps but modified by anthropogenic desertification of surrounding areas. Acute spatial variations in rates of soil erosion and deposition clearly exist and, given a tephrochronological framework containing numerous, well-dated and extensive isochrones, there is a remarkable potential for studying local patterns of geomorphological change in considerable and unusual detail.

Eyjafjallajökull is a small ice cap lying west of Mýrdalsjökull in S.W. Iceland (Fig.1). It is fringed by a gently sloping coastal plain consisting largely of sandur laid down by the rivers Markafljót, Irá, Holtsá, Kaldaklifsá and Jökulsá, and is backed by a former marine cliff. Above the coastal escarpment the land rises gently to the ice cap. To the east, jökulhlaup modified sandur lies in front of the valley glacier Sólheimajökull, but lush hayfields occupy much of the coastal plain west of Skógar. Grasslands extend a short way onto the lower slopes of Eyjafjöll, providing some grazing for sheep, which occasionally range over the poorly vegetated slopes to the edge of the ice cap.

2. METHODOLOGY

A combination of field mapping, remote sensing and careful detailed recording of the Holocene stratigraphy was employed to assess rates of accumulation of aeolian sediments and the spatial patterns of both erosion and deposition. Stratigraphic sections were measured at over 200 sites around Eyjafjallajökull and Sólheimajökull, and 148 were recorded as part of a wider study of Holocene environmental change (Dugmore, 1987, 1989). A subset of these profiles was selected on the basis of the completeness of the sedimentological record and approximately comparable geomorphological location. The tephrochronological dating was derived from the regional frameworks established by Thórarinsson and Larsen (Thórarinsson, 1944, 1967, 1980; Larsen, 1979, 1984; Einarsson et al. 1982), supported by additional chemical analyses and radiocarbon dates (Dugmore, 1987). The detailed stratigraphy of the complete soil profiles were recorded to an accuracy of 2mm; tephra layers were identified and their thickness was excluded from calculations of the aeolian sedimentation rate between tephra layers of known age.

3. RESULTS

Runolfsson (1978) has suggested that anthropogenically triggered erosion has removed approximately half of the soils of Iceland and it is informative to compare his estimate with actual figures for areas above the sandur plain around Eyjafjöll. Three natural landscape

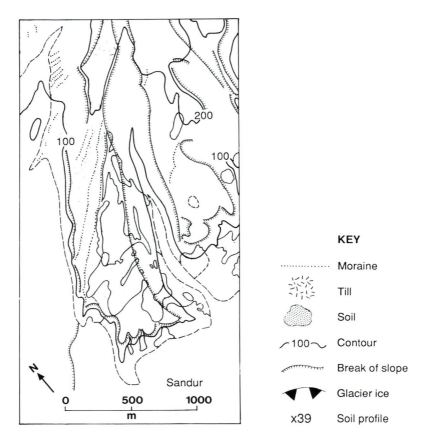

KEY

················ Moraine

Till

Soil

⌒100⌒ Contour

Break of slope

Glacier ice

x39 Soil profile

Fig.2 Geomorphology of Eystriheiði.

units have been selected, which cover the altitudinal range discussed in this paper, ranging between the 50m and 700m contours, and extending from the sandur edge to the limits of recent glaciation (Figs.2-4).

Aerial photograph interpretation combined with field mapping shows that between 60 and 90% of the landscape in these sample areas had been stripped of soil cover by 1985, and the process is continuing, if partially compensated for by reclamation on the sandur and reseeding elsewhere. Over 90% of the upland area between the 300m contour and the edge of Eyjafjallajökull has been stripped of soil (Fig.3). In contrast most of the low-lying ridge north of Lambafell is still covered in soil, although this area is actively eroding. The pattern is reinforced by the sites to the north and east (Figs.2 and 4).

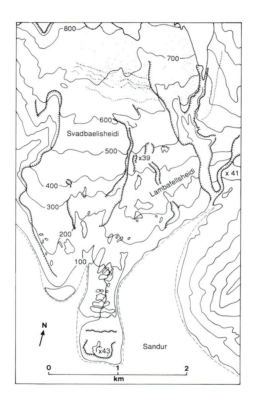

Fig.3 Geomorphology of Lambafellsheiði (same key as Fig.2).

Another approach to the documentation of soil erosion and accumulation is, as Thórarinsson (1961) adeptly showed, by examination of the vertical stratigraphy and this can be expanded by areal correlation by sections to provide a three dimensional picture of geomorphological processes. It is immediately apparent from the sections in Fig.5 that

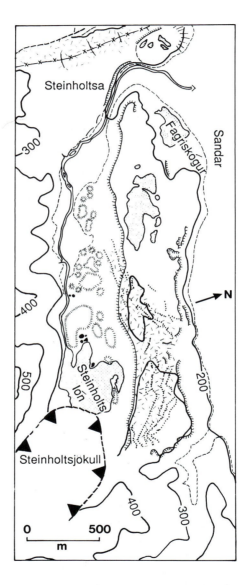

Fig.4 Geomorphology of Fagriskógur (key as Fig.2).

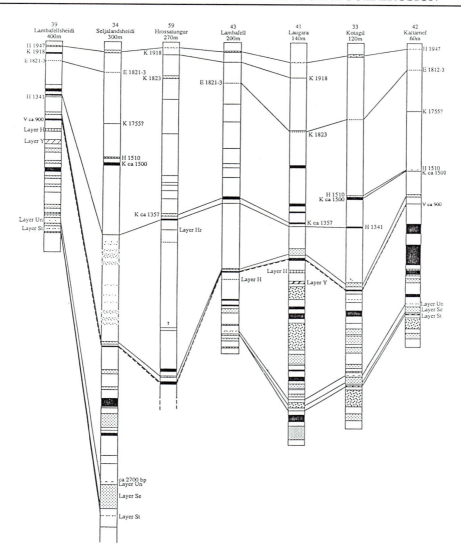

Fig.5 Tephra stratigraphy. Tephra layers are shown by horizontal lines, the blank areas are loessial soils, unconfined dots are fine gravel slope wash. Letters and calendar dates refer to known locations and ages of tephras: H-Hekla; K-Katla: E-Eyjafjallajökull; V-Veiðivötn. Vertical and diagonal lines refer to light coloured tephras, for the dark layers (olive (olive-brown, grey-brown, dark grey brown, blue grey and black), symbol size is proportional to particle size: solid <0.2mm; light dots 0.2-2mm; heavy dots >2mm. Profile numbers refer to locations on maps. Layer Un is a needle-grained tephra radiocarbon-dated in study area and to the north (Larsen, 1984; Dugmore, 1989).

deposition before Landnám is dominated by airfall tephra, a pattern which extends back approximately 8,000 years to the early Holocene. After Landnám, aeolian sedimentation rapidly comes to dominate, but the period at which maximum accumulation occurred varies both in timing and location (Fig.6). At Seljalandsheiði and a number of other localities, discrete lenses of coarser-grained sediment appear within the aeolian sequence recording episodes of slope wash.

The location of profiles selected for detailed consideration are shown in Figs.1 and 3, and the stratigraphic sections are shown in Fig.5.

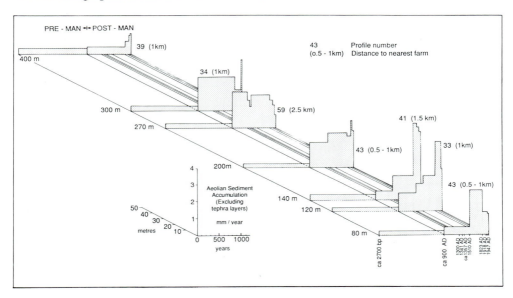

Fig.6 Rates of loess accumulation.

4. DISCUSSION

Two important assumptions are made in order to step from a study of aeolian sediment accumulation rates to models of regional soil erosion: i) it is assumed that rates of aeolian sediment accumulation are directly proportional to rates of local wind erosion, and, ii) that the intensity of this erosion reflects the general intensity of soil erosion affecting the immediate area. The interpretation of sediment accumulation rates has to be approached with some caution because of the practical difficulties of profile measurement and uncertainty about the significance of the depositional record.

Practical difficulties abound in badly eroded areas, few deposits have survived and those that have frequently contain a disturbed or incomplete sedimentological record. The older parts of successions show evidence of modification by diagenesis. In more recent sections, cryoturbation of profiles is common and principally associated with þúfur, or frost hummock, formation. The resulting distortion of tephra layers makes representative measurements of sediment accumulation difficult, and sometimes impossible.

The significance of the depositional record is complex and accumulation rates at a particular point probably vary through time solely as a result of position relative to areas of erosion. There is evidence that contemporary sediment accumulation rates increase towards the edge of vegetated areas because of the effectiveness of grassland in trapping the coarser particles set in motion on eroding slopes. Consequently, it could be predicted that accumulation rates may increase through time not because the intensity of erosion in the area is increasing, but because an eroding slope is moving closer. Indeed, against a trend of overall reduction in erosion rates, deposition at a particular site could increase because of the increasing proximity of sediment supply. Given the preceding caveats, all of which were considered during site selection, a preliminary model of regional soil erosion may be derived from the interpretation of the profiles displayed in Fig.6.

The generally similar rates of accumulation in pre-history can be interpreted as a reflection of regional fall-out over stable vegetated areas. Given an extensive pre-Landnám vegetation cover it is likely that this accumulation is of sediment eroded some distance away along the fluctuating ice cap margins. Lesser sources of silt may have been derived from unstable hillslopes, river and stream margins and perhaps the interior. At the site ca 400m a.s.l. (Fig.6, profile 39), there is little overall change in accumulation rates from pre-Landnám through the fall of the Hekla ash in A.D.1341, and up to the fall of the Eyjafjalla-jökull ash in 1821. Increases after 1821 may reflect an increase in supply of sediment from areas of nearby deglaciation (Dugmore, 1989), although increased stocking levels may also be relevant (Preusser, 1976). In contrast, the sites between ca 300m and 200m a.s.l. indicate abrupt increases in sediment accumulation rates in the early post-Landnám period; at the upper sites rates appear to then decline through time (Fig.6, profiles 34 and 59), at the lower site there is a general, though less steep increase (Fig.6, profile 43).

Early in the post-Landnám period vegetation cover on Seljalandsheiði (Fig.6, profile 34) is broken sufficiently to permit the initiation of intermittent slope wash and the deposition of discrete lenses of gravel. These lenses were excluded from calculations of aeolian sediment accumulation rates. The deposition of coarse-grained layers fades, returns and finally ceases shortly before 1341. The layers of fine gravel may indicate phases of increased local grazing pressure, although patterns of sheet flow would cross the emerging rofbard invariably leading to local variations in patterns of slope wash. Whilst rofbards continue to grow vertically by aeolian accumulation, their great extent is progressively reduced by undercutting, frequently exacerbated by the direct action of sheep. As area is reduced, the catchment leading to slope wash is curtailed and eventually ceases. These progressive changes in the hydrological regime and in vegetation cover from woodland to grass, may lead to first the inception of þúfur formation and eventually to their destruction (Fig.7). Short term fluctuations in sediment accumulation rates between closely-spaced volcanic eruptions are difficult to assess but there appears to be a significant decline in deposition at Hrossátunga after the Hekla eruption of 1341, when contemporary records note very high mortality rates amongst stock (Thórarinsson, 1967) presumably as a result of

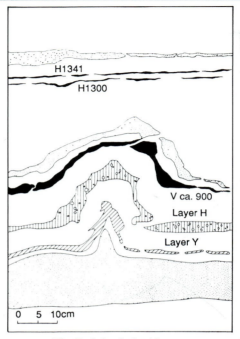

Fig.7 A buried þúfur.

fluorosis. Sites at still lower hillside locations indicate a delayed onset of rapid sediment accumulation (Fig.6, profiles 41, 33 and 42). Overall this pattern is interpreted to suggest an
early onset of soil erosion at about 300m and the gradual downslope migration of a
particularly intensive zone of instability.

This hypothesis may be summarised in a simple model of landscape instability (Fig.8)
- the earliest anthropogenically triggered episodes of acute soil erosion may have begun in
marginal upland areas at the limits of natural woodland. Through time, intensification and
concentration of anthropogenic, or rather ovigenic, impact would have spread instability
downhill into lands that were originally less marginal, but which, as a result of general
ecological degradation and climatic deterioration, became ever more vulnerable to erosion.
The impact of the 'Little Ice Age' when the length of the growing season in many years was
seriously reduced (Friðriksson, 1978; Grove, 1988), was to increase pressure upon 'middle
range' grazing lands. This should be evident in the sedimentary record as a phase of
accelerated deposition downslope of these areas. The highest absolute levels of aeolian
sediment accumulation do occur below 200m after 1823 (Fig.6, profiles 41 and 33).

Detailed archaeological survey work (Sveinbjarnardóttir, 1982) has highlighted the early
abandonment of the inland sites in the Þórsmörk area, where modern grazing pressure has
led to the complete denudation of sites over the last thirty years (Sveinbjarnardóttir et al.,
1982). There can be no doubt that sheep are the prime culprits in initiating soil erosion when

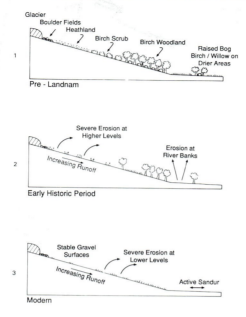

Fig.8 A model of soil erosion.

the boundary fence of the national park in Þórsmörk is examined. The survival of woodland in this area is a combination of isolation and the deliberate conservation of coppiced wood, considered to be a valuable natural resource. The model would predict a significant and early onset of soil erosion in this area, which may have been a contributory factor to the failure of the settlements. Later abandonment of farms in Eyjafjallasveit (Fig.1) appears to have been primarily the result of social and economic factors. The progressive loss of vegetation cover and resulting increased runoff may have led to the destabilistaion of the floodplains of local rivers, in particular Markafljót (Buckland et al., 1990), leading to additional pressures on the already reduced resources of Eyjafjallasveit's farmers.

5. CONCLUSION

The potential for further studies of soil erosion is considerable; established tephrochronological frameworks could be used to constrain detailed studies of small areas that focus not only on sediment accumulation rates but also on detailed particle size studies (cf. Guðbergsson, 1975, and Sigbjarnarson, 1969). Enhancement of the existing tephrochronologies could bring still greater rewards in terms of more detailed dating control. Further study could critically test existing ideas and perhaps bring detailed understanding of one of the most acute environmental changes to have occurred in Iceland in the postglacial period.

ACKNOWLEDGEMENTS

We wish to gratefully acknowledge the support of the Leverhulme Trust, the guidance and assistance of Guðrún Larsen, Guðrún Sveinbjarnardóttir and Þórður Tomasson. The National Research Council of Iceland kindly granted permission for this work to be undertaken. During part of the work AJD was in receipt of a NERC studentship.

REFERENCES

Árnalds, A. (1987) 'Ecosystem disturbance in Iceland', Arctic and Alpine Research 19, 508-513.

Árnolds, Ó. (1988) 'Jarðvegur á ogronu landi', Náttúrufræðingurinn 58, 101-116.

Árnalds, Ó., Aradóttir, A.L. and Thorsteinsson, I. (1987) 'The nature and restoration of denuded areas in Iceland', Arctic and Alpine Research 19, 518-525.

Ashwell, I.Y. (1966) 'Glacial control of wind and of soil erosion in Iceland', Annals of the Association of American Geographers 56, 539-540.

Ashwell, I.Y. (1972) 'Dust storms in an ice desert', Geographical Magazine 44, 322-327.

Bjarnarson, Á. H. (1942) 'Abuð og ortroð', Ársrit Skógraktarfélags Íslands 1942, 8-41.

Bjarnarson, Á. H. (1976) 'The history of the woodland in Fnjóskadalur', Acta Phytogeographica Suecica 68, 31-42.

Bjarnarson, Á. H. (1978) 'Erosion, tree growth and land regeneration in Iceland', in M.W.Holdgate & M.J.Woodman (eds.) The Breakdown and Restoration of Ecosystems, 241-248. Plenum, London.

Buckland, P.C., Dugmore, A.J., Perry, D.W., Savory, D. and Sveinbjarnardóttir, G. (1989) 'Holt in Eyjafjallasveit, Iceland: a palaeoecological study of the impact of Landnám', in T.H.McGovern & J.Bibelow (eds.) The Norse in the North Atlantic, British Archaeological Reports, Oxford, 97-111.

Dugmore, A.D. (1987) Holocene Glacier Fluctuations around Eyjafjallajökull, South Iceland: a Tephrochronological Study. Unpublished Ph.D. Thesis, University of Aberdeen.

Dugmore, A.D. (1989) 'Tephrochronological studies of Holocene glacier fluctuations in south Iceland', in J.Oerlemans (ed.) Glacier Fluctuations and Climatic Change, 37-55, Kluwer, Dordrecht.

Einarsson, Th. (1961) 'Pollenanalytische Untersuchungen zür spät- und postglazialen Klimageschichte Islands', Sonderveröffentlichungen des Geologisches Institutes der Universität Köln 6, 1-52.

Einarsson, E.H., Larsen, G. & Thórarinsson, S. (1980) 'The Sólheimar tephra layer and the Katla eruption of c.1357 A.D.', Acta Naturalia Islandica 28, 1-24.

Friðriksson, S. (1978) 'The degradation of Icelandic ecosystems', in M.W.Holdgate & M.J.Woodman (eds.) The Breakdown and Restoration of Ecosystems, 241-248. Plenum, London.

Grove, J.M. (1988) The Little Ice Age. Methuen, London.

Guðbergsson, G. (1975) 'Mýndun moajarðvegs í Skagafirði', Journal of the Agricultural Research Institute of Iceland 7, 20-45.

Hallsdóttir, M. (1987) Pollen analytical studies of human influence on vegetation in relation to the Landnám tephra layer in south-west Iceland, Lundqua Thesis 18.

Larsen, G. (1979) 'Um aldur Eldgjárhrauna', Náttúrufræðingurinn 49, 1-26.

Larsen, G. (1984) 'Recent volcanic activity of the Veiðivötn fissure swarm, southern Iceland: an approach to volcanic risk assessment', Journal of Volcanology and Geothermal Research 22, 33-58.

Preusser, H. (1976) The Landscapes of Iceland. Junk, The Hague.

Runolfsson, S. (1978) 'Soil conservation in Iceland', in M.W.Holdgate & M.J.Woodman (eds.) The Breakdown and Restoration of Ecosystems, 231-240. Plenum, London.

Runolfsson, S. (1987) ' Land reclamation in Iceland', Arctic and Alpine Research 19, 514-517.

Sigbjarnarson, G. (1969) 'Afok og uppblastur-pattie úr groðursögu Haukadalsheiðar', Náttúrufræðingurinn 39, 68-118.

Sveinbjarnardóttir, G. (1982) 'Farm abandonment in Eyjafjallasveit Southern Iceland', Department of Geography, University of Birmingham, Working Paper Series 14.

Sveinbjarnardóttir, G., Buckland, P.C. & Gerrard, A.J. (1982) 'Landscape change in Eyjafjallsasveit, Southern Iceland', Norsk Geografisk Tidsskrift 36, 75-88.

Thórarinsson, S. (1944) 'Tefrokronologiska studier på Island', Geografiska Annaler 26, 1-217.

Thórarinsson, S. (1961) 'Population change in Iceland', Geographical Review 51, 519-533.

Thórarinsson, S. (1981) 'The application of tephrochronology in Iceland', in S.Self & R.S.J.Sparks (eds.) Tephra Studies, 109-134. Reidel, Dordrecht.

Þórarinsson, S. (1980) 'Langleiðir gjosku úr premur Kötlugosum', Jökull 31, 65-81.

A STUDY OF FARM ABANDONMENT IN TWO REGIONS OF ICELAND

Guðrún Sveinbjarnardóttir
Department of Scandinavian Studies
University College London
Gower Street
London WC1E 6BT
UK

ABSTRACT. This article surveys the changing pattern of farm occupation and abandonment in two areas of Iceland, a southern coastal area and an inland area in the east, and examines the causes of change on the basis of documentary, archaeological and tephrochronological evidence. The two areas differ considerably in physical character, which is reflected in their different settlement patterns. Two main causes of abandonment are however evident in both areas, since both contain sites which have suffered from erosion, the most common cause of early abandonment, and sites known in later times to have been abandoned for economic and social reasons. The general pattern in both areas is for settlement to move away from the far inland. In the southern region, coastal erosion has also driven settlement away from the coast. In the past farm abandonment was commonly attributed to simplistic causes. A warning is issued against such generalisations.

1. INTRODUCTION

Abandoned farm-sites in the Icelandic landscape have for long intrigued scholars of many disciplines and a number of theories as to the cause of desertion have emerged. Of particular interest have been inland areas, uninhabited for a long time, where often whole valleys of deserted sites can be found. The main causes of abandonment were traditionally thought to have been climatic deterioration and epidemics, in particular the big epidemic of 1402-04, which in many sources has wrongly been termed as the Black Death, but is more likely to have been an influenza (on this subject see for example Jóhannesson, 1928; Bjarnadóttir, 1986). More recently studies of inland sites in particular have shown them to have suffered from erosion. Volcanic activity was in the past largely blamed for this (Guðmundsson, 1952, 1953), but more recently it has been suggested that the cause lies to a great extent in too heavy grazing (Þórarinsson, 1977; Friðriksson, 1972; Bjarnason, 1968).

161

J. K. Maizels and C. Caseldine (eds.), Environmental Change in Iceland: Past and Present, 161–177.
© 1991 *Kluwer Academic Publishers. Printed in the Netherlands.*

The methods of historians, most often adopted in settlement studies up to now, in tracing regional development of settlement on the basis of documentary sources, generally only give an accurate, detailed picture of later centuries, particularly after 1600 when sources become more abundant. For the period before that information consists of scanty documentary evidence, a small number of archaeological investigations and, recently, studies based on the application of tephrochronology as a means of dating many sites in an extensive area. Past regional studies of farm abandonment in Iceland have been either purely historical (Sveinsson, 1954; Þormóðsson, 1972; Helgason, 1969; Gissel et.al,1981) or based on tephrochronology (Þórarinsson, 1976; 1977). In this article aspects of a study where both methods were applied, are discussed, the latter with the aid of geologists Guðrún Larsen and Gunnar Ólafsson of the Nordic Volcanological Institute in Reykjavík. An archaeological survey was also carried out and some limited excavation.

The research began as part of the project 'Viking settlement, climate and environmental change around the North Atlantic', based at Birmingham University and funded by the Leverhulme Trust during the period 1979-1981. It covered the areas Eyjafjallasveit in the south and Austurdalur in Skagafjörður in the north, but was subsequently expanded to include Vesturdalur in Skagafjörður in the north and the Berufjörður area in the east (Fig.1). The project was an attempt at throwing light on farm abandonment by using a combination of methods belonging to many disciplines. The contributions of three other authors in this

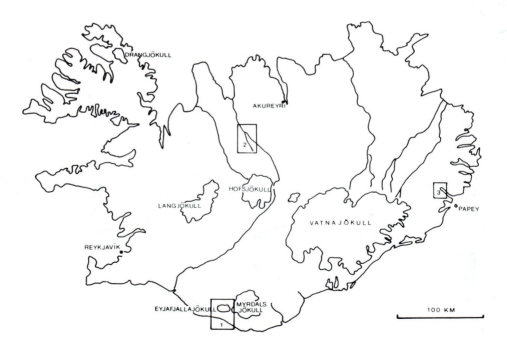

Fig. 1   Map of Iceland showing study areas.

volume, Paul Buckland, Andy Dugmore and John Gerrard, are also products of this project. A wealth of material on the history and development of settlement was collected from a variety of sources. In this present article material collected on Eyjafjallasveit and Beru-fjörður will be reviewed and discussed, providing some historical and archaeological background to other aspects of these areas covered in this volume. The whole study will be published in more detail in the near future.

2. EYJAFJALLASVEIT

The inhabitable area of the lowland, coastal parish of Eyjafjallasveit now lies on the stretch between the mountain and the sea, but formerly there were also settlements in the inland area of Þórsmörk, which lies north of the glacier (Fig. 2). Until flood barriers were built at the eastern side of the Markarfljót river early this century, the western half of the parish was frequently flooded by the river which changed its course to flow eastwards into the lagoon Holtsós, causing extensive damage to land and structures on the way. As a result of this the farms were located either close to the mountain edge or close to the sea-shore. Now this stretch of land consists of fertile grazing land and fields.

The earliest major documentary source of settlement in Iceland, the Book of Settlements (*Landnámabók*), surviving in a 13th century version, lists eight settlers in Eyjafjallasveit (ÍF I, 338-340, 342-346, 350). The exact location of the farms of all these pioneer settlers is not known, but as far as can be made out they were fairly evenly distributed throughout the parish, two in the eastern half and six in the western half, including two in the inland area Þórsmörk (Fig. 2). These two and two more sites in the western half are now lost to erosion, the other locations are still occupied.

The population density in the area increased rapidly. By the 14th century a further 38 farms are known from documentary sources, and when the Árni Magnússon survey was compiled in 1709, the first land survey to include all categories of farms, 78 farms are listed as occupied and a further 27 as abandoned in the area (Jarðabók 1, 27-102). Settlement seems to have been fairly stable in the lowland with periods of abandonment in the 11th/12th centuries, the 17th/18th centuries and then in the 20th century. The first period refers largely to inland and coastal abandonment, the second to that of small dependent farms either through erosion by natural forces or because of pressure on land-use; the last mentioned has largely social causes, the movement from rural to urban areas. All categories will now be discussed further.

2.1. The inland sites

Five sites are known in the Þórsmörk area (6-9 on Fig.2; the 5th site was not investigated by the author). A site belonging to the parish north of the Markarfljót river (10 on Fig. 2), is included in this discussion because it throws light on the desertion of the others. All these sites have been published elsewhere (Sveinbjarnardóttir, 1982, 1983). They will therefore

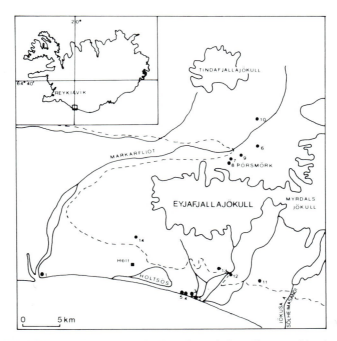

Fig. 2 Map of Eyjafjallasveit showing the location of sites discussed in the text. 11-14 are some of the shieling sites in the area.

Fig. 3 A view of the lower Þuríðarstaðir site in Þórsmörk, Eyjafjallasveit, showing its eroded nature. The site lies in the right centre part of the photo.

not be described in detail here, but discussed as cases of farm abandonment.

Þórsmörk served in the past as the mountain pasture for the farms in Eyjafjallasveit. Now, a large part of it has been fenced off and made into a National Park. It was in the past no doubt more heavily forested than it is today, as is implied in the place-name, mörk, meaning forest. The difference in vegetation in and outside the National Park boundaries, where sheep are still allowed to graze, is very noticable. Four of the known sites lie in exposed positions outside the National Park boundaries and are all badly eroded (Fig. 3). The fifth site (9 on Fig. 2), briefly reoccupied in the early 19th century, is well overgrown, lying at the bottom of a valley-floor within the Park.

No proper archaeological investigation was possible at the eroded sites. The position and orientation of structures was, however, detectable from the scatter of stones on the surface, and a number of artifacts have been picked up from the sites over the years. In Þórsmörk it is entirely on the basis of these that the occupation of the sites has been dated. At Steinfinnsstaðir, site no. 6, heathen graves were found, putting its occupation before the conversion of the country to Christianity in the year 1000. Artifacts found include a glass-eyed bead of a typical Viking type (museum no.: Þjms. 1974, 222), dated in Scandinavia to the latter half of the 10th century (Callmer, 1977, 77 & 85, Arbman, 1943, 444, ibid 1940, Taf. 121, 13a). At Þuríðarstaðir efri, site no. 8, finds included a bronze ringed pin (museum no.: Þjms. 1977; 52), undecorated and of a simple type, dated in Ireland to between ca the 7th and the 10th centuries, but in graves in Scotland and Scandinavia to the 9th and 10th centuries (Fanning, 1969, 6-11), and a bronze stud (museum no.: Þjms. 1974; 232), heart-shaped with palmett decor, dated to the 10th century in Scandinavia (Jansson, 1975, 7, figs. 243-244) (this object is wrongly dated to the 12th or 13th century in Sveinbjarnardóttir 1983, 46-7).

The objects found in Þórsmörk suggest that the area was settled by the 9th or 10th century. Evidence of human occupation found in a soil section close to the site Einhyrningsflatir, which lies north of the Markarfljót river (10 on Fig. 2), perhaps the site of the farm of one of the pioneer settlers (Sveinbjarnardóttir, 1982), supports this dating of occupation in the general area. The site is badly eroded, now consisting only of a scatter of stones (Fig. 4). None of the several artifacts picked up here is precisely datable. In the two soil sections studied nearby (Fig. 5) several volcanic ash layers were identified and dated with the aid of geologist Guðrún Larsen. In one of the sections (column 2 on Fig. 5) occupational debris was found on top of and partly dug into the so-called Landnám-layer (Landnámslag), dated to about 900 (Þórarinsson, 1944, 1968; Einarsson, 1963, 448 ff.; Larsen, 1982, 63; Larsen & Þórarinsson, 1984; Hallsdóttir, 1987), suggesting human activity in the area shortly after it fell.

As for the date of abandonment in this general area, there are two indications. In the soil section from Einhyrningsflatir (column 1 on Fig. 5) there is evidence of heavy erosion, dated by volcanic ash layers to between about 1000 and 1341. Erosion has clearly been a severe problem in the whole area. This is evident on the surface today, and has, in addition to the Einhyrningsflatir-sections, been demonstrated in a number of sections studied by John Gerrard close to three of the archaeologically investigated farm-sites in the Þórsmörk

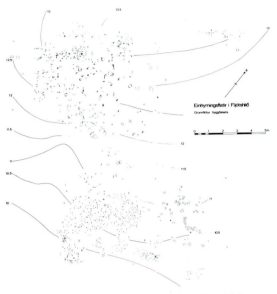

Fig. 4 A site-plan of Einhyrningsflatir in Fljótshlíð.

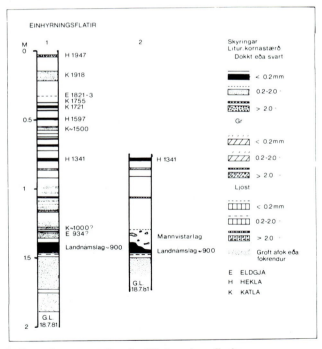

Fig.5 Soil sections near the site Einhyrningsflatir.

area (Gerrard, 1983) and in others done by Andy Dugmore in the same general area (Dugmore, 1987, Sheet 1, 28 & 31). It is suggested that the evidence from Einhyrningsflatir may throw light on the cause of abandonment of all the above mentioned sites, i.e. that the delicate inland vegetation cover could not take the grazing and other human activities for very long, erosion set in and rendered the areas uninhabitable. Reliable, contemporary sources suggest that the Þórsmörk area was already abandoned by the 12th century (Biskupa sögur I, 1858, 291; Sturlunga saga I, 1946, 532).

2.2. The coastal sites

A number of deserted farm-sites are known on the coast of Eyjafjallasveit (1-5 on Fig. 2). Remains of some can still be seen, others have been completely eroded away by the action of the sea and rivers, or have disappeared in the coastal sandur. Only one of these sites, Stóraborg (no. 2 on Fig 2) has been, and still is in the process of being excavated. All the coastal sites have been eroded in some way and it is clear that the reason for their abandonment lies in their closeness to the sea. It is also a fact that the farms were not abandoned as such but rather relocated to sites further inland, away from the encroachment of the sea. In some cases, like for example that of Stóraborg, a farm with that name is still occupied some distance further inland than the site under excavation, in others the farms have been given different names.

The farm-mound of Stóraborg, at present under excavation (Snæsdóttir, 1981, 1988; Sveinbjarnardóttir et.al., 1981), is a good example of these coastal sites. The mound, which at the outset was quite high, is placed at the edge of the sea, at the point where two rivers run into it. It has suffered erosion for a long time caused by the sea and changing course of the rivers and contained one building phase on top of another. It is thought to be the original site of the farm (Sigurðsson, 1886, 507), which is first mentioned with certainty in written records in 1332 (DI II, 678), but probably dates to further back. The excavation has rendered evidence of early occupation, but so far no firm dates have been published.

Accounts exist of coastal sites on both sides of Stóraborg being moved away from the sea more often than once (Sigurðsson, 1886). The original sites were in these cases placed further out towards the sea and not on as high and substantial stabilized dunes as the one the Stóraborg farm stood on. One of these is Miðbæli (3 on Fig. 2), just across the river to the west of Stóraborg. Here, the clear remains of farm-buildings can be seen at the edge of the sandy beach. The site was abandoned in 1867, but had, according to the 1709 land survey (Jarðabók 1, 47) then already been moved once from a site further out towards the sea. There is an account of building remains having been visible in 1862 on the beach, over 300 m to the south of the present site (Sigurðsson, 1865, 67-8, ibid. 1886, 507). Similar remains are in one of these accounts reported south of the next farm to the west of Miðbæli, and the next one to the east of Stóraborg (Sigurðsson, 1886, 507).

East of the mouth of the Markarfljót river, six sites have been identified (1 on Fig. 2), now all consisting of just scatters of stones among the blown out sand dunes (Fig. 6).

Fig. 6 One of the six Sandar sites east of the mouth of the river Markarfljót in Eyjafjallasveit.

They are placed fairly close together and are therefore likely not all to have been contemporary. A number of artifacts have been picked up at these sites over the years, few of which are precisely datable. All but one of these sites seem already to have been deserted in 1709 when the Árni Magnússon land survey for the area was compiled. In the survey one of the farms in that area is by then said to have been moved at least three times because of the encroachment of sand and the Markarfljót river (Jarðabók 1, 81), by which all the sites on the sandur were affected.

In *Landnámabók* (the Book of Settlements) it is suggested that one of the earliest settlers had his farm in this area (DF I, 340), but that it had already been abandoned by the time *Landnámabók* was written, perhaps as early as about 1100 (Rafnsson, 1974). Investigation of objects picked up at the known sites at the mouth of the Markarfljót suggests that none of them is the site of the pioneer farm. It looks as if it may have been located even closer to the sea, in an area long since eroded by sand and water.

2.3. Other sites in Eyjafjallasveit

In addition to the two very obvious areas of farm abandonment discussed above, a number of other deserted sites were discovered through the documentary and archaeological survey undertaken in the area. The bulk of these are the sites of small, so-called dependent farms, established on the land of larger holdings providing revenue for them through the payment of rent. A number of these, placed just east of the Markarfljót river, at the point where the

lowland spreads out to the south, were deserted in the 17th century as a result of the action
of the river (Jarðabók 1, 77 ff).   Others, often clustered around the church farms, were
abandoned because of pressure on land-use; the home-farm required the land. Thus, for
example the church seat and pioneer farm Holt had 12 dependent farms in 1709 (Jarðabók
1, 63-67), only one of which is still occupied. Some of these were first settled in the 17th
century, others earlier, some were abandoned already by the time the land survey was
compiled, others as late as the early 20th century. Most peoples livelihood in Iceland
depended entirely on the land and the sea until the rapid growth of towns after the 2nd World
War, when there was a further phase of small and often isolated farms being abandoned all
over the country.

3. THE BERUFJÖRÐUR AREA

Berufjörður in eastern Iceland (2 on Fig. 1) is physically very different from the Eyjafjallasveit area. It is a narrow fjord lined with steep mountains. The only lowland is a narrow
strip along the coast and in three valleys cutting into the inland from the south and west. One
of these is Fossárdalur (Fig. 7), now occupied by only one farm (Eyjófsstaðir).   Until
recently there were 3 farms there and more in the past. Tradition has it that there was once
a whole church community in the valley, with the suggested number of farms ranging from

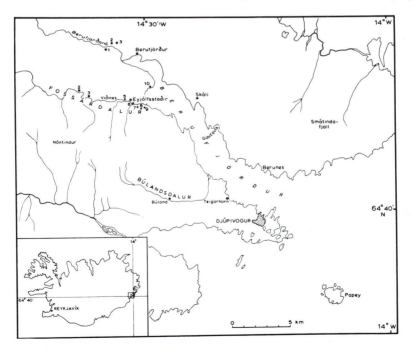

Fig. 7   Map of Berufjörður showing location of sites discussed in the text.

11 (Stefánsson, 1970, 186) to 15 (Skýrslur um Þjms. 30.7.1953), to 11-14 (Sveitir og jarðir, 406), depending on the source. Tradition also has it, treated with scepticism, however, by the above sources, that the valley was deserted, at least temporarily, as a result of the big plague epidemic of 1402-04.

Access to Fossárdalur is gained by way of a steep basalt ridge, the foremost of several lying accross the ascending valley. The difference in height between the lowest and the highest platform is about 235 m. The lowland lies between these and that is where the sites are. Nine sites have been identified in the valley itself and one at its mouth (10 on Fig. 7). Of the nine five are definitely those of farm-sites, two are likely to be (3-9 on Fig. 7).

The prime location in the valley is towards its mouth, where the now occupied farm, Eyjólfsstaðir, lies. Its former name was Fossárdalur, appearing in the 1703 census (Manntalið, 1703), but it is not known when it was first occupied. The name Eyjólfsstaðir first appears in sources in 1866 (Sveitir og jarðir, 407). Diagonally across the river from it are the sites of 3 small farms (7-9 on Fig. 7), occupied for a short time from the early 19th century to the early 20th century.

Further inland, on the third platform from the mouth of the valley, lies the site of Víðines on a dry grassy area which is now being eroded by the river (Fig. 8). It was first built, as far as is known, in 1845 on land belonging to the present Eyjólfsstaðir, and abandoned in 1944, because of isolation and difficult access to the main communication route, as well as, it is suggested, hard winters (Sveitir og jarðir, 409). The site lies 75 m higher above sea level than Eyjólfsstaðir. The occupation of the farm falls within a period

Fig. 8   Víðines in Fossárdalur, looking north.

when there was an increase in the number of settled farms generally, and not least in the east of Iceland (Stefánsson & Jónsson (eds), 1947, 167; Þormóðsson, 1972; Þórarinsson, 1977). This increase coincided roughly with the improvement in climatic conditions during the middle of the 19th century (Bergþórsson, 1969). Víðines was attractive as a farm-site because it offered easy access to natural hay-fields and meadows. But as farming practices changed and farms became more dependent on trading centres, its isolation became a major obstacle. Added to this was an increase in work available in towns as a result of the Second World War, drawing people there from the countryside.

Between Eyjólfsstaðir and Víðines is the site named Broddaskáli (5 on Fig. 7), abandoned well before 1362. It consists of a longhouse with a total external length of 16-17 m, divided into two rooms with the longside parallel to the river which lies less than 10 m to the south. Between the structure and the river there is about a 5 m drop. At present the site is surrounded by very little lowland. This was not always so. The river seems to have changed its course after the site was first settled, cutting it away from extensive lowland now to the south of the river (Fig. 9). It is suggested that this was the main cause of the abandonment of the site.

Fig. 9   Broddaskáli in Fossárdalur, looking east.

When a trial hole was dug into the structure in 1953 a floor laid with slabs and 3 objects were reported. The finds consisted of two loom-weights and a small, red stone with perforation (Stefánsson, 1970, 95-6; Skýrslur um Þjms. 30.7.1953). The section of this trial-hole was investigated in 1983 (left on Fig. 10). It showed an up to 5 cm thick floor-layer 10 cm above which was a 1-1.5 cm thick layer of volcanic ash, identified as deriving

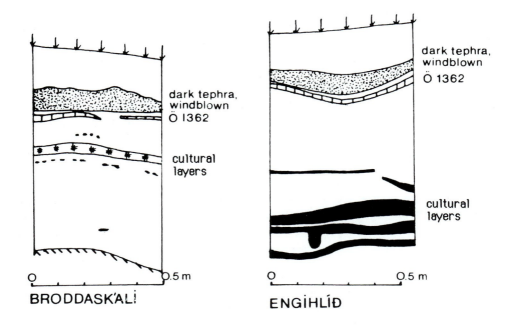

Fig. 10 Sections from structures at Broddaskáli and Engihlíð in Fossárdalur, Berufjörður.

from an eruption in Öræfajökull in the year 1362 (Thórarinsson, 1958). Between the two is windblown, erosion material. This stratigraphy shows that the site must have been abandoned already by the time the Ö1362 tephra fell. It is suggested that the main cause of the abandonment of the site is the changed course of the river leaving it without easy access to lowland hay- and grazing land.

On the 4th platform, the best pastureland in the valley is to be found. Here, most of the meadow-haymaking was formerly done from the farm Fossárdalur. At the top end of the meadow where the land is drier lies the site Engihlíð, at a height of 180 m.a.s.l. Two structures were found at the site in 1983, the larger of which may be divided into 3 or 4 rooms. A trial hole was also dug here in 1953, at which time were found (in the top cultural layer) 3 fragments of an oval brooch, dated to the 10th century (Eldjárn, 1956, 296 ff) and a bead carved in soft, green stone (Stefánsson, 1970, 95-6; Skýrslur um Þjms. 30.7.1953). The section of this trial hole was investigated in 1983 (right on Fig. 10). It shows 4 cultural layers, with 25-30 cm of windblown erosion material on top of the uppermost one, followed by a 1.5 cm thick, undisturbed layer of the aforementioned Ö1362 tephra layer. Here again the site was abandoned well before 1362.

In documentary sources it is assumed that this was a farm-site. The findings of a palaeoecological study carried out on some samples taken close to the site may, however, suggest otherwise. Preservation was poor, especially in the samples which were more likely to have been associated with the occupation at the site. An insect species list was compiled and analysed by Dr Paul Buckland and Jon Sadler, then at the University of Birmingham, now at Sheffield University. They have suggested, on the basis of the absence in the upper samples of insects linked with hay storage, an activity usually associated with farming in Iceland, that the site was, at least during its last phase, not occupied as a farm (Buckland & Sadler 1989). An alternative use of the site would be that of a shieling.

These findings throw doubt on the interpretation of a number of other sites as those of farms and serve to remind of the caution which has to be shown when dealing with material of this kind. No shieling has so far been excavated in Iceland. Whether by excavation a distinction could be made between a shieling and a full farm is not certain. In both people lived and cooked. A larger range of activities though took place in the latter, a fact one would hope would show in the excavation.

Regardless of whether Engihlíð was a farm or a shieling, it was clearly abandoned well before 1362. As for the reason for the abandonment of the site, it matters what type it was. The following hypothesis are suggested. If there was a farm here originally, its early desertion may have to do with its position so high above sea level, bearing in mind the comments made earlier about the contribution of hard winters to the abandonment of Viðines at a later stage. Engihlíð lies at a considerably higher altitude than Viðines. If the site was a shieling this argument does not hold. Shielings were only used in the summer and we know from later sources that the Engihlíð meadow was the main meadow hay-making area in the valley for a long time. The abandonment of the site as a shieling would have had its reasons in changes in the management of farming rather than anything else. The area was needed for hay-making for winter fodder, so the shieling had to give way, and perhaps be moved up onto the next platform, where building remains, definitely not those of a farm (2 on Fig. 7) were found.

3.1. Búland

In the next valley to the south, one site, Búland, is known. It was last used as a shieling, but is alleged to have been a farm in the past (Stefánsson, 1970, 186). The structures, which are quite substantial, were cored in 1983, without any success in establishing stratigraphy because of the core hitting stones all the time.

The valley Búlandsdalur lies at about 200 m a.s.l. It is wide and grassy, but has thin soil cover, a typical area to tempt early settlers, but turning out to be more suitable for summer grazing, i.e. shieling activity, than for permanent settlement. The element -land in the site-name is suggestive of a farm, being one of the more common ones in farm-names in Iceland (Jónsson, 1907-15, 418-19), with the valley and a large, dominating mountain in the general area, Búlandstindur, attributed to it.

4. DISCUSSION

The documentary and archaeological data on settlement gathered in two limited areas in the south and east of Iceland and discussed here gives some quite clear idea of trends in early settlement in these two areas. These will now be summarised.

It appears that the southern area, Eyjafjallasveit, suffered early medieval farm abandonment in two easily defined regions; the inland and the coastal ones. Both regions have suffered acutely from erosion and this seems to have been the cause of abandonment.

No farms were ever placed in the upland areas of Eyjafjallasveit proper (above the limit of the lowland shown on Fig. 2). These were reserved for shielings, of which several sites are known (11-14 on Fig. 2 represent only part of the known ones). Shielings were, however, not always restricted to the upland. There is considerable place-name evidence of early lowland shielings, of sites which after having served as shielings were turned into farms (Sveinbjarnardóttir, forthcoming). In at least one case, that of Seljavellir, the original shieling, placed within the limit of the lowland, was turned into a farm in 1774 (Sunnlenzkar byggðir IV, 90), and had by then already been replaced by another shieling up in the highland (13 on Fig 2).

Later farm abandonment in Eyjafjallasveit was not caused so exclusively by natural forces as that earlier. It mostly affected minor farms, established and abandoned at various times. Some of these were placed in unfavourable locations, suffered erosion and were hence abandoned. Others were placed too close to the home-farm and had to give way with changing farming practices and demands for higher living standards. Many of these small farms were only occupied for a short period of time in the 19th and early 20th centuries.

The main trends in farm abandonment in Eyjafjallasveit can be summed up as being early coastal and inland abandonment caused by natural forces, and later abandonment on the coast and in the lowland, caused either by erosion or by social effects.

In Fossárdalur, where the physical environment differs considerably from that of Eyjafjallasveit, being an inland valley in a narrow fjord, the picture is somewhat different. Although the sites have neither suffered coastal erosion nor deflation as in Eyjafjallasveit, some of them have been visibly eroded, not least by the river. This is in fact likely to have been the cause of abandonment of one of the sites (5 on Fig. 7). The valley ascends quite steeply inland, with a difference of 110 m in height a.s.l. between the lowest and highest known settlements, a difference which could be of consequence in site viability. The evidence, indeed, shows a tendency here like in Eyjafjallasveit for settlement to move from the inland towards the coast. In later times, when easy access to trading centres became more important, isolation and difficult access to main communication routes were the main factors affecting abandonment. At present only 1 site is occupied in the valley, until recently there were 5, at least for a short period, all but one located at the mouth of the valley. It is suggested that the total number of occupied farms in the valley at any one time never exceeded that number.

It is evident from the data discussed here that there has been a certain amount of movement of settlement in the areas under study and that the survival of settlement was affected by a whole number of factors. Environmental effects, so often stressed because of their visibility, have always played a large part, erosion being the most easily detected one. The effect of climatic change is not as easy to determine. It is clear that some of the sites to have caused a temporary total desertion of occupation in the valley. No documentary evidence exists for the abandonment of any specific farm in either of the studied areas as a result of any subsequent epidemic. Epidemics are, however, known to have affected the occupation of farms at various times in various parts of the country. Last but not least social and economic factors played a part in the survival of farms, not least in later times. All the above examples serve to remind of the complexity of factors affecting farm abandonment in Iceland and warn against generalisations and simplifications when discussing the topic.

REFERENCES

Arbman, H. (1940-43) Birka I: Die Gräber, Stockholm/Uppsala.

Bergþórsson, P. (1969) 'Hafís og hitastig á liðnum öldum', in M.Á. Einarsson (ed.),Hafísinn, Almenna bókafélagið, Reykjavík, pp. 333-345.

Biskupa sögur I, Hið Íslenzka bókmenntafélag, Kaupmannahöfn, 1858.

Bjarnadóttir, K. (1986) 'Drepsóttir á 15. öld', Sagnir, 7. árgangur, 57-64.

Bjarnason, Á. H. (1968) 'The History of Woodland in Fnjóskadalur', Acta Phytographica Suecica, 31-42.

Buckland, P. & Sadler, J. (forthcoming) 'Farm or shieling: an entomological approach', Appendix in G.Sveinbjarnardóttir, 'Shielings in Iceland - an archaeological and historical survey,' in The Norse North Atlantic Conference Proceedings, Acta Archaeologica, forthcoming.

Callmer, J. (1977) 'Trade Beads and Bead Trade in Scandinavia ca. 800-1000 A.D.', Acta Archaeologica Lundensia, Series in 4° Nr 11, Lund.

DI-Diplomatarium Islandicum, Íslenzkt fornbréfasafn. Kaupmannahöfn og Reykjavík 1857-.

Dugmore, A. (1987) Holocene glacier fluctuations around Eyjafjallajökull, South Iceland: a tephrochronological study. PhD University of Aberdeen, Department of Geography.

Einarsson, þ. (1963) 'Vitnisburður frjóreiningar um gróður, veðurfar og landnám á Íslandi', Saga 1962, 442-469.

Fanning, T. (1969) 'The bronze ringed pins in Limerick City Museum', North Munster Antiquaries Journal XII, 6-11.

Friðriksson, S. (1972) 'Grass and grass utilization in Iceland', Ecology, Vol 53, No 5, 785-796.

Gerrard, J. (1983) 'Contemporary soil erosion in Þórsmörk, Southern Iceland', Appendix I in G. Sveinbjarnardóttir, Byggðaleifar á Þórsmörk, Árbók hins íslenzka fornleifafélags 1982, pp.57-59.

Gissel, S. et.al. (eds.) (1981) Desertion and Land Colonozation in the Nordic Countries ca.1300-1600, Almquist & Wiksell International, Stockholm.

Guðmundsson, V. (1952) 'Eyðibýli og auðnir á Rangnárvöllum,' Árbók hins íslenzka fornleifafélags 1951-52, 91-164

Guðmundsson, V. (1953) 'Eyðibýli og auðnir á Rangnárvöllum' Árbók hins íslenzka fornleifafélags 1953, 5-79.

Hallsdóttir, M. (1987) 'Pollen Analytical Studies of Human Influence on Vegetation in Relation to the Landnám Tephra Layer in Southwest Iceland', Lundqua thesis, Volume 18, Lund University, Department of Quaternary Geology.

Helgason, Ö (1969) 'Bæjanöfn og byggð í Hryggjardal og Víðidal, Skagafjarðarsýslu', Saga VII, 196-220.

ÍF- Íslenzk fornrit I Íslendingabók, Landnámabók, Jakob Benediktsson (ed.), Reykjavík 1968.

Jansson, I (1975) 'Ett rembeslag av orientalsk typ funnet på Island. Vikingetidens orientalska balten och deras eurasiska sammanhang', Tor, Vol. XVII, 383-419.

Jarðabók Árna Magnússonar og Páls Vídalín, Hið Íslenzka fræðafélag í Kaupmannahöfn, 1913-43.

Jóhannesson, Þ. (1928) 'Plágan mikla 1402-1404', Skírnir, 73-95.

Jónsson, F. (1907-15) 'Bæjanöfn á Íslandi', Safn til sögu Íslands IV, 413-584.

Larsen, G. (1982) 'Gjóskutímatal Jökulsdals og nágrennis', in Eldur er í Norðri, afmælisrit helgað Sigurði Þórarinssyni sjötugum 8.janúar 1982, Sögufélag, Reykjavík, pp. 51-65.

Larsen, G. & Þórarinsson, S. (1984) 'Kumlateigur í Hrífunesi í Skaftártungu IV', Árbók hins íslenzka fornleifafélags 1983, 31-47.

Manntalið 1703. Hagskýrslur Íslands, Reykjavík 1960.

Rafnsson, S. (1974) Studier i Landnámabók, Lund.

Sigurðsson, P. (1865) 'Bæjafundurinn undir Eyjafjöllum', Þjóðólfur 16.-17.tbl. 1865, 67-68.

Sigurðsson, P. (1886) 'Um örnefni, goðorðaskipan og fornmenjar í Rangárþingi', Safn til sögu Íslands II, 498-557.

Skýrslur um Þjms. 30.7.1953, manuscript at The National Museum of Iceland.

Snæsdóttir, M. (1981) 'Anna á mig. Um snældusnúð frá Stóruborg', Árbók hins íslenzka fornleifafélags 1980, 51-57.

Snæsdóttir, M. (1988)˹ Kirkjugarður að Stóruborg undir Eyjafjöllum', Árbók hins íslenzka fornleifafélags 1987, 5-40.

Stefánsson, H. (1970) 'Fornbýli og eyðibýli í Múlasýslum', Múlaþing 5, 172-187.

Stefánsson, H. & Jónsson, Þ.M. (eds.) (1947) Austurland. Safn austfirskra fræða I, Akureyri.

Sturlunga saga I, J. Jóhannesson, M. Finnbogason & K. Eldjárn (eds.), Reykjavík 1946.

Sunnlenzkar byggðir IV. Rangnárþing austan Eystri Rangnár. Búnaðarsamband Suðurlands 1982.

Sveinbjarnadóttir, G. (1982) 'Byggðaleifar við Einhyrningsflatir í Fljótshlíð', in Eldur er í Norðri, afmælisrit helgað Sigurði Þórarinssyni sjötugum 8.janúar 1982, Sögufélag,Reykjavík, pp. 67-77

Sveinbjarnardóttir, G. (1983) 'Byggðaleifar á Þórsmörk', Árbók hins íslenzka fornleifafélags 1982, 20-61.

Sveinbjarnardóttir, G. et.al. (1981) 'Excavation at Stóraborg: a palaeoecological approach', Árbók hins íslenzka fornleifélags 1980, 113-129.

Sveinbjarnardóttir, G. (forthcoming) 'Shielings in Iceland: an archaeological and historical survey', The Norse North Atlantic Conference Proceedings, Bowdoin College Maine, April 1988, Acta Archaeologica, forthcoming.

Sveinsson, Þ. (1954) 'Bæjatalið í Auðunarmáldögum', Árbók hins íslenzka fornleifafélags 1954, 26-47.

Sveitir og jarðir í Múlaþingi 3, Á. Halldórsson (ed.), Búnaðarsamband Austurlands 1976.

Thórarinsson, S. (1958) 'The Öræfajökull eruption of 1362', Acta Naturalia Islandica, Vol. II, No. 2, Reykjavík.

Þórarinsson, S. (1944) 'Tefrokronologiska studier på Island', Köbenhavn.

Þórarinsson, S. (1968) Heklueldar, Reykjavík.

Þórarinsson, S. (1976) 'Þáttur af Þegjandadal', in Minjar og menntir. Afmælisrit helgað Kristjáni Eldjárn, Reykjavík, pp. 461-470.

Þórarinsson, S. (1977) 'Gjóskulög og gamlar rústir, Árbók hins íslenzka fornleifafélags 1976, 5-38.

Þormóðsson, E. (1972) 'Byggð í Þistilfirði', Saga 1972, 92-133.

PART 2

ENVIRONMENTAL CHANGES DURING THE POSTGLACIAL

2.2 Glacier Fluctuations

NEW OBSERVATIONS ON THE POSTGLACIAL GLACIAL HISTORY OF TRÖLLASKAGI, NORTHERN ICELAND

Johann Stötter
Institute of Geography
University of Munich
8000 Munich 2
FRG

ABSTRACT. Early in the Lateglacial period the glaciers of the Tröllaskagi peninsula became independent. These small alpine glaciers were able to react to climatic influences in a sensitive and direct way and preserve more information about palaeoenvironmental changes. Thus the landscape of the Tröllaskagi mountains provides a valuable potential source area for research on the environmental history of Iceland. By means of moraine analysis a first attempt is made towards developing a chronology of the postglacial environmental history of this area. Two glacial advances can be traced and dated to the periods of 6,000-4,800 B.P. and 3,400-2,800 B.P. The local terms Vatnsdalur I and Vatnsdalur II are suggested for these advances. Beside these there are remnants of older advances during the early postglacial period.

1. INTRODUCTION

An increasing number of scientific projects concerning climate and its fluctuations since the end of the last ice age have taken place throughout the world, reflecting the growing importance of this area of research. Most of these investigations are focused on modelling climate, the goal of which is to develop an understanding of the influence of changing frame conditions on climate, and to make climatic variations more predictable. On the one hand present climatic parameters are documented by a series of actual measurements, on the other hand data relating to palaeoclimatic conditions are used as analogous states for future climatic variations (Frenzel, 1989).

Regular climatic measurements only reach back to the late 17th century (Rudloff, 1965), but additional sources of information on climate can be found in early documents which, in Iceland, record aspects of environmental conditions back to the original settlement (Ogilvie, 1981). For the pre-settlement period it is necessary to develop the potential information to be found in the landscape itself.

J. K. Maizels and C. Caseldine (eds.), Environmental Change in Iceland: Past and Present, 181–192.

Because of the clear relationships that exist between climate, mass and energy balance and changes in length, small glaciers of alpine type like those in Tröllaskagi provide a very valuable source of palaeoclimatic data. The development of a secure recent glacial history for such glaciers which is well documented by moraines and associated deposits, should therefore be an important contribution to the understanding of palaeoclimatic conditions.

Iceland provides a unique field laboratory for such an approach to landscape research, as it is possible to distinguish temporally between the period of natural impact on the environmental and climatic system, and the later period of human influence following settlement. Within Iceland the Tröllaskagi peninsula, with its alpine type of landscape, is the only area with a range of small corrie glaciers which provides the sort of potential information discussed above.

2. PREVIOUS WORK ON THE GLACIAL HISTORY OF ICELAND

Iceland has a long tradition of research on both glacial environments and glacial history. Around the beginning of the 20th century Thoroddsen (1892, 1906) was the first to develop a broad knowledge of Icelandic glaciers. He described 139 glaciers and collected a wide variety of information about them, especially about the advances of these glaciers during historical time. In 1909 Pjeturss demonstrated the existence of different ice-ages, some years before the theory of polyglaciation became established by a wide range of geoscientists in the Alps (Ampferer, 1914).

Since this early work further studies have been made to enlarge knowledge of Iceland's glaciers and their fluctuations. Both Bárðarsson (1934) and Thórarinsson (1943) summarised again the history of glaciers, based on a large number of historical documents. Before the 1930's no work was carried out on glacial oscillations during the period between the maximum of the last ice age and historical times: the time between 18,000 B.P. (Einarsson, 1968) or 23,000 B.P. (Norðdahl, 1983) and the beginning of the so called 'Little Ice Age' around A.D. 1570 (Grove, 1988).

A first subdivision of this whole period was made by the definition of the Búði stadial in southern Iceland (Kjartansson, 1939, 1943), equivalent to the Hólkot stadial in northern Iceland (Thórarinsson, 1951), and the earlier Álftanes stadial (Einarsson, 1960, 1961). The Búði/Hólkot stadial was correlated with the Sâlpausselka stadial in Scandinavia and put into the Younger Dryas (Einarsson, 1968), whilst the Álftanes stadial was equated with the Older Dryas. New investigations based on radiocarbon dating now show that the Búði advance took place during the Preboreal period (Hjartarson and Ingólfsson, 1988).

3. THE MODEL FOR THE POSTGLACIAL PERIOD

The development of the model for the postglacial period in Iceland has been based on rather limited evidence. Comparison of the pattern for postglacial climatic variations in

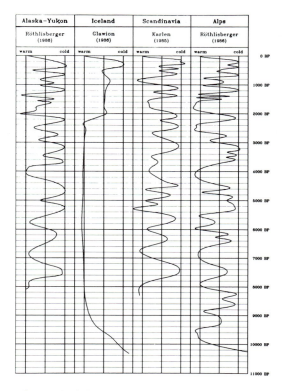

Fig.1 Patterns of postglacial climatic fluctuations for Iceland compared with Alaska,
Scandinavia and the Alps.

Iceland (Glawion, 1986) with curves for North America (Alaska/Yukon) and Europe
(Scandinavia/Alps), as summarised by Röthlisberger (1986) shows hardly any similarity
(Fig.1). There are two principal possible reasons for this:

    (i) Either climatic conditions in Iceland developed totally independently and were
therefore very different to other countries;

    (ii) Or the results of the research in Iceland are for some reasons, which have yet
to be understood, of a character which are not equivalent to those of other countries.
A short review of the genesis of the model may help to find an answer to this problem.

    Thórarinsson (1956) suggested that Kvíárjökull, Skaftafellsjökull and Svínafellsjökull
had advanced over their historical maximum at about 1,500 B.P. In a profile from the
Stóralda moraine at Svínafellsjökull he found the Ö1362 tephra (tephra of the Öræfi
eruption of 1362) but not the rhyolitic, 6,000 year old tephra, which is widespread over
the Öræfi district. His conclusion (Thórarinsson, 1956 p.7) was ' in my opinion it (i.e.
the moraine) was most likely formed as a result of the climatic deterioration during the

first centuries of the Subatlantic Time, that began at ab. 600 B.C.'.

This is the origin of the so-called Subatlantic advance, which has been referred to by many scientists when they described moraines outside the gletschervorfeld, the area covered by the glacial deposits and moraines of advances since the beginning of the 17th century A.D. (Kinzl, 1949). Such a Subatlantic advance has never been dated absolutely, for instance by the use of radiocarbon.

North of Hofsjökull Kaldal (1978) described eight moraine ridges. As their course is at right angles to the edge of Hofsjökull they were not connected with the glacier but with the readvances or stationary phases of a more extensive Vatnajökull ice cap, which was wasting down on the whole. Compared with the Subatlantic moraines of Thórarinsson they were declared to be older and therefore of Preboreal age. Further on it is said that all ice caps disappeared during the postglacial climatic optimum and, ' like all the large ice-caps in Iceland, the present Hofsjökull is supposed to have started to form during the cold period that began 2500 years ago, i.e. during Subatlantic times' (Kaldal, 1978 p.29).

These ideas of Kaldal more or less followed the subdivision of the Holocene of Iceland proposed by Einarsson (1961). Based on pollen-analytical studies of bog profiles he distinguished four pollen zones (Fig.2). In his curve of temperature changes he interpreted a postglacial warm period lasting from about 9,000 B.P. (around the end of the Preboreal) until about 2,500 B.P. (the beginning of the Subatlantic period).

This system has remained unchanged rather in the form of a dogma and until recently no evidence to the contrary has been published. This was rather like the situation in the Alps where up until the 1960's results were fitted in to an accepted scheme. According to the glacial readvances of Penck and Brückner (1901/1909) knowledge of the lateglacial period in the Alps had been developed by several authors (Kinzl, 1929; Klebelsberg, 1950) and for the postglacial period a long lasting climatic optimum (based on pollen-analytical studies) was believed. However Heuberger (1966), Patzelt (1972, 1973), Bortenschlager

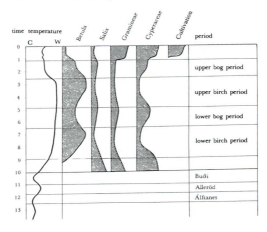

Fig.2 Subdivision of the Holocene of Iceland following Einarsson (1961).

(1970, 1972) and others found evidence for several glacial advances and periods of climatic deterioration, and established a more subtly differentiated curve for the postglacial period with multiple changes of warm and cold phases.

4. THE RESEARCH AREA

The research area of Svarfaðardalur-Skíðadalur is one of the main valleys on the eastern side of the Tröllaskagi peninsula (Fig.3). Both the Svarfaðardalsá and Skíðadalsá have their origin in glaciers at the divide. After their confluence the Svarfaðardalsá drains into Eyjafjörður at Dalvík. The catchment area is 451.5 km², and the highest point is

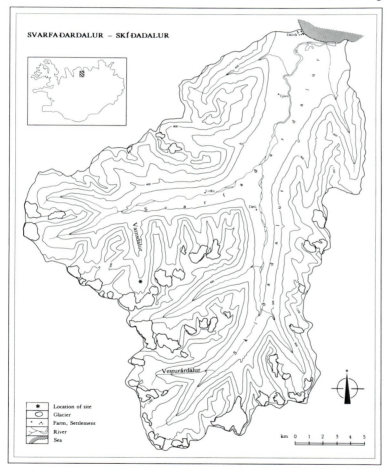

Fig.3 Research area of Svarfarðadalur/Skíðadalur.

Dyafjallshnjúkur at 1456m a.s.l. The valley system is cut into a plateau of Tertiary basalts, which have been dated to 9-10.5 ma by Sæmundsson et al. (1980). During the Quaternary the area was covered by ice caps or filled by ice streams. At present 48 small corrie glaciers with a total area of 30.7 km² (Stötter, 1989) can be found. Thus almost all geomorphological features are the result of both glacial and periglacial processes, many of which are still very active.

Previous research has been concentrated mostly on the glacial history of the area. Caseldine (1983, 1985a, 1987) and Kugelmann (1989) reported on recent glacier oscillations, and Hjartarson (1973), Müller et al. (1984) and Meyer and Venzke (1986) on the late- and postglacial history.

Both Svarfaðardalur and Skiðadalur have a number of tributary valleys, of which the biggest is Vatnsdalur, a tributary valley of Svarfaðardalur with an area of 20 km². In Vatnsdalur there are at present 6 glaciers with a total area of 3.72 km². The glaciation rate of 18.5% is one of the highest in the valley.

4. THE MORAINE COMPLEX IN VATNSDALUR

4.1 Description

In front of the gletschervorfeld of Vatnsdalsjökull, which is the biggest glacier (1.39 km²) in Vatnsdalur, there is a moraine complex (Fig.4). A well preserved terminal moraine is covered by a dense grassy vegetation and can easily be distinguished from the gletschervorfeld area, which is almost free from vegetation. Due to the lack of fine material only traces of soil development can be found here.

The moraine complex can be found down to 688 m a.s.l. On the eastern part the ridge is two- or sometimes threefold and runs up almost to the end of a large rock glacier at 718m a.s.l. On the western side the moraine ridge is not so well preserved. Next to the glacial river a section was dug through the moraine complex (Fig.5). The base of the exposure shows till, which can be interpreted as the basal till of an older, advanced glacier, or morainic debris from a retreating glacier. Above the till there are remnants of a fossil soil with traces of tephra. Above this there is a terminal moraine, which can be divided into two, an upper and a lower part. The central part of the complex is filled by a V-shaped section of peat material. On the inner side another terminal moraine completes the complex. The whole complex is covered by a well developed brownish soil, which is asymmetrically spread. Due to dominating katabatic winds blowing down from the glacier the soil thickness on the inner side is thinner than on the outer side where a colluvium has accumulated on the leeward side of the moraine complex.

4.2 Interpretation of Genesis and Age

In the period between 6,000 B.P. and 4700±205 B.P. (GSF-IS -287) Vatnsdalsjökull advanced and deposited a terminal moraine. This is the outermost part of the moraine

Fig.4 Gletschervorfeld of Vatnsdalsjökull.

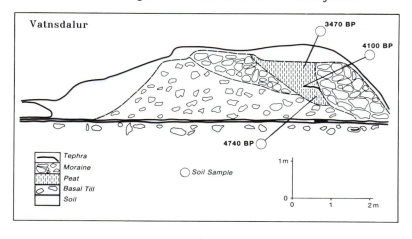

Fig.5 Section through the outer moraine ridge, Vatnsdalur.

complex described above. The maximum age is given by the H5 tephra layer that could be found in the fossil soil beneath the whole moraine complex. The soil itself has not yet been dated but the minimum age is derived from the lower part of the peat section, which was only partly disturbed by the later glacial advance. It is suggested that this outer part of the moraine complex is the remnant of two glacial advances or of a twofold event. An upper, outer part with a high proportion of fine-grained material can be distinguished from a lower, inner part with a lack of fine material, thus indicating two separate advances. Between these two parts there is no trace of any fossil soil. This lack of a separating layer implies that there was only a short period of time between the two phases of deposition.

When the glacier retreated from this advanced position a period of considerably improved climate began. A bog developed on the inner side of the moraine using it as a kind of dam, but only seems to have developed close to the moraine; there is no evidence of widespread bog formation. This growth of the bog was almost totally undisturbed. Within the peat it is possible to find the H4 tephra layer. Peat material in contact with this tephra gave an age of 4,100±190 B.P. (UZ-2415:ETH 4115). No indication of any climatic deterioration can be traced within this part. The period from 4,740±205 B.P. to 3,470±160 B.P. (GSF-IS -187), during which bog formation took place, seemed to represent very good climatic conditions.

More evidence for this pattern can be found in the Svarfaðardalur-Skíðadalur area. A peat profile in Vesturárdalur shows a layer of birch twigs immediately beneath the H4 tephra. This section lies at about 500m a.s.l., about 150m higher than the present potential tree-line in this area (Krístinsson, pers. comm.), and today represents the uppermost site of bog formation. There is again a difference of about 150m between present day vegetation boundaries and those during the earlier postglacial, as was indicated in Vatnsdalur.

All these observations can be interpreted as indications of good climatic conditions. It is quite within the bounds of probability that the period between about 4,800 B.P. and 3,400 B.P. reflected the best climatic conditions during the whole Holocene and thus in this part of Iceland may represent the rest of the climatic optimum. It is very likely that almost all glaciers in the Tröllaskagi disappeared during this period. If it can be assumed that both the tree-line and snowline react in parallel to climatic changes, then the snowline on glaciers reached a level higher than the critical height mentioned by Björnsson (1971). This is also supported by reconstruction of the snowline following Gross et al. (1976). An increase of about 150m causes the whole of the theoretical snowline to rise above the upper end of most of the glaciers in Tröllaskagi.

Vatnsdalsjökull readvanced after 3,470 B.P., but this readvance did not reach the size of the older event. The bog was only influenced slightly, there are a few traces of disturbance, such as compression of the peat by boulders of the new moraine. As there is almost a total lack of fine-grained material this advance seems to have been quite a rapid event, perhaps even a glacier surge. There is not yet a minimum age for this readvance. The lack of H3 tephra in the peat suggests that bog formation ceased before 2,800 B.P. The climatic deterioration which preceded the glacier advance and the subsequent proximity of the glacier terminated peat formation. In Barkárdalur Häberle (1989) has a minimum age of 3,000±95 B.P. for the beginning of peat formation inside a supposedly Holocene moraine. Further, Häberle describes a moraine in front of the gletschervorfeld of Bægisárjökull, put into the 18th century by Sigurjónsson (1963), on which the soil contains H3 tephra (2,800 B.P.). These points may be used to indicate improving climate around/after 3,000 B.P. in the whole of the Tröllaskagi area. There is no evidence for any further advance of Vatnsdalsjökull between about 3,000 B.P. and historical times, nor are there any indications of significantly improved climatic conditions. The next readvance of Vatnsdalsjökull took place in the beginning of the 19th century, as dated lichenometrically by Kugelmann (1989).

The only other accurately dated Holocene advances in Iceland are those described by Dugmore (1989) at Sólheimajökull, where he defined advances between 7,000-4,100 B.P., and just before 3,000 B.P. Dugmore relates these oscillations to changed conditions in the accumulation area of the glacier, but the correspondence between the results from north and south Iceland may indicate, at least in part, a climatic influence. Nevertheless these results cast more doubt upon the original accepted chronology for the course of climatic variability during the postglacial period in Iceland.

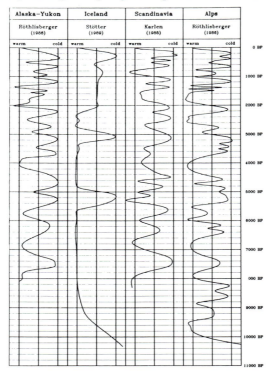

Fig.6 Revised pattern of postglacial climatic fluctuations for Iceland compared with Alaska, Scandinavia and the Alps.

6. CONCLUSIONS

This study shows that during the Icelandic postglacial 'climatic optimum' climatic conditions were not always as optimal as supposed by the traditional interpretation (Fig.6). Two glacial advances have been identified and demonstrate a greater variability in the climatic record.It is proposed to use the local terms Vatnsdalur I and Vatnsdalur II for these advances.

This revision should be seen as an initial impulse to develop a more sensitive chronology to that originally proposed. There must now be renewed investigation of the whole of the postglacial period using all available modern methods. The landscape of Iceland provides a lot of information which has to be adequately exploited and interpreted to develop a new postglacial history comparable to that either of North America or Europe.

ACKNOWLEDGEMENTS

Permission to work in Svarfaðardalur-Skíðadalur was kindly given by the Icelandic Research Council (permits 52/87 and 46/88). I am grateful to Oskar Gunnarsson and Lene Zachariassen at Dæli and Hjörtur Þórarinsson and Sigriður Hafstad at Tjörn for their hospitality and help. Assistance in the field was provided by Ottmar Kugelmann, Margret Thomas and Dieter Graser of the Arbeitsgruppe Island. Prof. dr. F.Wilhelm, Dr. Chris Caseldine, Thomas Häberle and Ottmar Kugelmann gave generously comments on many points in this paper. Radiocarbon analyses were supported by the Deutsche Forschungsgemeinschaft (Wi 159: 24-3) and gratefully done in the laboratories of the Gesellschaft für Strahlen- und Umweltforschung, Institut für Hydrologie, in München-Neuherberg and of the Institut für Geographie der Universität Zürich.

REFERENCES

Ampferer, O. (1914) 'Über die Aufschließung der Liegendmoräne unter der Höttinger Breccie im östlichen Weiherburggraben bei Innsbruck', Zeitschrift für Gletscherkunde 8, 145-159.
Bárðarsson, B. (1934) 'Islands Gletscher', Societas Scientiarum Islandica XVI, Reykjavík.
Caseldine, C.J. (1983) 'Resurvey of the margins of Gljúfurárjökull and the chronology of recent deglaciation', Jökull 33, 111-118.
Caseldine, C.J. (1985a) 'The extent of some glaciers in Northern Iceland during the Little Ice Age and the nature of recent deglaciation', Geographical Journal 151, 215-227.
Caseldine, C.J. (1985b) 'Survey of Gljúfurárjökull and features associated with a glacier burst in Gljúfurárdalur', Jökull 35, 61-68.
Caseldine, C.J. (1987) 'Neoglacial glacier variations in northern Iceland: examples from the Eyjafjörður area', Arctic and Alpine Research 19, 296-304.

Dugmore, A.J. (1989) 'Tephrochronological studies of Holocene glacier fluctuations in South Iceland', in J.Oerlemans (ed.) Glacier Fluctuations and Climatic Change, Kluwer Academic Press, Dordrecht, pp. 37-56.

Einarsson, Th. (1960) 'Geologie von Hellisheiði', Sonderveröffentlichung des Geologischen Instituts der Universität Köln 5.

Einarsson, Th. (1961) 'Pollenanalytische Untersuchungen zür spät- und postglazialen Klimageschichte Islands', Sonderveröffentlichung des Geologischen Instituts der Universität Köln 6.

Einarsson, Þ. (1968) Jarðfræði, Mál og Menning, Reykjavík.

Frenzel, B. (1989) 'Klimasituation und Klimabläufe', Resultate der terrestrischen Paläoklimatologie, in GSF (ed.) Klimaforschungsprogramm Statusseminar 10.1-12.1.1989, München.

Glawion, R. (1986) 'Rezente Klimaschwankungen und Vegetationsänderungen', Geowissenschaft in unsere Zeit 4, 141-150.

Grove, J. (1988) The Little Ice Age, Methuen, London.

Gross, G., et al. (1976) 'Methodische Untersuchungen über die Schneegrenze in alpinen Gletschergebieten', Zeitschrift für Gletscherkunde und Glazialgeologie XII, 223-251.

Häberle, T. (1989) 'Spät- und postglaziale Gletschergeschichte des Hörgárdalur-Gebiets, Tröllaskagi, Nordisland', Dissertation, Universität Zürich.

Heuberger, H. (1966) 'Gletschergeschichtliche Untersuchungen in den Zentralalpen zwischen Sellrain- und Ötztal', Wissenschaftliche Alpenvereinshefte 20.

Hjartarson, Á. (1973) 'Rof jarðlagastaflans milli Eyjafjarðar og Skagafjarðar og ísaldarmenjar við utanverdan Eyjafjörð', unpublished B.Sc. thesis, University of Iceland, Reykjavík.

Hjartarson, Á. and Ingólfsson, Ó. (1988) 'Preboreal glaciation of southern Iceland', Jökull 38, 1-16.

Kaldal, I. (1978) 'The deglaciation of the area north and northeast of Hofsjökull', Jökull 28, 18-31.

Kinzl, H. (1929) 'Beitrage zür Geschichte der Gletscherschwankungen in den Ostalpen', Zeitschrift für Gletscherkunde XVII (1-3), 66-121.

Kinzl. H. (1949) 'Formenkundliche Untersuchungen im Vorfeld der Alpengletscher', Veröffentlichung des Museums Ferdinandeum 26, Innsbruck.

Kjartansson, G. (1939) 'Stadier i isens tilbag erykning fra det sydvestislanske lavland. En isdammaet sö. Marine dannelser. Postglacial tektonik', Meddelelser fra Dansk Geologisk Forening 9, 426-458.

Kjartansson, G. (1943) Árnesinga saga, Reykjavík.

Klebelsberg, R. v. (1950) 'Das Silltal bei Matrei', Schlern-Schriften 84, 76-86.

Kugelmann, O. (1989) 'Gletschergeschichtiche Untersuchungen im Svarfaðardalur und Skíðadalur, Tröllaskagi, Nordisland', Diplomarbeit, München.

Müller, H.-N. (1984) Spätglaziale Gletscherschwankungen in den westliche Schweizer Alpen (Simplon-Süd und Val de Nendaz, Wallis) und im nordisländischen Tröllaskagi-Gebirge (Skíðadalur), Dissertation, Näfels.

Müller, H.-N. (1984) 'Glazial- und Periglazialuntersuchungen im Skíðadalur, Tröllaskagi (N-Island, Polarforschung 54, 95-109.

Norðdahl, H. (1983) 'Late Quaternary stratigraphy of Fnjóskadalur, Central North Iceland, a study of sediments, ice-lake strandlines, glacial isostasy and ice-free areas', Lundqua Thesis 12, Lund.

Ogilvie, A.E.J. (1981) Climate and Society in Iceland from the Medieval Period to the late 18th Century, Unpublished Ph.D Thesis, University of East Anglia, Norwich.

Patzelt, G. (1972) 'Die spätglazialen Stadien und postglazialen Schwankungen von Ostalpengletschern', Berichte der Deutschen Botanische Gesellschaft 85.

Patzelt, G. (1973) 'Die postglazialen Gletscher- und Klimaschwankungen in der Venedigergruppe (Hohe Tauern, Ostalpen) Zeitschrift fur Geomorphologie, Neue Folge Supplementband 16, 25-72.

Penck, A. and Brückner, H. (1901/1909) Die Alpen im Eiszeitalter, Leipzig.

Pjeturss, H. (1909) 'Über marines Interglazial in Südwest-Island', Zeitschrift der Deutschen Geologischen Gesellschaft 61, 274-287.

Röthlisberger, F. (1986) 10000 Jahre Gletschergeschichte der Erde, Aarau.

Rudloff, H. v. (1965) Die Schwankungen und Pendelungen des Klimas in Europa seit Beginn der regelmäßigen Instrumentum (1670), Braunschweig.

Sæmundsson, K., Krístjánsson, L., McDougall, I. and Watkins, N.D. (1980) 'K-Ar dating, geological and palaeomagnetic study of 5km lava succession in Northern Iceland', Journal of Geophysical Research 85, 3628-3646.

Sigurjónsson, J. (1963) Bægisá, Nord-Island, Oslo.

Stötter, J. (1989) 'Geomorphologische und landschaftsgeschichtliche Untersuchungen im Svarfaðardalur-Skíðadalur, N.-Island', Dissertation, München.

Thoroddsen, Th. (1892) 'Islands Jökler i Fortid og Nutid', Geografisk Tidsskrift 11, 111-146.

Thoroddsen, Th. (1906) 'Island. Grundriß der Geographie und Geologie', Petermanns Geographische Mitteilungen, Ergänzungshefte 153, 162-358.

Thórarinsson, S. (1943) 'Oscillation of the Icelandic glaciers in the last 250 years', Geografiska Annaler 21, 1-54.

Thórarinsson, S. (1951) 'Laxárgljúfur and Laxáhraun. A tephrochronological study', Geofrafiska Annaler 33A, 1-88.

Thórarinsson, S. (1956) 'On the variations of Svínafellsjökull, Skaftafellsjökull and Kvíárjökull', Jökull 6, 1-15.

Venzke, J.-F. and Meyer, H.-H. (1986) 'Remarks on the late glacial and early Holocene deglaciation of the Svarfaðardalur and Skíðadalur valley system, Tröllaskagi, Northern Iceland', Berichte der Forschungsstelle Neðri Ás 45.

HOLOCENE GLACIAL HISTORY OF THE HÖRGÁRDALUR AREA, TRÖLLASKAGI, NORTHERN ICELAND

Thomas Häberle
Egg 373
5026 Densbüren AG
Switzerland

ABSTRACT. Radiocarbon dating, tephrochronology and lichenometry are used to determine the Holocene glacial history of the Hörgárdalur area, Tröllaskagi. Glacial advances at around 4,275B.P. and 1,000 B.P. have been identified in Bægisárdalur, whereas in Barkárdalur pre-'Little Ice Age' advances are recognised between 2,240-1,835 B.P. and shortly before 1,555 B.P. During the 'Little Ice Age' there were a series of advances in both valleys between the mid-18th century and the first half of the 20th century, with a further advance in the 1970's and 1980's.

1. INTRODUCTION

The evidence presented in this paper is part of a wider study of the glacial history of the region to be presented at the University of Zürich as a doctoral thesis. The main aim of the work described here is to examine the record of Holocene glacial change and to infer a measure of climatic variability from this and associated evidence. The research area is situated on the eastern side of the Tröllaskagi peninsula in mid-northern Iceland and covers the Hörgárdalur valley system, an area in total of ca 770 km² (Fig.1).

2. PREBOREAL/HOLOCENE GLACIAL LIMITS

The location and age of Preboreal glacial limits in Iceland has recently been the subject of considerable reappraisal with many moraines of supposed Búði or Younger Dryas age being assigned to the Preboreal period (e.g. Hjartarson and Ingólfsson, 1989). Radiocarbon-dating of the base of the peat at Ytri-Bægisá, at the mouth of the easternmost southern tributary valley of Öxnadalur (Fig.1), shows that this location was ice-free by 8,850±120 B.P. (NPL-160: Bartley, 1973). Reconstruction of the former glacier which would have

J. K. Maizels and C. Caseldine (eds.), Environmental Change in Iceland: Past and Present, 193–202.

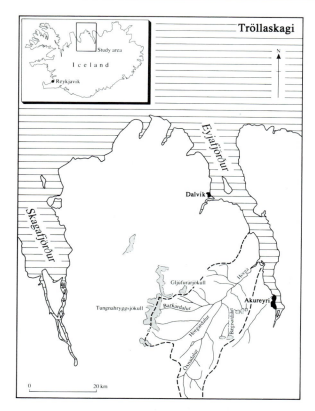

Fig.1 Tröllaskagi peninsula, location of sites mentioned in the text.

and which terminated close to the dated site gives a depression of the Equilibrium Line
Altitude (E.L.A.) relative to today of 350m. To the east of Eyjafjörður end moraines
associated with valley glaciers from Ytri- and Syðri-Brettingsstaðadalur show a depression
of only 220m. These end moraines were thought to be of Búði age, and dated to prior to
9,650±120 B.P. (Lu-1433: Norðdahl, 1979). In Barkárdalur the major Sörlatunga moraine
which lies at the end of the valley where it joins Hörgárdalur (Fig.2) defines a former valley
glacier with an E.L.A. depression of 535m. Sections through this moraine have not yet
provided organic material suitable for dating so the age of this limit remains uncertain.

    The amount of E.L.A. depression implied by the Sörlatunga moraine suggests that
it must represent a much earlier stage than either the Younger Dryas (Búði) or Preboreal,
unless this glacier behaved very differently to those close by, and hence marks an, as yet,
undated limit earlier in the Late Weichselian. It would appear that glaciers were probably
withdrawn into the tributary valleys of Öxnadalur by the Younger Dryas (Búði) stage and

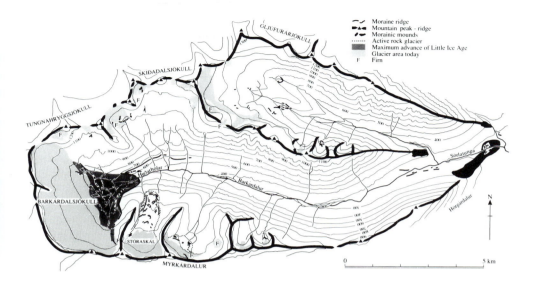

Fig.2 Major morphological features in Barkárdalur.

that there is very little morphological evidence for this stage in the area. An exception to this is Skriðudalur, a tributary valley of Hörgárdalur (Fig.1), where end moraines correspond to depression values of 160m and 180m respectively, the sort of values perhaps expected for limits of Younger Dryas age.

Higher in the valleys of Bægisárdalur and Barkárdalur moraine evidence of later Holocene advances, thought to be older than 'Little Ice Age', can be found. These are so close to present glacial limits that, at most, they represent a depression of the E.L.A. of only 5m. Thus any depression greater than this is likely to indicate glacier expansion of at the latest Preboreal age. Radiocarbon dating of wood and soil found in association with pre-'Little Ice Age' moraines has not provided a secure chronology as yet for Barkárdalur, although in Bægisárdalur dates indicate moraine formation around 4,275 B.P. and 1,000 B.P. (Häberle, in prep.).

The results of a number of radiocarbon determinations from sites in Barkárdalur indicate some of the problems encountered in dating Holocene advances in this area. Fig.3 illustrates the location and stratigraphies of a series of pits excavated in association with four moraines found at Húðarhólar immediately in front of Barkárdalsjökull. (Details of all the radiocarbon dates from this location are presented in Table 1.) Profile G includes tephra layers H3, H4 and H5 and a date from immediately below H5 shows that tephra deposition took place after 5,950±110 B.P. Radiocarbon dating of H5 in Iceland has yet to provide a consistent age, with dates from northern Iceland generally being younger than those previously published (Table 2). Nevertheless, the profile at G clearly shows that the area

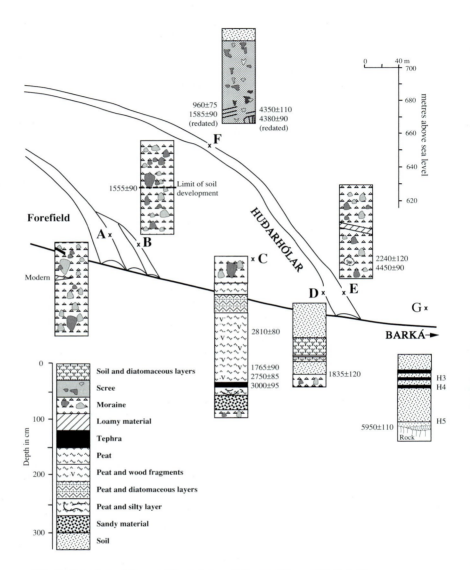

Fig.3 Stratigraphies and locations of the profiles at Huðahólar.

immediately outside the outermost moraine has been ice free since at least ca 6,000 B.P.

Radiocarbon dates from both wood and soil at D, E and F on the moraine immediately inside G present a range from 4,450±90 B.P. to 960±75 B.P. Site F is a complex profile involving a combination of scree and morainic material. As the stratigraphic implications of the sequence are not clear the dates from here are not used for dating the moraine. Separate pieces of wood found underneath the same moraine at E were dated to 4,450±90 B.P. and 2,240±120 B.P. respectively using accelerator dating. This spread of dates is interpreted as representing a period of relatively warm conditions with shrubs able to grow up to ca 560m,

Table 1. ¹⁴C determinations from Upper Barkárdalur (* denotes AMS).

Site	Age ¹⁴C yrs.	δ¹³C	Laboratory Reference No	Material dated
G	5,950±110	-24.6	UZ 2361=ETH 3565*	Soil
F	4,350±110	-	UZ 1199	Soil
	4,380±90	-20.5	UZ 2400=ETH 3981*	Soil
	960±75	-	UZ 1198	Soil
	1,585±90	-20.7	UZ 2399±ETH 3980*	Soil
E	4,450±120	-29.3	UZ 2467±ETH 4570	Wood
	2,240±120	-33.9	UZ 2431±ETH 4251*	Wood
D	1,835±120	-24.9	UZ 2271	Soil
C	3,000±95	-23.2	UZ 2330±ETH 3400*	Wood
	2,750±85	-	UZ 1218	Peat
	1,765±90	-23.8	UZ 2329±ETH 3399*	Peat
	2,810±80	-	UZ 1222	Wood
B	1,555±90	-22.8	UZ 2369±ETH 3570*	Soil

an altitude close to the limit proposed by Glawion (1985). In view of the temperatures needed to sustain such growth it is unlikely that Barkárdalsjökull would have survived during this time. The lack of continuity within the profile at E cannot confirm the continued existence of warm conditions throughout the period between 4,450 and 2,240 B.P., but the dates for fragments of wood in the peat in profile C inside the outer moraines suggest that shrubs were growing there between at least 2,750±85 B.P. and 3,000±95 B.P. It is interesting that below the 3,000 B.P. level the sediments are sandy or intercalated silts and peat.

Table 2. Radiocarbon ages of the tephra layer H5 according to different authors.

Author	S.Þórarinsson (1971)	Þ.Einarsson (1968)	E.Vilmundardóttir & I.Kaldal (1982)	T.Häberle (average of two dates)
H5 (yrs B.P.)	6610±170	6400±170	6100	5785±70

This may indicate some snowmelt or even glacial effect but there is no real evidence for the history of the valley between 3,000 and 4,450 B.P., thus the redevelopment of some ice cannot be ruled out. The advance which produced the moraine with the profile at E is likely therefore to have been produced sometime before 2,240 B.P. A profile at D on the inner slope of the same moraine, produced a date from the base of the soil developed on the moraine of 1,835±120 B.P. As this was a bulk soil sample it is likely to be a maximum age for soil development (Matthews, 1981) and hence brackets the period for the formation of the moraine to between 2,240 and 1,835 B.P. (allowing for 95% errors on the dates this would extend to between 2,480 and 1,595 B.P.).

At C, a buried peat inside the two outer moraines, there are clearly problems with the radiocarbon dates as two dates appear to be out of sequence. To some extent this may be due to combining both conventional and AMS dates, but in the case of the date of 1,765±90 this would seem to be clearly in error. At present the reasons for this are unknown. The basal date of 1,555±90 from the soil on the moraine at B, the outermost of a series of three closely juxtaposed moraines, implies that the formation of this moraine took place shortly before this date and it may therefore represent either a gradual reduction from the extent marked by the previous limits or a period of renewed but less extensive glaciation. Dating of material at A provided only a modern date.

3. 'LITTLE ICE AGE' LIMITS AND RECENT DEGLACIATION

Dating of moraines lying inside those described above which were thought to be of 'Little Ice Age' date on the basis of geomorphological and biogeographical criteria was carried out mainly by lichenometry. Growth curves for *Rhizocarpon geographicum* agg. were constructed for different valleys in the study area using a wide range of calibration surfaces including farm ruins, tombstones, bridges and a single moraine of documented age. For Barkárdalur (Figs.4A and 4B) data is presented using both longest axis and largest inscribed circle, the former producing a significant correlation between size and age of r= +0.81. For Skriðudalur, Fossárdalur and Myrkárdalur results from the use of the largest inscribed circle are used, but present a similar level of correlation (Fig.4C). In Fig.5A all the results are combined and show a much lower level of correlation, whereas in Fig.5B the use only of farm ruins represents a relatively high level of correlation of +0.74. The problems of using such calibration data in this area have been discussed elsewhere (Häberle, in press; Kugelmann, in press) but overall the calibrations demonstrate some degree of variability between valleys. For the moraines discussed here an overall growth rate of 0.41 mm a^{-1} has been derived, a figure that agrees well with the rate of 0.44 mm a^{-1} used by Kugelmann immediately to the north in Skíðadalur and Svarfaðardalur.

Results from the application of the separate lichen curves for the moraines in five areas are presented in Table 3; also included are historical and recent observational results. The earliest 'Little Ice Age' date of the mid-18th century for Barkárdalur is earlier than any previous determination in the area as a whole but is open to question as it lies beyond the calibrated area of the curve. Nevertheless the pattern and dates of the advances generally correlate well with those to the north (Caseldine, in press; Kugelmann, in press) with particularly noticeable periods of formation in the mid-19th century and in the early 20th

Table 3. Lichenometric and other dating of moraines of 'Little Ice Age' date in the Hörgárdalur area.

Glacier	Year or period of advance/stillstand	Dating Method
Barkárdalsjökull	1744/1763	Lichenometry
	1827/1846	As above
	1858/1863	As above
	1900	Historical record (E.Guðmundsson)
	1945/1951	Lichenometry, aerial photography
	1975-1985	Observation
Bægisárjökull	1801	Historical record (T.Thoroddsen)
	1977-1988	Observation
Skriðudalur (corrie glacier)	1815	Lichenometry
Vindheimajökull	1855	As above
	1922	As above
	1952	As above
Myrkárjökull	1840	As above

century. Recent advances between 1975 and 1988 have been observed at Bægisárjökull and Barkárdalsjökull and equate with the results of Caseldine (1985) at Gljúfurárjökull.

4. CONCLUSION

Studies of the glacial history of Hörgárdalur using a variety of dating techniques have underlined some of the problems involved in establishing chronologies of glacier variations for the Holocene and 'Little Ice Age' in this part of Iceland. Radiocarbon dating and tephrochronology have indicated the existence of pre-'Little Ice Age' Holocene glacier advances in Barkárdalur between ca 1,835 B.P. and 2,240 B.P. and around 1,555 B.P. Lichenometric dating has shown a number of periods of moraine formation within the last 200 years in a pattern which conforms with previously published results for the Tröllaskagi peninsula. The results achieved so far therefore confirm the existence of pre-'Little Ice Age' glacier advances in northern Iceland, but to positions which only imply a depression of 5m

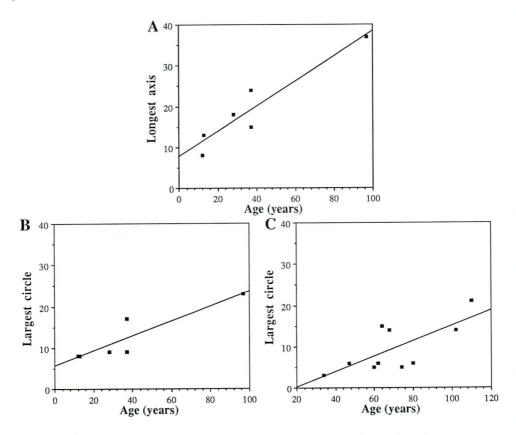

Fig.4 Barkárdalur lichen calibration curves, see text for explanation.

for the E.L.A. There still remains a lot of work to be done both in the improvement of the dating methods and in widening the range of areas covered before a clear pattern of the complete Holocene pattern of glaciation in Tröllaskagi can be established.

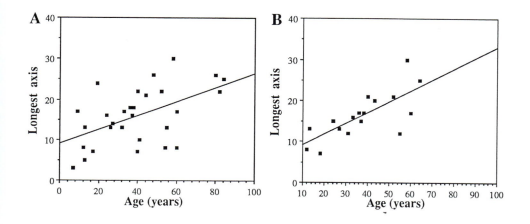

Fig.5 Lichen calibration data, see text for explanation.

ACKNOWLEDGEMENTS

Financial support for the work discussed here was provided by my mother and by the Geographical Institute of the University of Zürich, the latter also financed the radiocarbon dating. To all these people I offer my sincere thanks, as well as those with whom I had useful discussions: Chris Caseldine, Halldór Pétursson and Hans Stötter.

REFERENCES

Bartley, D.D. (1973) 'The stratigraphy and pollen analysis of peat deposits at Ytri-Bægisá near Akureyri, Iceland', Geologiska Föreningens i Stockholm Förhandlingar 95, 410-414.
Bergþórsson, P. (1956) 'Barkárjökull', Jökull 6, 29.
Caseldine, C.J. (1985) 'The extent of some glaciers in Northern Iceland during the Little Ice Age and the nature of recent deglaciation', The Geographical Journal 151, 215-227.

Caseldine, C.J. (1990) 'A review of Holocene dating methods and their application in the development of a Holocene glacial chronology for Northern Iceland', Münchener Geographische Abhandlungen Reihe B (in press).

Einarsson, Þ. (1968) Jarðfræði: Saga bergs og lands, Mál og Menning, Reykjavík.

Eyþorsson, J. (1957) 'Frá Norðurlandsjöklum. Úr dagbókum 1939 og 1957 (framhald)', Jökull 7, 52-58.

Glawion, R. (1985) 'Die natürliche Vegetation Islands als Ausdruck des ökologischen Raumpotentials', Bochumer Geographische Arbeiten H 45, 208 pp.

Häberle, T. (1990) 'Spät- und postglaziale Gletschergeschichte des Hörgárdalur Gebietes, Tröllaskagi, Nordisland', Dissertation an der Universität Zürich, in press.

Hjartarson, Á. and Ingólfsson, Ó. (1988) 'Preboreal glaciation of southern Iceland', Jökull 38, 1-16.

Kugelmann, O. (1990) 'Datierungen neuzeitlicher Gletscherschwankungen im Svarfaðardalur/Skíðadalur mittels einer neuen Flechten-Eichkurve', Münchener Geographische Abhandlungen Reihe B (in press).

Matthews, J.A. (1981) 'Natural ^{14}C age/depth gradient in a buried soil', Naturwissenschaften 68, 472-474.

Norðdahl, H. (1979) 'The last glaciation in Flateyjardalur, central north Iceland, a preliminary report', University of Lund Department of Quaternary Geology Report 18.

Vilmundardóttir, E. and Kaldal, I. (1982) 'Holocene sedimentary sequence at Trjávidarlækur Basin, Þjórsardalur, southern Iceland', Jökull 32, 49-59.

Þórarinsson, S. (1971) 'Aldur ljósu gjóskulaganna úr Heklu samkvæleiðréttu geislakolstímatali', Náttúrufræðingurinn 41, 99-105.

DATING RECENT GLACIER ADVANCES IN THE SVARFADARDALUR-SKÍÐADALUR AREA OF NORTHERN ICELAND BY MEANS OF A NEW LICHEN CURVE

Ottmar Kugelmann
Institute of Geography
University of Munich
8000 Munich 2
Germany

ABSTRACT. A new lichenometric curve is presented for the Svarfaðardalur-Skíðadalur area of north Iceland developed on the basis of 19 dated surfaces, including abandoned farmsteads or ruins, grave stones, memorial stones, bridges and a mudflow. The growth rate for *Rhizocarpon geographicum* agg. is found to be 0.44mm a^{-1}, a lower figure than that previously adopted in the area and in southern Iceland. Errors in the construction of earlier lichen growth curves in Iceland are outlined and the new curve used to date moraines of recent age in a number of valleys. Use of 10-year running means on these results show glacier advances at ca A.D. 1810, 1850, 1870-80, 1890-1900, 1920s and 1940s, and comparison with similarly treated sea-ice data shows possible correlations for the late 19th century and possibly the 1920s.

1. INTRODUCTION

The nature of recent climatic variations has been attracting increasing interest, and attention has been focused on the potential of past sequences of glacier variations as a source of palaeoclimatic information. The location of Iceland in an area of the North Atlantic which has a significant influence on the climate of North West Europe makes it an important location for the study of glacier variations. In this paper a new lichenometric dating curve for *Rhizocarpon geographicum* agg. is presented for the Svarfaðardalur-Skíðadalur area of northern Iceland which is then used to date a series of moraine sequences. These moraine sequences occur in front of either small valley glaciers or cirque glaciers which are characteristic of the area and which are very sensitive and react quickly to climatic changes.

2. PREVIOUS WORK ON RECENT GLACIER HISTORY IN ICELAND

Many of the earlier studies of recent glacier fluctuations concentrated on outlet glaciers from the major Icelandic ice caps and refer to written or oral sources, especially when glaciers advanced into cultivated land and damaged meadows, buildings or routeways

203

J. K. Maizels and C. Caseldine (eds.), Environmental Change in Iceland: Past and Present, 203–217.
© 1991 *Kluwer Academic Publishers. Printed in the Netherlands.*

(Grove, 1988). At the end of the last century Thoroddsen visited a large number of Icelandic glaciers and collected information on their former positions. These results were published in two extensive papers in 1895 and 1906, and were specifically referred to in later papers by Barðarson (1934), Ahlmann (1937) and Thórarinsson (1943). Critical consideration of these papers poses certain problems. Results were not always presented unequivocally and later authors have tended to refer to these papers uncritically. As an example, Thoroddsen was not always referring to the maximum extent of a glacier, only to an advanced position. Despite such difficulties a number of phases of general recent glacier advances can be identified, notably around A.D. 1750, 1850, 1870 and 1890, with most glaciers reaching their recent maximum extent around 1850. In North West Iceland Eyþorsson (1935) investigated outlet glaciers of Drangajökull and interpreted advances in the 18th century, around 1840-1850, 1860 -1870 and in the 20th century (1914, 1920, 1925).

Without explanation he also added in his summary advances at the end of the 16th century and in the first half of the 18th century. A further advance of Kaldalonsjökull after the 1840s was found by John and Sugden (1962). Also in North West Iceland Hjort et al. (1985) studied several cirque glaciers which they believed reached their recent maximum around 1860, or slightly earlier. Now only four cirque glaciers and a few snowfields remain. The results of the dating of advances of small valley glaciers and cirque glaciers in Tröll-askagi by Caseldine (1983, 1985, 1987) will be discussed below.

3. LICHENOMETRY AND ITS APPLICATION IN ICELAND

Lichenometry was first developed and applied by Beschel (1950, 1954, 1957, 1961, 1965) utilising the slow growth of epipetric lichens. Following the deposition of material on the recession of a glacier, colonisation by lichens soon ('asbald') begins (Beschel, 1950 p.152). After a short time the thallus can be observed macroscopically and then follows a short phase of, at most, a few decades of accelerated growth (Beschel, 1950). This phase was previously identified by Nienburg (1919) and called the great period. There then follows a constant size increase over a period of often a few hundred years (Beschel, 1950). This growth curve is illustrated in Fig.1. There is however still some confusion about the form of the lichen growth curve and the different phases of growth. The idea of the great period as defined by Beschel (1950) has been misinterpreted, with the identification of 'colonisa-tion lag' which is not mentioned by Beschel. Innes (1985a), in a summary of lichenometry, defines three periods of lichen thallus growth - prelinear, linear and postlinear. He refers to Beschel but equates the great period with the linear phase. He then cites Armstrong (1974) to confirm the tripartite curve, but Armstrong only talks about this division in relative growth. Concerning the radial growth which is important for lichenometry he states, 'There is no evidence for a postlinear phase in the growth of a lichen thallus' (Armstrong, 1976 p.309). Large differences exist about the duration of the phase of growth before the linear phase. Beschel (1950, 1958) talks about 4-8 years to at least a few decades, Webber and Andrews (1973 pps.298-299) mention values as high as 500 years and Locke et al. (1980 p.28) 200-400 years. In a paper in English in 1961 Beschel was unclear as to the size par-

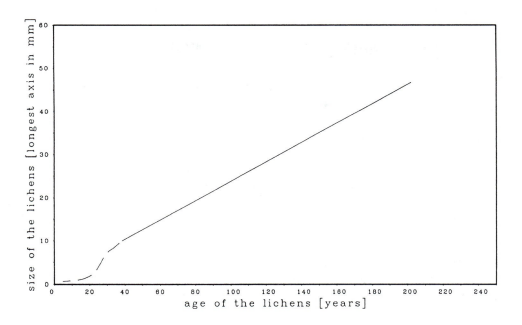

Fig.1 Lichen growth curve after Beschel (1950).

parameter of a lichen to be measured. Referring to this Locke et al. (1980) proposed the use of the shortest axis and this was adopted in Iceland by Caseldine (1983, 1985) and Gordon and Sharp (1983), although Caseldine (1987, in press) has also used longest axis measurement. Application of lichenometry thus requires several of these problems to be considered and comparison of lichenometric results also necessitates an awareness of methodological differences. Some scientists have seriously questioned the validity of the technique (Jochimsen, 1966, 1973; Worsley, 1981) but it has proved itself a valuable dating method. Beschel (1957) estimated its accuracy as 5%, and a comparison of lichenometric results with radiocarbon dates by Curry (in Heuberger, 1971) gave variations of less than 10%.

Several workers have carried out lichenometric studies in Iceland. Jaksch (1970, 1975, 1984), Gordon and Sharp (1983) and Maizels and Dugmore (1985) computed growth rates on the basis of largest lichens found on moraine ridges of believed known age. All the growth curves however relied on a small number of dated points, some of which were uncertain as will be shown below. Jaksch (1970) carried out pioneering work on lichenometry at Sólheimajökull. He found a largest lichen of 25 mm on a ridge he believed to have been deposited in 1930 and used this to derive an age of 80-90 years for the outermost ridge on which he found lichens up to 50 mm, thus placing the maximum advance within the latter half of the 19th century. On a later visit he was able to find measurable lichens on a ridge only 15 years old and refined his overall dating to produce an age for the outermost moraine of 1890 (Jaksch, 1975 p.35). However Thoroddsen (1906) wrote that in 1860 the glacier extended to the outermost moraines and retreated afterwards, and Barðarson (1934) stated that in 1860 the glacier covered the rocks of the 'Jökulhöfuð', touching a moraine the glacier lay 100 m from in 1893.

At Fláajökull Jaksch (1975) also put the maximum advance to 1890 but this is not possible according to Thoroddsen (1906 p.198). Barðarson (1934) cites accounts that in 1860-1870 the glacier was in contact with the neighbouring Heinabergsjökull and by 1903 the map of the Danish General Survey shows the distance between the two glaciers to be 1800 m. Although Barðarson (1934) refers to a report in Thoroddsen that in 1894 the glacier extended to the moraines it is not clear which moraine was meant. At Siðujökull Jaksch (1970, 1984) assumed a similar dating framework to Sólheimajökull because of a lack of direct dating evidence and placed the maximum advance in the 20th century. The results of Jaksch were used by Maizels and Dugmore (1985) at Sólheimajökull, who adopted a growth rate of 0.73 mm a $^{-1}$ based on a largest lichen of 50 mm and a colonisation lag of 15 years, this despite Jaksch's advocacy of a growth rate of 0.59 mm a^{-1} based on a regression through four dated points and his observation of measurable lichens on a moraine only 15 years old.

The growth rate for *Rhizocarpon geographicum* derived by Gordon and Sharp (1983) for Breiðamerkurjökull relies on a date for the outermost moraine of 1894 but this too can be queried. In 1894 Thoroddsen visited the glacier and wrote: 'Der kurzeste Abstand von den Moränen bis zum Strandwall betrug hier nur 213 m und weiter bis zur außersten Gletscherspitze 43 m,...' (Thoroddsen, 1906 p.196) (The shortest distance between the beach ridge and the moraines is here only 213 m, and from the moraines to the glacier edge 43m,...). Thus by 1894 the glacier was already more than 40 m from the outer moraine. In his summary of glacier variations in Iceland over the last 250 years Thórarinsson (1943 p.29) wrote, '1894˝ Shortest distance glacier - beach 256 m,˝...' and this has often been taken as defining the recent maximal extent, clearly an error in view of the observations by Thoroddsen. By using the 1894 date Gordon and Sharp (1983) computed a growth rate of 0.673 mm a $^{-1}$ for the shortest axis at Breiðamerkurjökull, and at Skalafellsjökull, assuming a similar moraine age and with a colonisation lag of 15 years, they derived rates of 0.769 mm a^{-1} (shortest axis) and 0.987 mm a $^{-1}$ (longest axis). Although the distribution of their measured points showed a linear relationship they assumed a decreasing rate of growth following colonisation. They also commented that their values fitted well with those published by Webber and Andrews (1973) but the highest rates found in their summary are 0.90 mm a^{-1} and 0.93 mm a^{-1} and there is little distinction between longest and shortest axis measurements. In Skiðadalur Caseldine (1983) used abandoned farmsteads as a source of dated surfaces assuming various possible lag times between abandonment and lichen colonisation. His work was based on a much smaller number of observations than those presented below and his results are discussed here in the light if the much larger data set now available.

4. LICHENOMETRIC RESEARCH IN SKÍÐADALUR -SVARFAÐARDALUR

For the research undertaken in the study area of Skíðadalur-Svarfaðardalur (Fig.2) a number of specific problems were identified:

    i) The establishment of a new lichen growth curve for *Rhizocarpon geographicum* agg. by the use of as large a number of dated surfaces as possible.

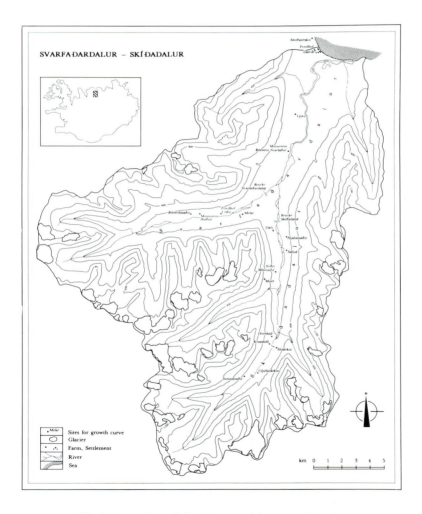

Fig.2 Location of the sites used in the calibration curve.

ii) Use of the new growth curve to date moraine ridges over a range of glacier fore-lands.

iii) Comparison of the pattern of glacial advances with known climatic parameters, especially temperature and precipitation, and with the distribution of sea ice around Iceland.

4.1 Establishment of a new lichen growth curve

Within the study area a range of sites were used to provide surfaces of known age for the growth curve of *Rhizocarpon geographicum* agg. (Innes, 1985b). These comprised 11 abandoned farmsteads or ruins, 5 grave or memorial stones, 2 old bridges and 1 mudflow. These sites were distributed throughout Skíðadalur and Svarfaðardalur from sea level to an altitude of 300m 25 km from the sea (Sveinsstaðir) (Fig.2). At farmsteads and ruins only lichens on the upper and inner sides of stone foundations were used. Because it takes time after abandonment for the loss of the roof and lichen colonisation to begin, it proved necessary to establish the likely duration of such a process. Local inhabitants commented that the time lag varied depending upon the quality and stability of the buildings. Roof construction for an inhabited building was much better than that of a stable or shed, and in some cases buildings would have already been in decay before the main buildings were abandoned. On the basis of this information a colonisation lag of 10 years was assumed for observation of the largest lichens on farmsteads. The actual date of abandonment is usually found on a sign on the ruins but detailed information was derived from a history of the valley and its farmsteads with a complete list of dates compiled by H.E.Þórarinsson (1973). The use of gravestones is also open to the problem of late emplacement of the stone, cleaning of the stone or the influence of a polished surface. Despite these problems the time of death was assumed to be the age of the surface and lichen development appeared to have been undisturbed.

The growth curve for *Rhizocarpon geographicum* agg. was therefore derived for the longest axis of the largest lichen based on observations of the above features assuming time of death for gravestones and a 10 year colonisation lag for farmsteads. It was derived by regression analysis between lichen size (x) and age of the surface (y), and produced a curve of the following form:

$$y = -1.26 + 0.44x \quad \text{where} \quad y = \text{age of lichen (surface)}$$
$$x = \text{longest axis of the largest lichen}$$

The equation has an r value of +0.987 which is significant at the 0.01% level and clearly indicates a linear relationship (Fig.3).

The early part of the curve still presents problems, for assumption of a linear form to the whole curve would give a colonisation lag of only 3 years. From observation of lichens on the memorial stones of Audnir and Þorstein Svarfaðar, and following various authors

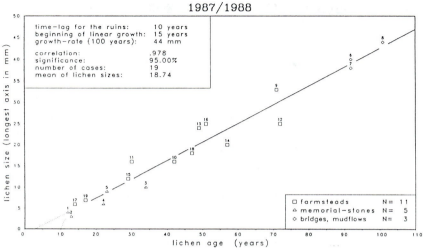

Fig.3 Lichen growth curve for *Rhizocarpon geographicum* agg. for Skíðadalur/ Svarfaðardalur.

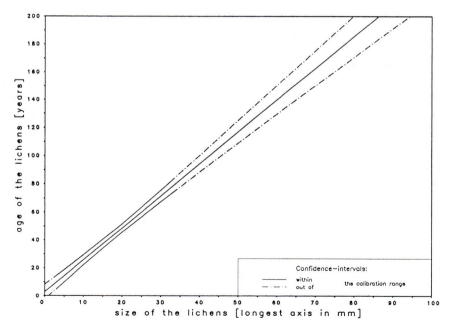

Fig.4 Lichen growth curve with 95% confidence bands.

(Beschel, 1950, 1957; Armstrong, 1974, 1976; Schröder-Lanz, 1983) lichens may normally be visible macroscopically after about 10 years. Thus the linear growth curve should only be considered after about this length of time, with a rate of growth of 0.44 mm a^{-1}.

For moraine dating, age (y) was predicted from thallus size (x) within a confidence interval computed from:

$$y \pm t_{(n-2)} S_y$$

where S_y = standard error of the mean
t = Student's t statistic

The graph used for predicting age within these confidence intervals is shown in Fig.4. It should be noted that these confidence intervals are statistically valid for the calibrated part of the curve but they do indicate increasing uncertainty of age for older moraines. The use of regression through points which demonstrate lichen thalli of sizes both greater than and less than predicted age i.e. deviating from optimal growth points, is believed justified due to the problems of the dated surfaces, especially the possibility of building abandonment earlier than overall farm abandonment and the small size of the surface area available for colonisation on the calibration surfaces. The former would account for lichens greater than the optimal size and the latter for sub-optimal sizes. The lichen curves derived independently by Häberle (in press) from the Barkárdalur area immediately south of Skíðadalur have generated very similar growth rates and this further supports the validity of the estimated figures used here.

4.2 Application of the growth curve - Þverdalur

Application of the new lichen growth curve is demonstrated for a small tributary valley of Vatnsdalur named Þverdalur (H.E.Þórarinsson, pers. comm.) where a small glacier lies within a series of well defined moraines (Fig.5). The ridge nearest to the ice is closely associated with the glacier and has no visible thalli present. Five ridges outside this moraine were dated lichenometrically (Table 1, Fig.6).

Table 1.   Moraine age estimates for Þverdalur.

Ridge Number	Age of deposition (years A.D.)	Confidence interval (95% - years)
1	Recent	
2	1943	±3
3	1922	±4
	1920	±4
4	1895	±6
5	1881	±7
6	1872	±8
	1870	±8

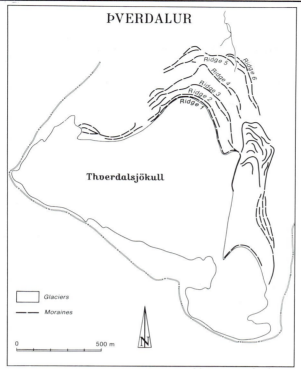

Fig.5 Location of moraines in Þverdalur.

4.3 Wider application

Altogether 8 glacier forelands were studied (5 by Kugelmann, 3 by Stötter) and the results of these are presented in Fig.7. The graph in Fig.8 represents a smoothing of this data whereby each date is expressed ±5 years and then each individual year represented by as many moraines as fall within the age band. This method of averaging demonstrates the existence of six phases of glacier advance: ca A.D.1810; 1850; 1870-1880; 1890-1900; 1920s and 1940s - although not all moraines can be placed within these phases. A number of reasons may be put forward as to why all the moraines do not follow the same pattern - differing response times of glaciers, nature of moraine material and their suitability for lichen colonisation, limits of accuracy within lichenometry and subjective errors such as not locating the largest lichen. Nevertheless the results do define a series of phases of glacier advance and moraine deposition throughout Skíðadalur-Svarfaðardalur in the later 'Little Ice Age'.

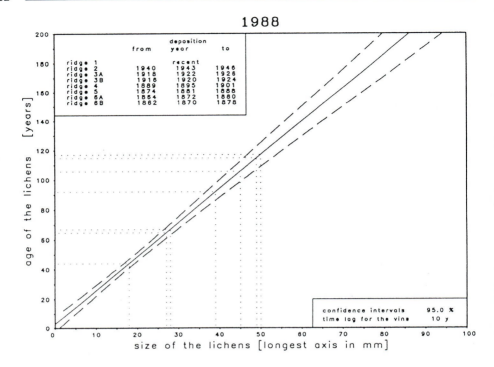

Fig.6 Dates of moraines in Þverdalur.

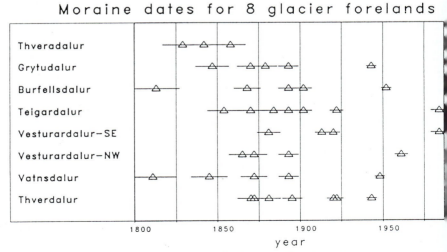

Fig.7 Moraine dates for eight glaciers in Skıðadalur/Svarfaðardalur.

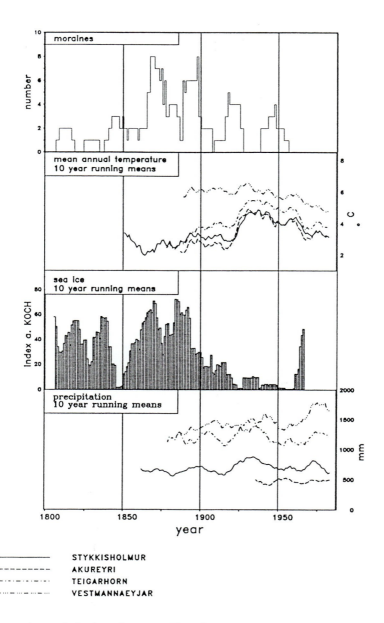

Fig.8 Comparison of glacier advances, climatic parameters and the incidence of sea ice (10 year running means).

5. COMPARISON OF GLACIER ADVANCES, CLIMATIC PARAMETERS AND THE PRESENCE OF SEA ICE

Regular temperature readings were first taken in Iceland at Stykkishólmur in 1845-46. Later in the century measurements were begun at Teigarhorn (1873), Grimsey (1875), Akureyri (1882) and on Vestmanneyjar (1884). Precipitation was first recorded at Stykkishólmur in 1857, Teigarhorn in 1873, Grimsey in 1874, Vestmanneyjar in 1884, but at Akureyri not until 1930. For the study area the absence of values from Akureyri before 1930 and the numerous gaps in the Grimsey record are disappointing, particularly in view of the number of glacier advances noted in the latter half of the last century. Topographic variability between the study area and the stations with good records makes extrapolation of values to fill the gaps unwise.

Sea ice conditions were investigated by Koch (1945) who computed a sea ice index using the duration of the appearance of ice in ten contiguous sections covering the entire Icelandic coastline. These studies were used by Bergþórsson (1969) to reconstruct temperature. This sea ice data was converted to 10 year running means for comparison with the glacial evidence (Fig.8). There is a good fit between the two records with low temperatures and a high incidence of sea ice corresponding to glacier advances in the last half of the 19th century and possibly in the 1920s. The relationship between sea ice and climate in northern Iceland is complex and the significance of precipitation in determining the level of glacier activity is as yet unclear.

6. CONCLUSIONS

The new lichen growth curve for *Rhizocarpon geographicum* agg. developed for Skíðadalur-Svarfaðardalur has been used to determine the occurrence of six phases of glacier advance and moraine deposition in recent, late 'Little Ice Age' time. The most recent period of glacial advance in the 1970s and 1980s cannot be dated lichenometrically. It is disappointing that there are no moraines of known age in the research area to prove the validity of the estimated growth curve, and the limited timescale of the calibrated section of the curve affects the resolution level of the results. However, the new curve provides an acceptable dating basis for the last 100 years in this part of Tröllaskagi, and in view of the absence of other dating methods (Caseldine, in press) is an important contribution to studies of recent glacial history in northern Iceland. The relationship between glacier response and climatic parameters in northern Iceland is an area requiring further study. Although it is possible to demonstrate the influence of temperature, especially summer temperature, the role of precipitation is unknown, and the lack of suitable precipitation records makes it difficult to estimate the likely relationship that exists.

Methodological problems of lichenometry encountered in Iceland have been discussed and points made concerning the errors involved in incorrect use of German language material in previous studies. The absence of lichen thalli apparently older than about 1800 in northern Iceland is a further area for future study, especially the possible influence of the huge volcanic eruption of Lakigígar in 1783 on lichen growth. The results from this area of northern Iceland are still preliminary due to the recent development of a reasonable lichen curve. They need to be seen in the context of results from other glacial areas in Iceland, especially the outlet glaciers from the big ice caps in southern Iceland and the small glaciers of the north west peninsula. It is only by developing a comprehensive and sound data set for recent glacier fluctuations that the nature of climatic variation across the country can be demonstrated.

REFERENCES

Ahlmann, H.W. (1937) 'Oscillations of the other outlet-glaciers from Vatnajökull', H.W.Ahlmann and S.Thórarinsson (eds.), Scientific results of the Swedish-Icelandic investigations 1936-1937, Geografiska Annaler 19, 195-200.

Armstrong, R.A. (1974) 'Growth phases in the life of a lichen thallus', New Phytologist 73, 913-918.

Armstrong, R.A. (1976) 'Studies on the growth rates of lichens ', in D.H.Brown, D.L.Hawksworth and R.H.Bailey (eds.), Lichenology Progress and Problems, Academic Press, London, pp. 309-322.

Barðarsson, G.G. (1934) 'Islands Gletscher. Beiträge zur Kenntnis der Gletscherbewegungen und Schwankungen auf Grund alter Quellenschriften und neuster Forschung', Vísindafélag Íslendinga 16, 1-60.

Bergþórsson, P. (1956) 'Barkárjökull', Jökull 6, 29.

Bergþórsson, P. (1969) 'An estimate of drift ice and temperature in Iceland in 1000 years, Jökull 19, 94-101.

Beschel, R. (1950) 'Flechten als Altermaßstab rezenter Flechten', Zeitschrift für Gletscherkunde und Glazialgeologie 1, 303-309.

Beschel, R. (1954) 'Growth of lichens, a mathematical indicator of climate (Lichenometry)', Rapport Communication 8me Congresse Botanie International Section 78, 148.

Beschel, R. (1957) 'Lichenometrie im Gletschervorfeld', Jahrbuch der Vereinigung zum Schütze der Alpenpflanzen und Alpentiere 22, 164-185.

Beschel, R. (1961) 'Dating rock surfaces by lichen growth and its application to glaciology and physiograph-lichenometry', in G.O.Raasch (ed.), Geology of the Arctic, Volume 2, University of Toronto, Toronto, pp.1044-1062.

Beschel, R. (1965) 'Epipetric succession and lichen growth rates in the eastern Nearctic, '7th International Quaternary Congress, Boulder, 25-26.

Caseldine, C.J. (1983) 'Resurvey of the margins of Gljúfurárjökull and the chronology of recent deglaciation', Jökull 33, 111-118.

Caseldine, C.J. (1985) 'The extent of some glaciers in northern Iceland and the nature of recent deglaciation', Geographical Journal 151, 215-227

Caseldine, C.J. (1987) 'Neoglacial glacier variations in northern Iceland: examples from the Eyjafjörður area', Arctic and Alpine Research 19, 296-304.

Caseldine, C.J. (1990) 'A review of Holocene dating methods and their application in the development of a Holocene glacial chronology for Northern Iceland', Münchener Geographische Abhandlungen Reihe B, in press.

Eyþorsson, J. (1935) 'On the variations of glaciers in Iceland.Some studies made in 1931', Geografiska Annaler 17, 121-137.

Gordon, J.E. and Sharp, M. (1983) 'Lichenometry in dating recent landforms and deposits, southeast Iceland', Boreas 12, 191-200.

Grove, J.T. (1988) The Little Ice Age. Methuen, London.

Häberle, T.H. (1990) 'Spät- und Postglazial Gletschergeschichte des Hörgárdalur Gebietes Tröllaskagi, Nordisland ', Dissertation an der Universität Zürich, in press.

Heuberger, H. (1971) 'Roland Beschel und der Lichenometrie', Zeitschrift für Gletscherkunde und Glazialgeologie 7, 175-184.

Hjort, C., Ingólfsson, Ó. and Norðdahl, H. (1985) 'Late Quaternary geology and glacial history of Hornstrandir, Northwest Iceland: a reconnaissance study', Jökull 35, 9-28.

Innes, J.L. (1985a) 'Lichenometry', Progress in Physical Geography 9, 187-225.

Innes, J.L. (1985b) 'A standard Rhizocarpon nomenclature for lichenometry', Boreas 14, 83-85.

Jaksch, K. (1970) ' Beobachtungen in den Gletschervorfeldern des Sólheima- und Siðujökull im Sommer 1979' Jökull 20, 45-48.

Jaksch, K. (1975) 'Das Gletschervorfeld des Sólheimajökull', Jökull 25, 34-38.

Jaksch, K. (1984) 'Das Gletschervorfeld des Vatnajökull am Oberlauf der Djúpa, Sudisland', Jökull 34, 97-103.

Jochimsen, M. (1966) 'Ist die Große des Flechtenthallus wirklich ein brauchbar Maßstab zur Datierung von glazialmorphologischen Relikten ?', Geografiska Annaler 48A, 157-164.

Jochimsen, M. (1973) 'Does the size of lichen thalli really constitute a valid measure for dating glacial deposits', Arctic and Alpine Research 5, 417-424.

John, B.S. and Sugden, D.E. (1962) 'The morphology of Kaldalon, a recently deglaciated valley in Iceland', Geografiska Annaler 44A, 347-365.

Koch, L. (1945) 'The East Greenland ice', Meddelelser om Grønland 130, 1-375.

Locke, W.W., Andrews, J.T. and Webber, P.J. (1980) 'A manual for lichenometry', British Geomorphological Research Group Technical Bulletin 26.

Maizels, J.K. and Dugmore, A.J. (1985) 'Lichenometric dating and tephrochronology of sandur deposits, Sólheimajökull area, southern Iceland', Jökull 35, 69-77.

Meyer, H.H. and Venzke, J.F. (1985) 'Der Klængshóll-Kargletscher in Nord-Island', Natur und Museum 115, 29-64.

Nienburg, W. (1919) 'Studien zur Biologie der Flechten, I-III', Zeitschrift für Botanik 11, 1-38.

Schröder-Lanz, H. (1983) 'Establishing lichen growth curves by repeated size (diameter) measurements of lichen individuals in a test area - A mathematical approach ', in H.Schröder-Lanz (ed.) Late and postglacial oscillations of glaciers : Glacial and periglacial forms, A.A.Balkema, Rotterdam, pp.393-409.

Thórarinsson, S. (1943) 'Vatnajökull - Scientific results of the Swedish-Icelandic investigations 1936-1937-1938. Chapter IX: Oscillations of the Icelandic glaciers in the last 250 years', Geografiska Annaler 25, 1-54.

Thoroddsen, T. (1895) 'Fra det sydöstlige Island. Rejseberetning fra sommeren 1894', Geografisk Tiddskrift 13, 167-234.

Thoroddsen, T. (1906) 'Island. Grundriß der Geographie und Geologie II', Petermanns Geographische Mitteilungen 153, 162-358.

Webber, P.J., and Andrews, J.T. (1973) 'Lichenometry : a commentary', Arctic and Alpine Research 5, 295-302.

Worsley, P.J. (1981) 'Lichenometry', in A.Goudie (ed.), Geomorphological Techniques, George Allen and Unwin, pp. 302-305.

Þórarinsson, H.E. (1973) 'Svarfaðardalur og gönguleiðir um fjöllin', Ferðafélag Íslands Árbok 1973, Reykjavík.

LICHENOMETRIC DATING, LICHEN POPULATION STUDIES AND HOLOCENE GLACIAL HISTORY IN TRÖLLASKAGI, NORTHERN ICELAND

Chris Caseldine
Department of Geography
University of Exeter
Amory Building
Rennes Drive
Exeter EX4 4RJ
UK

ABSTRACT. In view of the problems of lichenometric dating as applied in northern Iceland, studies of populations of *Rhizocarpon geographicum* s.l. are advocated in addition to using largest thalli alone as a measure of age. It is concluded that such studies can provide a supporting dating method for the last two centuries and also indicate the presence of disturbed lichen populations. 'Traditional' lichenometric dating of recent end moraines in Tröllaskagi confirms that there were five periods of moraine abandonment : A.D.1810-1820; 1845-1875; late 1880s-early 1890s; 1915-1925; late 1930s-early 1940s. In order to extend moraine dating beyond the lichenometric timescale (pre-A.D. 1800) it is argued that greater emphasis will have to be placed on surrogate climatic records from peats and lake sediments which can be dated by other methods.

1.INTRODUCTION

Glacier variations provide a valuable source of terrestrial information concerning Holocene climatic variability and studies of Holocene glacier variations have taken place throughout the world in glaciated and formerly glaciated regions (Grove, 1979, 1988; Röthlisberger, 1986). With the increasing development and use of models to examine relationships between climatic variability, glacier response and postulated forcing functions (Porter, 1986; Oerlemans, 1988), it is important to derive chronologies of Holocene glacier variations that are as complete and accurate as possible. The success of any modelling exercise is often strongly dependent upon the accuracy and detail of the empirical data against which the model can be tested, and in the case of regional or wider scale climatic modelling, the spatial spread of the data is of crucial importance. Northern Iceland lies in

J. K. Maizels and C. Caseldine (eds.), Environmental Change in Iceland: Past and Present, 219–233.

a significant location for the climate of North West Europe, close to both major atmospheric and oceanic boundaries, yet neither Porter nor Oerlemans were able to include data from the area, and only Oerlemans included results from southern Iceland. In northern Iceland the glacial record for the Holocene is extremely limited, both in spatial and temporal terms (Björnsson, 1979; Caseldine, 1985, in press). It is therefore of considerable importance that a data base suitable for inclusion in climatic models is developed.

Recognition of former glacier limits can be made on morphological grounds and the nature of such evidence is well documented (Porter, 1981). It is, however, the dating of the limits that can present most problems. A wide range of dating techniques is available for the Holocene timescale (Mahaney, 1984), but many are not applicable in all environments. The most widely applicable is probably lichenometry and in northern Iceland this has proved a particularly valuable technique, especially on a timescale of the last 150 years. In view of the importance of lichenometry for establishing a recent glacial chronology in northern Iceland, this paper concentrates on the problems and potential of its application in the Tröllaskagi peninsula.

2. LICHENOMETRY

2.1 Problems of Application in Northern Iceland

Because of the widespread use of lichenometry there have been a number of general reviews which outline the basis of the technique and discuss constraints and underlying assumptions (Locke, Andrews and Webber, 1979; Worsley, 1981; Innes, 1985). Curves for the growth rate of *Rhizocarpon geographicum* s.l. have been published for a number of locations in Iceland (Jaksch, 1970, 1975, 1984; Caseldine, 1983; Gordon and Sharp, 1983; Thompson and Jones, 1986; Kugelmann, in press). That by Kugelmann (in press) is the most detailed and has been calibrated using the largest number of surfaces of known age. It is possible that this curve may allow a re-evaluation of other lichen dating curves in Iceland as the present range of growth rates between the north and south of the island is considerable, and the validity of the assigned ages of some of the fixed points is in question.

The application of lichenometry in Iceland has encountered two recurrent problems which have been commented on by several authors:

(i) Lack of calibration surfaces - the earliest surface of believed known age used so far in Iceland is the outermost moraine at Skaftafellsjökull believed to date to A.D. 1870 (Thompson and Jones, 1986). Thompson and Jones (1986) also located a feature of similar age at Svínafellsjökull, but not with the same degree of certainty morphologically, although it was used to construct their lichen growth curve. Attempts to extend the record further back in time by using lichens growing on the Kotá fan, which was produced by a jökulhlaup in A.D. 1727, failed. In this case the problem was the small size of the available clasts and competition from mosses preventing the survival of large lichens. Dating of surfaces which have lichen thalli larger than the largest calibrated thallus necessarily has to assume an extrapolated growth rate. Several empirical studies in a range of environments have emphasised the curvilinear nature of the overall growth curve (Worsley, 1981) but for Iceland only the final curve of Thompson and Jones (1986) suggests that this slowing of the growth rate occurs, within 60-80 years for *Rhizocarpon geographicum* s.l. In Tröllaskagi the oldest surfaces of known age are generally found either in abandoned farmsteads, of which the oldest that has been used is A.D. 1905 (Caseldine, 1983; Kugelmann, in press), or on grave and memorial stones and bridges which have provided a date of A.D. 1887, although in Barkárdalur a moraine has been identified from A.D. 1900 on documentary evidence (Bergþórsson, 1956; Häberle, in press). The lack of earlier dated surfaces, however, is very restrictive in this area of northern Iceland and all earlier lichenometric ages remain estimates, probably minimum estimates.

(ii) Absence of large thalli of *Rhizocarpon geographicum* s.l. in Iceland - as noted above several workers have commented upon the absence of large thalli of this lichen group and the same is true for northern as for southern Iceland. Several reasons have been put forward for this phenomenon as well as those proposed by Thompson and Jones (1986), and many of these have been discussed by Maizels and Dugmore (1985). In Iceland an important consideration is likely to be the parent material which is commonly volcanic in origin and highly susceptible to frost shattering and rapid weathering, but both snowkill and burial by tephra are possible influences, especially on basaltic substrates away from active volcanic areas.

Thus the application of lichenometry in Iceland appears appropriate only for the last two centuries at most with, as yet, uncertainties over both the true lichen growth rates that are applicable in different areas, and over age estimates that are beyond the calibrated growth curves. Lichenometric research therefore clearly needs to address these two areas and it is the examination of lichen population characteristics that has been adopted here for this purpose.

2.2 Introduction to Size-Frequency Studies

Benedict (1985), working in the Arapaho Pass area of the Colorado Front Range, constructed size-frequency curves for populations of *Rhizocarpon geographicum* s.l. in order to derive a lichen growth curve based on the gradients of the population curves. He was also able to use the structure of the population to identify disturbed or composite lichen populations influenced by some disturbing factor, in this case disturbance of sites by American indian communities. The construction of a growth curve based on characteristics of the population as a whole, independent of a curve based on largest thalli sizes, proved successful over a long timescale i.e. 4000-5000 years, but Benedict considered it unlikely to be applicable over shorter timescales 'in temperate or maritime-tundra environments, where lichen growth is rapid' (1985, p. 106). Lichen growth rates in northern Iceland are high relative to his study but even if such an approach could not produce a new and longer chronology, it still has potential for identifying populations older than those which developed during the last 100-200 years, and also for identifying disturbed populations. The concept of using the gradient of the population curve also provides a potential mechanism whereby recent populations can be assigned to very similar age ranges, and a relative series of ages confidently determined.A major assumption of the work of Benedict was that the size-frequency distribution of lichen thalli approximated to a log-normal model (more correctly a log-linear form as pointed out in Locke (1983)). Benedict discussed the arguments against such an assumption as developed in Locke, Andrews and Webber (1979) and Innes (1983), but supported his belief by referring to a number of empirical studies, including his own, which have shown best fits to a negative log-linear distribution. There is, understandably, some variation in the nature of the size-frequency distribution of populations from different environments (Innes, 1983, 1986; Haines-Young, 1988), but where negative log-linear curves can be consistently fitted to sample data and provide the best estimate of the mathematical distribution represented, it is reasonable to apply such a method. A further complication may be the aggregation of different species of *Rhizocarpon* but this is not thought to be a problem over the size ranges covered in this study. It should be pointed out that in both the work of Benedict and that described here the term population does not always refer to the total lichen population. In most cases large samples were used (i.e. ca 1000), a large proportion of the populations.

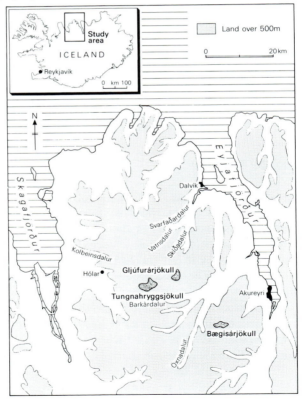

Fig.1   Location map of the Tröllaskagi peninsula.

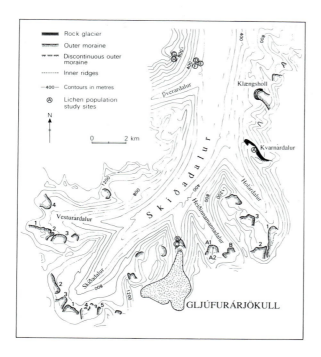

Fig. 2   Location map of the sites in Skíðadalur. Numbers refer to glaciers with moraines dated lichenometrically which are tabulated in Table 1.

In the present study the approach developed by Benedict has been applied to a series of sites in Skíðadalur, on the eastern side of the Tröllaskagi peninsula, northern Iceland (Fig. 1), with two principal aims. Firstly, to evaluate the use of the largest lichen approach to lichenometric dating, and, secondly, to identify disturbed or composite lichen populations. At eight sites in Skíðadalur (Fig. 2) the long axes of 1000 thalli of *Rhizocarpon geograph-icum* s.l. >10mm in diameter, were measured to within 0.5 mm and the results graphed using varying size intervals. Statistical analysis of this data revealed that in all cases a negative log-linear distribution best described the linear relationship between size class and frequency, and variations in the class intervals used in the analysis made no significant difference to the results down to a minimum class interval of 2 mm, as similarly found by Innes (1986). For most of the results presented here a class interval of 4.5 mm has been adopted (effectively 5mm but rounded to 4.5mm as measurements were only made to the nearest 0.5mm). The eight sites examined comprised: A - Kvarnárdalur rock glacier surface; B - Gljúfurárjökull outer moraine; C - Gljúfurárjökull inner moraine; D, E and F - Skíðadalur debris flows; G & H - Þverardalur debris flows. At all sites care was taken to examine surfaces of relatively uniform position and aspect, avoiding positions likely to experience 'green zone' effects (Innes, 1986).

2.3 Lichen Population Size-Frequency Curves and Dating

In Tröllaskagi a lichen growth curve has been developed based on single largest lichen thalli observed on calibrated surfaces (Kugelmann, in press), so the lichen population data provides a means of evaluating the validity of this curve by another lichenometric dating technique. Size-frequency studies have been used elsewhere to interpret a '1 in 1000' thallus which is then used for dating surfaces in a similar way to the largest observed lichen thallus, but the validity of this approach has been questioned by several authors (e.g. Benedict, 1985). Instead of using an observed single largest thallus to date a surface, the size-frequency curve is used to extrapolate a '1 in 1000' thallus size which is then used for dating. This approach has been criticised on the grounds that the eventual size used for dating is a function of the class intervals employed and not a measured value from the population itself. In proposing using the slopes of size-frequency distribution curves to construct a dating curve, Benedict argues that its strength is that this feature is 'an intrinsic characteristic of the lichen population' (1985, p. 94). Thus in Skíðadalur the development of a curve based on the '1 in 1000' thallus has not been adopted. It is possible, however, to use the gradients of the population curves as measures of relative age and a start has also been made to calibrate the gradients such that it will be possible in the future to date accurately surfaces from their population characteristics.

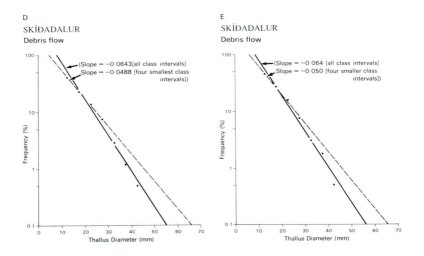

Fig. 3   Lichen size-frequency distributions for older debris flows in Skíðadalur (see text for explanation of the letters).

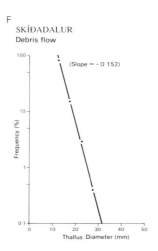

F
SKÍÐADALUR
Debris flow

Fig. 4   Lichen size-frequency distribution for youngest debris flow (A.D. 1920's) in Skíðadalur.

It can be shown that populations from features of the same age have virtually identical size frequency distributions. This can be seen in Fig. 3 where populations from adjacent debris flows (D and E) in Skíðadalur are plotted. There is documentary evidence for the flows occurring during a period of intense debris flow activity in the early 1880's, and although there is some slight difference in the size of the observed largest lichens the use of the Kugelmann growth rate would indicate a date of between 1883-1886. A more recent debris flow from the same location (F), originating in a period of enhanced debris flow activity documented in 1929 (with a lichenometric estimation of 1926), is clearly differentiated by a much steeper gradient to the curve of the frequency distribution (Fig. 4). The gradients for the two 1880s flows are statistically indistinguishable from each other, and are also indistinguishable from the gradient found on an inner moraine ridge in front of Gljúfurárjökull (C) (Fig. 5), which was assigned a lichenometric age of 1877 on the basis of the largest thallus.

Ideally the gradients should be plotted against known surface ages to provide a dating curve independent of calibration by a largest lichen curve as carried out by Benedict (1985). At present, on the data available, it is not possible to construct a fully independent curve. However, a preliminary analysis was carried out to demonstrate the potential of this approach. The gradients do plot as an exponential curve best described by a log-linear function with a high degree of correlation ($r = -0.97$), and provide age estimates with a 95% confidence interval of ±7 years over the last 100 years. It is also noticeable that the form of the curve is similar to that found by Benedict (1985) although the timescales covered vary

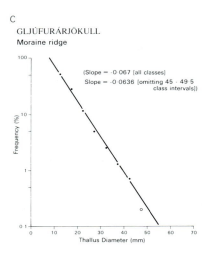

Fig. 5   Lichen size-frequency distribution for Gljúfurárjökull inner moraine.

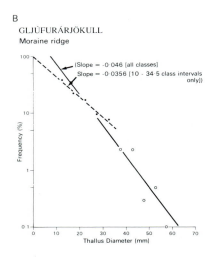

Fig. 6   Lichen size-frequency distribution for Gljúfurárjökull outer moraine.

by an order of magnitude. However the curve is based on only eight points and only two of those are calibrated by documentary evidence; all the rest are based on observed largest lichen sizes. Until more independently dated surfaces can be used covering a wider time range over the last two centuries, there is an obvious danger of circularity in the dating approach. Nevertheless it is believed that clear relative age separation can be achieved and that in time close age estimates will be possible. If the mathematical form of the curve can be closely approximated then extrapolation beyond the calibrated range may be more confidently carried out.

A further implication for dating arises where the population data reveal a scatter in the upper size ranges of the lichen thalli. This is well illustrated in Fig. 6 where there is clearly considerable variation in the upper size classes on the outer moraine ridge in front of Gljúfurárjökull (B), in contrast to the distribution on the inner moraine ridge. Whereas use of the largest lichen on the inner ridge probably provides a good estimate of age, this is not the case on the outer moraine, for there is little relationship between the largest observed thallus and the overall size-frequency distribution. Indiscriminate use of the single largest lichen with no examination of the population could therefore provide a misleading estimate of age for the feature. Because of the time involved in measuring a representative sample from each lichen population (i.e. at least 1000 thalli), and in some cases the lack of a large enough population to produce such a figure, it is unlikely that this detailed approach will be undertaken. It would however be of considerable value for some attempt to be made to evaluate the likely character of the population, even by the use of relatively small samples, to check for possible errors, especially where examining older moraines with very variable lichen sizes.

2.4 Size-Frequency Distribution Curves and Disturbed Populations

The results from the Gljúfurárjökull outer moraine discussed above demonstrate the existence of composite lichen populations on features such as moraines and debris flows. At Arapaho Pass Benedict (1985) was able to show how populations had been disturbed with larger lichen thalli representing the remnants of older populations. From the Tröll-askagi data, e.g. Fig. 6, the disturbed populations are characterised by an under-represen-tation of older thalli with a close linear fit for small size ranges producing a low gradient to the curve, i.e. an 'old' age. This suggests that not only are the older lichens somehow selectively removed or prevented from developing, as for instance by competition with other lichens or higher plants, but that there is also disturbance of thalli of varying sizes, not just over an individual class. Disturbance is not universal for some relatively old sites with lichen thalli in the higher size ranges are not affected (e.g. the debris flows in Fig. 3). Although 'a priori' it was the surface of the rock glacier (A) that might have been expected to show most disturbance, this produced a very close fit to a negative log-linear distribution (Fig. 7). Apart from the Gljúfurárjökull outer moraine, one of the debris flows from Þverardalur (G) located at a higher altitude to the Skíðadalur debris flows, showed the most erratic distribution (Fig. 8) and probably represents debris flow activity on more than one occasion on the same feature.

The close fits to a negative log-linear distribution for the majority of the populations suggests that over the time period there has been no widespread disruption of lichen growth. Severe ash falls due to Icelandic volcanic activity have not been recorded.in the north over the last 150 years but there is evidence for climatic severity, if not as prolonged as during earlier centuries, during the 19th century (Sigfússdóttir, 1969; Sigtryggsson, 1972). The even development of populations of *Rhizocarpon geographicum* s.l. over at least the last 150 years at a range of sites and altitudes (200-800m) could be interpreted as pointing to a non-climatic cause for the absence of large thalli. This inference is further strengthened by the selective nature of the loss of such thalli, their under-representation pointing to influences within the population structure itself. While it remains unclear whether volcanic activity (either through ash or gases) or internal population dynamics is the underlying cause of lichen mortality over a long timescale, the close approximation of maximum lichen ages to a period of high volcanic activity remains an intriguing coincidence.

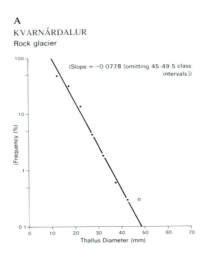

Fig. 7   Lichen size-frequency distribution for Kvarnárdalur rock glacier.

3. HOLOCENE GLACIER VARIATIONS IN TRÖLLASKAGI

3.1 'Little Ice Age' Limits

Measurements of single largest lichens (longest axis) on outermost Holocene (i.e. post-Weichselian ice sheet disappearance) moraines at 18 glaciers in Skíðadalur and Kolbe-insdalur (Caseldine, 1987) have revealed that the majority of these moraines mark the limits

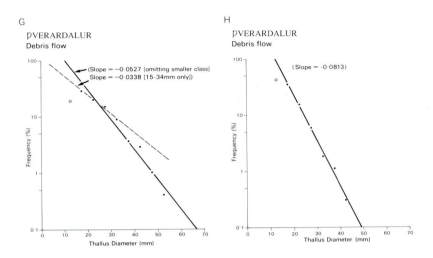

Fig. 8   Lichen size-frequency distribution for Þverardalur debris flows.

of glacial advances that culminated in the late 'Little Ice Age'. These are certainly the most extensive Neoglacial limits and probably mark Holocene glacial maxima after the remnants of Weichselian ice finally disappeared in the earliest part of the Holocene. Revision of the dating of these moraines based on the curve derived by Kugelmann (in press) identifies five main periods of moraine abandonment : A.D. 1810-1820; 1845-1875; late 1880s-early 1890s; 1915-1925; late 1930s-early 1940s (Table 1). These periods agree well with the wider geographical range of sites examined by Kugelmann, in which the majority of moraines date from the period between 1865-1875 and from the 1890's. The possible reasons for this distribution of ages are discussed in Kugelmann (in press). Accurate dating of 'Little Ice Age' glacier limits preceding the calibrated lichen curve i.e. pre-1887, is still problematical, although it is thought that the 'true' ages are not radically different from those estimated. Dating of the outer moraine at Gljúfurárjökull using the gradients of the population curve would not, for instance, give a date earlier than 1845 on the present evidence available. In this context it is worth noting that in north west Iceland Hjort, Ingólfsson and Norðdahl (1985) similarly found maximum Neoglacial ice limits of late 'Little Ice Age' date and reinterpretation of their data using the Kugelmann growth rate places the majority of the moraines in the mid-19th century.

3.2 Pre-'Little Ice Age' Holocene Limits

Evidence in Iceland for Holocene glacial advances beyond 'Little Ice Age' limits is relatively scarce. In Tröllaskagi a small number of glaciers appear to have extended beyond

Ages of moraines in N. Iceland (Skíðadalur-Kolbeinsdalur)

A	B	C	D	E
A.D.1810-1820	1845-1875	Late1880's-1890's	1915-1925	Late1930's-1940's
Tungnahryggsjökull	Klængshóll	Holárdalur 3	Holárdalur 2	Holárdalur
	Kvarnárdalur	Skíðadalur 2	Skíðadalur 1	Gljúfurárjökull
	Heiðinnamannadalur A1	Skíðadalur 5	Skíðadalur 3	Tungnahryggsjökull
	Gljúfurárjökull	Vesturárdalur 2	Skíðadalur 4	
	Vesturárdalur 1	Vesturárdalur 3	Gljúfurárjökull	
	Holárdalur 1	Vesturárdalur 4	Kvarnárdalur?	
	Tungnahryggsjökull	Heiðinnamannadalur A2		
		Heiðinnamannadalur B		
	———— Outer Moraine	Gljúfurárjökull		
	– – – – Inner Moraine	Tungnahryggsjökull		

Table 1   Dates for moraine sequences in Skíðadalur-Kolbeinsdalur.

their 19th century positions as both Stötter (in press) and Häberle (in press) have shown. In Vatnsdalur (Fig. 1) Stötter has evidence for advances in the period 5,800-4,740 BP and between 3,500-3,000 BP, whereas Häberle has identified features in Barkárdalur indicating a possible advance at ca 1,800 B.P. Meyer and Venzke (1985) identified two moraine ridges in front of the Klængshóll glacier (Fig. 2) which they defined as 18th and 19th century positions, but these have been interpreted by Caseldine (1987) as marking earlier Holocene limits on the basis of weathering characteristics of large clasts. Furthermore Caseldine (1987) has also argued that several of the moraines in Skíðadalur, especially the major terminal-lateral ridge at Klængshóll and features in upper Skíðadalur and Holárdalur, are of such a size that they were formed by more than one advance and thus also mark pre-'Little Ice Age' limits. Sites with evidence for pre-'Little Ice Age' limits are, therefore, in the minority in Tröllaskagi, although a large number of glaciers still have to be studied.

Elsewhere in Iceland more extensive pre-'Little Ice Age' glacier positions have been identified at Halsájökull in eastern Iceland (Thórarinsson, 1964), around the southern edge of Vatnajökull at Svínafellsjökull and Kvíárjökull (Thórarinsson, 1956), at Gigjökull, a northern outlet of Eyjafjallajökull, and at Sólheimajökull, an outlet glacier of the Mýrdalsjökull ice cap (Maizels and Dugmore, 1985; Dugmore, 1989). In the latter case where earlier Holocene moraines lie up to 4km outside the 'Little Ice Age' limits, Dugmore has convincingly argued for a change in the catchment area of the glacier as being responsible for the behaviour of the glacier, which is in contrast to all the other glaciers, except perhaps Gigjökull, that he studied around Mýrdalsjökull and Eyjafjallajökull. As in northern Iceland the evidence from the major Icelandic ice caps is for maximum Holocene development of ice in the 'Little Ice Age', usually in the 19th century.

3.3 Potential for Improved Chronologies

At present there is still only a relatively poor understanding of both the sequence and extent of Holocene glaciation in the Tröllaskagi peninsula of northern Iceland, especially when compared with information available from other glaciated areas such as the Alps or Scandinavia. Whereas considerable progress has been made in unravelling the pattern of the last two centuries, information about earlier periods, and even the earlier part of the 'Little Ice Age', is at best piecemeal. Improvement of lichenometric dating is clearly feasible as demonstrated above and should eventually allow both precise and accurate age estimates for the last two centuries so long as suitable calibration surfaces can be found. For the rest of the Holocene it is likely that, apart from occasional finds of organic material in contexts suitable for defining former glacier margins, it will be necessary to look for other sources of surrogate evidence for glacier fluctuations. This topic has been discussed in greater length by Caseldine (in press) where it is argued that palaeoecological evidence from peat deposits close to former glacier margin positions and combined palaeoecological/sedimentological studies of suitable lake sediments offer the best hope for improving our understanding of Holocene glaciation. In Tröllaskagi the rather small number of potentially suitable sites for the latter approach is likely to prove a major obstacle.

4. CONCLUSION

Studies of Holocene glacial history rest heavily upon an adequate chronological framework if they are to provide information of value to the understanding of climatic variability. Increasingly detailed information from northern Iceland is now becoming available, and the area of the Tröllaskagi peninsula offers considerable opportunities for future research. Many of the problems posed by the nature of the glacial record and the difficulties encountered in applying 'traditional' dating techniques will provide a stimulus to the refinement of well established techniques and the development of alternative sources of information. The 'new sense of urgency' noted by Wood (1988, p. 404) behind studies of the behaviour of glaciers in response to the potential impact of trace-gas induced climatic changes, applies just as much to studies on the longer timescale of the Holocene. Without detailed understanding of the complete Holocene sequence it will not be possible to evaluate adequately the current state of glacier activity and the links between changes in glacier mass balance, extent and climatic variables.

ACKNOWLEDGEMENTS

I am most grateful to the many people who have helped with fieldwork for this paper and in particular to the University of Exeter and Royal Geographical Society for financial assistance. The diagrams were prepared in Exeter by Terry Bacon and the text prepared by

Rosemary Dempster. Permission to carry out work in 1988 was granted by the National Research Council to which I am grateful for for their support.

REFERENCES

Benedict, J.B. (1985) 'Arapaho Pass: Glacial Geology and Archaeology at the Crest of the Colorado Front Range', Center for Mountain Archaeology, Research Report No. 3, Colorado.

Bergþórsson, P. (1956) 'Barkárjökull', Jökull 6, 29.

Björnsson, H.J. (1979) 'Glaciers in Iceland', Jökull 29, 74-80.

Caseldine, C.J. (1983) 'Resurvey of the margins of Gljúfurárjökull and the chronology of recent deglaciation', Jökull 33, 111-118.

Caseldine, C.J. (1985) 'The extent of some glaciers in Northern Iceland during the Little Ice Age and the nature of recent deglaciation', Geographical Journal 151, 215-227.

Caseldine, C.J. (1987) 'Neoglacial glacier variations in Northern Iceland: Examples from the Eyjafjörður area', Arctic and Alpine Research 19, 296-304.

Caseldine, C.J. (in press) 'A review of dating methods and their application in the development of a chronology of Holocene glacier variations in Northern Iceland', Münchener Geographische Abhandlungen, Reihe B.

Dugmore, A.J. (1989) 'Tephrochronological studies of Holocene glacier fluctuations in south Iceland', in J. Oerlemans (ed.), Glacier Fluctuations and Climatic Change, Kluwer Academic Publishers, Dordrecht, pp. 37-56.

Gordon, J.E. and Sharp, M. (1983) 'Lichenometry in dating recent landforms and deposits southwest Iceland', Boreas 12, 191-200.

Grove, J.M. (1979) 'The glacial history of the Holocene', Progress in Physical Geography 3, 1-54.

Grove, J.M. (1988) The Little Ice Age. Methuen, London.

Häberle, T.H. (in press) ' Gletschergeschichtliche Untersuchungen im Barkárdalur, Tröllaskagi', Münchener Geographische Abhandlungen, Reihe B.

Haines-Young, R.H. (1988) 'Size-frequency and size-density relationships in populations from the Rhizocarpon sub-genus Cern on morainic slopes in southern Norway', Journal of Biogeography 15, 863-878.

Hjort, C., Ingólfsson, Ó., and Norðdahl, H. (1985) 'Late Quaternary geology and glacial history of Hornstrandir, Northwest Iceland: A reconnaissance study', Jökull 35, 9-28.

Innes, J.L. (1983) 'Size frequency distribution as a lichenometric technique: an assessment', Arctic and Alpine Research 15, 285-294.

Innes, J.L. (1985) 'Lichenometry', Progress in Physical Geography 9, 187-254.

Innes, J.L. (1986) 'Influence of sampling design on lichen size-frequency distributions and its effect on derived lichenometric indices', Arctic and Alpine Research 18, 201-208.

Jaksch, K. (1970) 'Beobachtungen in den Gletschervorfeldern des Sólheima- und Siðujökull in Sommer 1970', Jökull 20, 45-49.

Jaksch, K. (1975) 'Das Gletschervorfeld des Sólheimajökull', Jökull 25, 34-38.

Jaksch, K. (1984) 'Das Gletschervorfeld des Vatnajökull am oberlauf der Djupá, Südisland', Jökull 34, 97-103.

Kugelmann, O. (in press) 'Datierungen neuzeitlicher Gletscherschwankungen im Svarfaðardalur/Skiðadalur mittels einer neuen Flechteneich Kurve', Münchener Geographische Abhandlungen, Reihe 13.

Locke, W.W. (1983) 'Discussion of " Size Frequency Distributions as a Lichenometric Technique: An Assessment " by J.L.Innes', Arctic and Alpine Research 15, 419.

Locke, W.W., Andrews, J.T. and Webber, P.J. (1979) 'A manual for lichenometry', British Geomorphological Research Group Technical Bulletin No 26, 47pp.

Mahaney, W.C. (1984) Quaternary Dating Methods. Elsevier, Rotterdam.

Maizels, J.A. and Dugmore, A.J. (1985) 'Lichenometric dating and tephrochronology of sandur deposits, Sólheimajökull area, southern Iceland', Jökull 35, 69-77.

Meyer, H.H. and Venzke, J.F. (1985) 'Der Klængshóll-Kargletscher in Nordisland', Natur und Museum 115, 29-46.

Oerlemans, J. (1988) 'Simulation of historic glacier variations with a simple climate-glacier model', Journal of Glaciology 34, 333-341.

Porter, S.C. (1981) 'Glaciological evidence of Holocene climatic change', in T.M.L. Wigley, M.J. Ingram and G. Farmer (eds.), Climate and History, Cambridge University Press, Cambridge, 82-113.

Porter, S.C. (1986) 'Pattern and forcing of northern hemisphere glacier variations during the last millennium', Quaternary Research 26, 27-48.

Röthlisberger, F. (1986) 10,000 Jahre Gletschergeschichte der Erde. Verlag Sauerländer, Aarau.

Sigfúsdóttir, A.B. (1969) 'Temperature in Stykkishólmur 1846-1968', Jökull 19, 7-10.

Sigtryggsson, H. (1972) 'An outline of sea ice conditions in the vicinity of Iceland', Jökull 22, 1-11.

Thompson, A. and Jones, A. (1986) 'Rates and causes of proglacial river terrace formation in south east Iceland; an application of lichenometric dating techniques', Boreas 15, 231-246.

Thórarinsson, S. (1956) 'On the variations of Svínafellsjökull, Skaftafellsjökull and Kvíárjökull, in Öraefi', Jökull 6, 1-15.

Thórarinsson, S. (1964) 'On the age of the terminal moraines of Bruárjökull and Halsájökull', Jökull 14, 67-75.

Wood, F.B. (1988) 'Global alpine trends, 1960s to 1980s', Arctic and Alpine Research 20, 404-413.

Worsley, P. (1987) 'Lichenometry' in A.Goudie (ed.), Geomorphological Techniques, Allen and Unwin, London, pp. 302-305.

PART 3

RECENT LANDSCAPE CHANGE

AN ASSESSMENT OF SOME OF THE FACTORS INVOLVED IN RECENT LANDSCAPE CHANGE IN ICELAND

John Gerrard
Department of Geography
University of Birmingham
Edgbaston,
Birmingham B15 2TT
UK

ABSTRACT. The Icelandic landscape shows abundant evidence of recent phases of land-sliding and soil erosion. A number of factors, such as climatic change, overpopulation and grazing pressure, and volcanic ash falls, have been suggested as causes of such instability. A framework is devised to enable the impact of such factors to be assessed. This framework is then tested by examination of slope exposures from several parts of Iceland, as well as by analysis of processes currently in action. Use is made of tephrochronology to deduce past landscape change. The general conclusion is that all the major factors mentioned earlier have been significant in initiating landscape change but that their relative effect depends very much on environmental factors. The major influence appears to be human pressure but climatic changes reinforce such pressures. At present the results must remain tentative until subjected to a wider analysis.

1. INTRODUCTION

It is being recognised increasingly that many landscapes around the world are in a delicate state of balance and it may only take a small disturbance to initiate major changes. The Icelandic landscape would fit into that category with abundant evidence of recent phases of landsliding and soil erosion. A number of factors such as climatic change, overpopulation and grazing pressure, and volcanic tephra falls have been suggested as causes of such instability. External shocks to landscape systems can be conceived as either pulsed or ramped events (Brunsden and Thornes, 1979). With pulsed inputs, the imposed disturbance is short in relation to the time scale being considered. This type of change is typical of

237

J. K. Maizels and C. Caseldine (eds.), Environmental Change in Iceland: Past and Present, 237–253.
© 1991 *Kluwer Academic Publishers. Printed in the Netherlands.*

extreme, episodic events. In Iceland such events would include rainstorms, sudden snow melt, tephra falls, jökulhlaups and so on. The impact of such extreme events will vary in different geomorphological zones depending on the relative efficiency of more frequent events. Impact is also a function of reinforcing or restorative processes. In the ramped type of disturbance the changes in inputs are sustained at the new level as a result of permanent shifts in the controlling variables such as climatic or land use changes.

The response of the landscape will also depend on its sensitivity. Some parts of the landscape will be more sensitive than others. Mobile, fast responding systems have a high sensitivity to externally generated pulses. Slow responding insensitive areas tend to be far removed from the zones where changes are propagated. Thus, interfluves and plateaux, in general, are insensitive areas. Sensitivity is the result of many processes, often enhanced by human action. Iceland, with steep slopes, unstable materials and high energy processes, possesses many sensitive landscapes.

Using the concepts proposed by Gignon (1983) and Winiger (1983) landscapes may be classified as:

1. Stable - where they are subject to little change, are capable of sustaining the present use and of returning to equilibrium after having been disrupted.
2. Conditionally unstable - where they are liable to irreversible change.
3. Unstable - where they are easily disturbed and not capable of returning to equilibrium following disturbance.

In order to assess the way in which landscapes respond, research must concentrate on:

1. The evaluation of environmental controls capable of causing change.
2. The identification and measurement of processes in order to establish variability of effect.
3. The time required for characteristic forms to become established in relation to the frequency intensity of pulses of change.

These important concepts are now examined with respect to selected Icelandic landscapes.

2. MAJOR INFLUENCES ON ICELANDIC LANDSCAPES

During the last thousand years Iceland has been subjected to a number of external influences, many of which have been instrumental in creating the present landforms. Major, sustained disturbances have been largely the result of changes of climate and grazing pressure. Episodic, pulsed inputs are more difficult to assess but frequent tephra falls must have had a significant, if local, impact on surface processes and landform development. Extreme, but short-lived, climatic events are also difficult to assess but a few examples will be presented to illustrate their importance. An attempt will also be made to determine annual and seasonal cycles of activity upon which these more extreme events are superimposed. But before this it is important to review the major influences on the Icelandic landscape.

Settlement and grazing pressure. It is extremely difficult to estimate the sort of pressure on land created by settlement but there is little doubt that as population increased, more remote and marginal areas came under such pressure. The earliest Norse settlers arrived about A.D.874 and were essentially farmers with animal husbandry their main occupation. An idea of the early population pressure can be gained from the Landnámabók, the Book of Settlements, which mentions about 540 farms. However, as Landnámabók was written sometime in the 12th-13th centuries A.D., this figure can only be an approximate estimate. There are no complete records of farms in Iceland until 1703 when the first census was taken but it is clear that the number of farms must soon have become considerably larger than that mentioned in the Landnámabók. All known farm numbers for a series of years are listed in Table 1. It is not always clear in the land surveys which categories of farms are included but it seems that the number of farms has been fairly constant since the beginning of the 18th century, with a decline in 1785 as a result of the Laki eruption of 1783-84.

Table 1   Number of farms in Iceland at different times (from Sveinbjarnardóttir 1988).

Year	Independent and dependent farms
Landnám	ca 540
1097	ca 4500
1311	ca 3800
1695-97	4033
1703	5456
1759	4252
1785	4456
1795	5356
1802-06	5286
1843-46	5611
1848-49	5675
1861	4343
1915	5701

A more realistic picture of pressure on the land is provided by the numbers of sheep and cattle. Accurate figures are not available before the 18th century, but after 1800 figures exist for almost every year. A selection of animal numbers are listed in Table 2. Numbers have fluctuated considerably, often as a result of epidemics. The drastic drop after 1783 is the result of fluorine poisoning caused by the Laki eruption. The figures indicate a major increase in animal numbers after about 1800 which must have resulted in considerable pressure being put on the land. Early descriptions indicate that the woodlands were being seriously depleted. Hooker (1813), in the early nineteenth century, describes Iceland as being extremely bare with attempts at cultivating trees being ineffectual. It has been estimated that over half of the vegetation cover present at the time of settlement has been destroyed (Bjarnason, 1978), with the natural climax vegetation of birch forest being reduced to about 1000km^2 and about half the area below 400m being practically devoid of soil (Thórarinsson, 1970).

Thórarinsson (1962) has produced convincing evidence of increasing erosion since settlement times. The rate of soil thickening by wind blown material has been on average 4-5 times greater after A.D.1104 than before. Climatic deterioration during the 'Little Ice Age' and volcanic activity may have contributed to the increase in soil erosion but Thórarinsson (1962) is convinced that the main cause of the erosion is the grazing of animals.

Table 2 Total number of sheep and cattle in Iceland from the 18th to
20th centuries (based on Thoroddsen, 1919; Bergþórsson, 1985;
Sveinbjarnardóttir, 1988).

Year	Sheep	Cattle
1703-14	278,994	35,860
1760	356,927	
1770	140,056	30,096
1783	236,251	20,067
1784	49,613	9,804
1785	64,459	17,592
1791	153,551	20,670
1795	241,171	22,488
1800	304,198	23,296
1801	225,306	
1802	152,947	22,247
1810	309,524	21,855
1830	542,200	28,010
1889	402,264	18,546
1893	519,298	19,948
1900	469,477	23,569
1903	486,347	26,992
1914	585,022	25,380
1981-82	750,000	65,000

Climatic fluctuations. The evidence that has been used to estimate climatic fluctuations since the time of settlement has been admirably summarised by Grove (1988). Climatic reconstructions have been based on the severity of sea ice incidence, deep ice cores and documentary sources. Some of the reconstructions are shown in Figure 1. Figure 1A shows the estimated mean annual temperature from A.D.900 based on information from ice cores (Dansgaard et al., 1975). This indicates marked fluctuations in temperature up to A.D.1200 with a few comparatively warm phases. The period 1500-1900 is characterised by generally below average temperatures ('The Little Ice Age'), before a considerable warming in the 20th century.

The ice core data can be compared with sea ice incidence. The first major study of sea ice incidence was that provided by Thoroddsen (1916-17) in his book 'The Climate of Iceland in a Thousand Years'. Koch (1945), relying extensively on Thoroddsen's evidence, has reconstructed sea ice variations since about A.D.800 (Fig.1B). The high incidence of

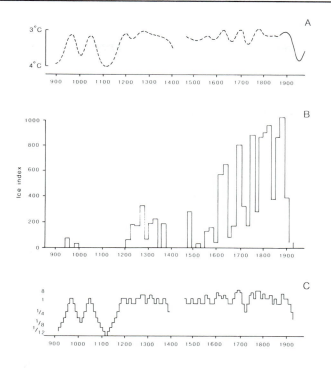

Fig.1   Various indices of Icelandic climate from A.D.900: (A)
Temperatures from ice core evidence (after Dansgaard
et al., 1975); (B) Sea ice variations (after Koch,
1945); (C) Decadal sea ice index (after Bergþórsson,
1969).

ice from 1600-1900 is a very conspicuous feature of this reconstruction, although it must
be stressed that cold periods are not necessarily associated with the presence of sea ice.
There are major similarities between Koch's reconstruction and that of Bergþórsson's graph
(Fig.1C). The graph also indicates significant fluctuations between A.D.900 and A.D.1300
reinforcing the curve produced by Dansgaard et al. (1975). The cold period from A.D.1600
to A.D.1900 is also a conspicuous feature.

   An interesting reconstruction of the climatic record based on documentary sources has
been provided by Ogilvie (1984). This work suggests that the climate was comparatively
mild at the time of Norse settlement and it was not until the 1180's that the climate
deteriorated. This accords well with the general conclusions of Haraldsson (1981) from his
detailed sedimentological and stratigraphical studies in the Markafljót sandur area. Cold
periods occurred sporadically until more harsh conditions in the 1280's and 1290's. The first
two decades of the 14th century were mild but there were then a number of cold periods,

especially after 1365. Mild conditions returned at the end of the fourteenth century up to about 1430. It seems that harsh conditions occurred in the late 16th century followed by a warm period from 1640-1660. A cold phase then set in and continued throughout most of the 18th and 19th centuries with a few warmer periods.

Volcanic activity. Volcanic tephra falls can have a tremendous effect on the natural balance of slopes. Thick falls will not only be redistributed by surface processes but will kill the vegetation thus initiating longer term erosion. Evidence from a bog at Ketilstaðir, southern Iceland, suggests that tephra from the 1357 eruption of Katla may have resulted in an increase in local erosion (Buckland et al., 1986). There are indications that the tephra layer was locally 200mm thick. The oldest tephra accurately dated through written sources is that produced by the Hekla eruption of A.D.1104. Many tephra falls since then have been dated reasonably accurately although there is some dispute over a few dates. The south of Iceland has been most affected with the volcanoes Hekla and Katla particularly active. One of the most important tephra layers is that known as the Landnám layer, which is thought to have originated in the Vatnaöldur-Hrafntinnuhraun crater row at about A.D.900, near the time of early settlement (Einarsson, 1963; Thórarinsson, 1967). It is a conspicuous marker horizon and separates pre-settlement from post-settlement slope activity.

This brief account of the potential disturbing forces in the Icelandic landscape demonstrates that by the 16th and 17th centuries population and grazing pressure were building up and there are signs that the climate was deteriorating. It might be expected that increasing instability was occurring and that erosion of one form or another was increasing. But the important questions relate to whether the landscape has been able to recover from these pressures or whether instability is the norm. These issues are now examined by using a number of specific examples.

3. SLOPE INSTABILITY

Instability of slopes occurs in a number of ways. Mass movements of one form or another are common and often lead to further erosion and the development of gullies. The plateau areas of Seljaland, southern Iceland, provide good examples of such instability. The soil and vegetation cover is being destroyed by a series of shallow mass movements. Fig. 2 shows 21 such movements on a 15° slope in an area 150m square. Movements vary from small rotational slips to debris slides and earth flows. Many of these features are retrogressing rapidly, coalescing with other features and developing into gully systems by overland flow and spring sapping. In this respect water flow is concentrated along cemented tephra layers, with piping being common. Once soil is exposed wind erosion is rapid and the landscape is clearly being degraded.

Using the terminology discussed earlier, the general impression is that such slopes are at least in a state of conditional instability if not instability. But it would be wrong to pre-

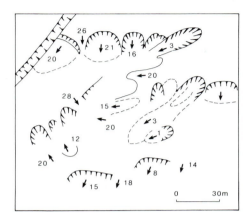

Fig.2 Soil movements in a 150m square area, above Seljaland,
south Iceland.

judge the situation as evidence from elsewhere demonstrates that stability can return. A good example of this is provided by a remarkable series of deposits exposed in a gully near the river Krossá, in Þórsmörk, southern Iceland (Gerrard, 1985). Material 2.6m thick occurs above the pale ash of the 1821 eruption of Eyjafjallajökull (Fig.3). This material consists of 74 different sedimentary units, mostly alternations of grit or sand and silts. Some of the grit layers, when traced upslope, merge with coarse particles and cobbles. The pebble layers appear to represent a number of infilled channels on a series of fans. The alternations of sands and silts high up the sequence may represent annual layers of slope wash and aeolian deposition. Such a sequence of deposits is being created today on neighbouring slopes where slope wash in spring and early summer is replaced by wind action during the main summer period. The sequence is, thus, a mixture of processes operating periodically with occasional disturbances of a more catastrophic and episodic nature. The interesting aspect of the sequence is that in the recent past it became stabilised with the development of a soil and vegetation cover.

Many other slope processes exhibit this mixture of the periodic and the episodic. This can be demonstrated by analysing the summary of landslides, rockfalls and mudflows for Iceland in the period 1958-1970 inclusive produced by Jónsson (1974). Although movements occurred in all months, there are peaks of activity in February and September-October which suggest that a relationship might exist with annual climatic patterns (Fig.4). Analysis of summary climatic statistics does not immediately suggest reasons for these

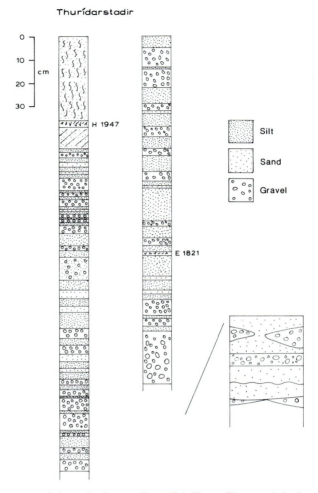

Fig.3 Sequence of deposits in a gully at Þúriðarstaðir, south Iceland.

peaks; probably because of the inappropriateness of standard climatic parameters. Frost activity is strongest in the period January to March, while September and October are the wettest months and have the highest precipitation intensities (Fig.5). Landslides are more frequent in the west, north-west and east of Iceland, all areas characterised by deeply-incised, glaciated valleys with oversteepened hillslopes. Slopes are mantled with a variety of superficial materials such as glacial drift, solifluction debris and coarse, angular screes; materials which are susceptible to movement given the right conditions. Topography is more subdued in south Iceland, apart from the former marine cliff, where a number of movements have occurred. Thus, the south is less sensitive to such movements.

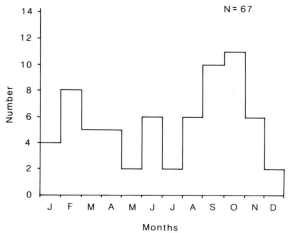

Fig.4 Frequency of landslides throughout the year in Iceland for the period 1958-1970 inclusive (data from Jónsson, 1974).

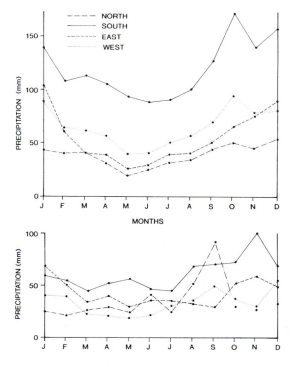

Fig.5 Mean monthly precipitation and maximum precipitation in 24 hour for stations representative of different parts of Iceland for the period 1931-66 inclusive.

The observations reported by Jónsson (1974) suggest that heavy rainfall, perhaps also associated wih the sudden thawing of snow, is the factor which triggers many movements. Many slides occurred in the north-west peninsula on 20 November, 1958 following 53-56mm of rain, with warm westerly winds raising the temperature to 10°C, thawing much of the lying snow. In the same area, extensive slipping occurred on 13-14 November, 1961 following a 10-day period in which 102mm of rain fell, and also on 21 October, 1962 following 96mm of rain in less than 24 hours. These relationships reinforce an eye witness account of the effects of heavy rain in the spring of 1922 in Austurdalur, north Iceland, when 40 screes were active at the same time. These observations endorse the pattern noted by Bjerrum and Jorstad (1968) and by Rapp and Stromqvist (1976) in Scandinavia.

The three examples reported here demonstrate that many Icelandic slopes are being subjected to a variety of processes, some of which are relatively low key but which occur frequently and regularly, whilst others are more akin to pulsed inputs as defined by Brunsden and Thornes (1979). But it is not possible to assess, with these examples, whether there has been a long term change in the operation of these processes. However, it may be possible to gain such information by a detailed examination of the build up of wind blown material.

4. RATES OF SOIL THICKENING

Extensive surveys by Thórarinsson (1962) and other workers have consistently produced soil thickening curves as indicated in Figure 6. Such a diagram is highly generalised and seems to indicate an increased but relatively constant rate of thickening after Landnám times. However, such a diagram conceals a great deal of variation. It is this local variation which enables the response of specific landscapes to be assessed. Peat bogs are excellent natural sinks and should provide a reasonable record of the variation with time of wind-

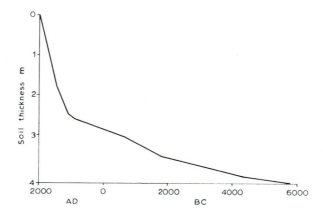

Fig.6 General soil thickening curves for south Iceland.

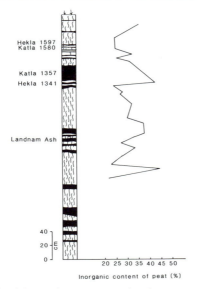

Fig.7 Loss-on-ignition values on samples from a peat bog at Ketilstaðir,
south Iceland (Buckland et al., 1986).

blown, inorganic material. A series of loss-on-ignition tests on samples from a peat bog at
Ketilstaðir, southern Iceland, produces highly variable results (Fig.7). There is a very slight
upward trend in inorganic content from pre- to post-Landnám times but this is masked by
considerable variability. This evidence suggests cyclic variability with periods of sudden
increase followed by equally sudden decline and other periods of more gradual increase
followed by a gradual decline. The landscape is clearly not responding in a simple manner
throughout the last thousand years.

Two exposures on the plateau above Skógar, in south Iceland, demonstrate that other
complications exist (Fig.8). The exposures occur in a fossil alluvial fan and an erosion scar
(rofbard) respectively, and ash layers allow depositional rates to be calculated for different
time periods. Both sections exhibit considerable variations in deposition rates and also in
the processes involved in that deposition. The alluvial fan section is dominated by slope
wash until 1597 when there is a change to aeolian input and a general reduction in deposition
rates. In the rofbard, rates of deposition until 1486 are extremely low. There is then a change
of process to slope wash deposition and an increase in deposition rates. So thresholds have
been crossed both in terms of rate of operation of processes and in the nature of those
processes. Thus, on the alluvial fan there has been high deposition of slope wash material
before 1597 and low rates of aeolian deposition after, and in the area represented by the
rofbard, there have been low rates of aeolian deposition before 1486 and high rates of slope
wash after 1486.

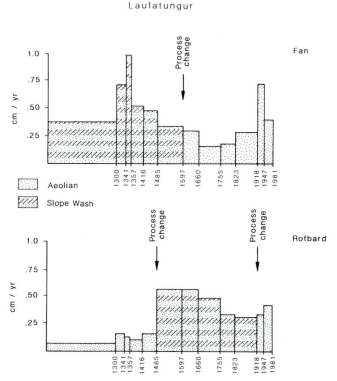

Fig.8 Variation in processes and rates of aeolian deposition on a plateau
surface above Skógar.

In each case, once the process has established itself there is a fairly constant rate of deposition. This is not always the case, as a variety of exposures from south Iceland show (Fig.9). The conspicuous ash from the 1362 eruption of Öræfajökull provides a base line for comparison. Each section can be equated with one of the three standard models of the way in which landscapes respond (Fig.10). Figure 10A is a model for landscapes that exhibit consistent stability over time with occasional rapid changes in activity being quickly compensated. Sections (A) and possibly (F) (Fig.9) correlate with this behaviour pattern. Figure 10C is a model for landscapes which are subjected to a disturbance which leads to a sudden, but sustained increase in rates of deposition. Stability is achieved but at a higher level. Section (D) (Fig.9) seems to fit this model with a sudden increase in deposition rates after 1625, with these increased rates being maintained for the next 350 years. The third possibility (Fig.10B) is that of a landscape in which equilibrium is upset, rates of operation of processes fail to stabilise and the slopes become rapidly degraded. Sections (B), (C) and (E) (Fig.9) indicate this type of behaviour. It is significant that in four of the sections the change of conditions occurs after about 1625, the period when grazing pressure was at a high

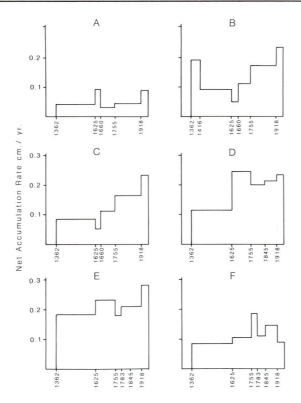

Fig.9 Variation in rates of aeolian deposition from a number of sites in
south Iceland.

level and climate was beginning to deteriorate.

5. DISCUSSION

The majority of the examples discussed so far have been taken from the south of Iceland
where rates of erosion are high. Thus, it must not be assumed that landscapes in other areas
of Iceland have reacted in similar ways. An interesting comparison is provided by the
valleyside slopes of Fossárdalur, a narrow valley leading off Berufjörður, northwest of
Djúpivogur, eastern Iceland. Although there may have been up to 9 farms in the valley,
settlement pressure has not been as great as in many southern areas. Also, slopes have rarely
been affected by tephra falls. Only three important tephras, two black and one light-
coloured, can be identified. It is assumed that the younger black tephra layer derives from
the 1755 eruption of Katla (Larsen, 1982). The other dark-coloured tephra was first desribed
by Thórarinsson (1958) and was produced by an eruption from Vatnajökull, probably in

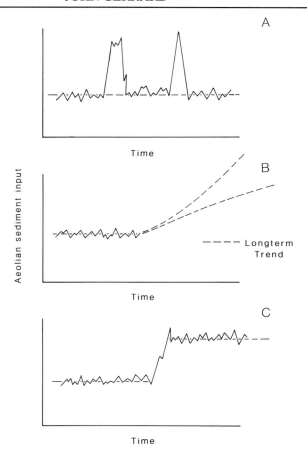

Fig.10 Possible models for rates of aeolian deposition.

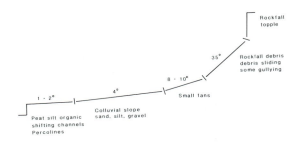

Fig.11 General arrangement of slope form, processes and materials on
valleyside slopes in Fossárdalur, east Iceland.

1477. The yellow-white tephra layer, which is very conspicuous in soil sections in the area, results from the 1362 eruption of Öræfajökull.

The valley of Fossárdalur is dominated by sidewalls of massive basalt which have been weathered and eroded into a series of benches and steps. Slope forms in the interior part of the valley are relatively simple whereas in the wider valley sections more complicated slope forms with a greater number of facets occur (Fig.11). Analysis of surface processes and exposures enables preliminary statements to be made concerning long term development of such slopes. Rock fall is an active process with the debris coming to rest on the 35° debris slope. But a cursory examination of lichen cover suggests that current rock fall activity is less than previously experienced.

The debris slope is active periodically with sliding and gullying processes moving material towards 8-10° angled fans. Some of the debris slopes are stabilised by soil and vegetation and there are small areas of stunted birch growth. But considerable erosion has also taken place. Late-lying snow patches contribute water to many of the gullies. Fan slopes are composed of alternating sequences of sandy silts and gravels and/or cobbles demonstrating periodic inundation by water interrupting aeolian input and soil formation. However, the number of sedimentary units and their thickness are much less than those described in south Iceland. On the lower slopes, deposits become more organic in nature resting on a silt or gravel substrate. This substrate grades upwards into fine sandy silt, all below the 1362 Öræfajökull ash layer. The succession above the ash is essentially silty with occasional coarse sandy lenses. Many of the silt layers possess old root holes with small rootlets indicating former soils. A number of infilled channels also occur indicating sudden changes in drainage. These sequences suggest a comparatively stable environment, with good soil and vegetation cover, and intermittent flooding by streams flowing off the valley sides constantly changing direction.

There is little doubt that slope processes have been intermittently active and that changes to slope form have occurred over the last thousand years. But there is very little evidence of long term instability. Of the three possible models represented in Fig.10, both model A and model C can be applied to slopes in Fossárdalur, but none of the slopes appears to possess a behaviour pattern typified by model B. Many of the changes simply represent differences in spatial activity such as the migration and shift of channels on alluvial fans. There is also little indication of major phases of activity associated with climatic fluctuations or pressures caused by settlement.

6. CONCLUSIONS

Each of the situations examined here has demonstrated that major changes have been occurring on Icelandic slopes. But the examples have also shown that considerable variability exists between different parts of Iceland. Although the use of depositional sequences has a number of pitfalls, the evidence can be used to show the way in which different slopes have reacted to different external pressures. The amount of material deposited must, in a simple way, relate to the amount of erosion going on elsewhere, although not necessarily on the same slope. Some slopes have shown major changes since the time of settlement whereas others seem to have been largely unaffected. Other slopes have exhibited major disturbances in the past but are now stable. Although it is clear that changes of climate and land· use pressures have both been involved in creating slope instability it is not easy to state, categorically, that one factor has been more influential than another. But, in general, the major influence over the last thousand years appears to be human pressure. The evidence also indicates that it is not easy to predict how a particular slope will respond to changing circumstances, although it is hoped that concepts introduced here will enable further work to go some way to solving this problem.

REFERENCES

Bergþórsson, P. (1969) 'An estimate of drift ice and temperature in Iceland in 1000 years', Jökull 19, 94-101.

Bergþórsson, P. (1985) 'Sensitivity of Icelandic agriculture to climatic variations', Climatic Change 7, 111-127.

Bjarnasson, Á. H. (1978) 'Erosion, tree growth and land regeneration in Iceland', in M.W.Holdgate and M.J.Woodman (eds.), The Breakdown and Restoration of Ecosystems, Plenum Press, New York, 241-248.

Bjerrum, and Jorstad (1968) 'Stability of rock slopes in Norway', Norwegian Geotechnical Institute Publication 79, 1-11.

Buckland, P.C., Gerrard, A.J., Larsen, G., Perry, D.W., Savory, D.R. and Sveinbjarnardóttir, G. (1986) 'Late Holocene palaeoecology at Ketilstaðir in Mýrdalur, South Iceland', Jökull 36, 41-55.

Dansgaard, W., Johnsen, N., Reeh, N., Gundestrup, N., Clausen, H.B. and Hammer, C.U. (1975) 'Climatic changes, Norsemen and modern man', Nature 255, 24-28.

Einarsson, Th. (1963) 'Pollen analytical studies on the vegetation and climate history of Iceland in late and postglacial times', in Á.Löve and D.Löve (eds.) North Atlantic Biota and their History, Pergamon, Oxford, 355-365.

Gerrard, A.J. (1985) 'Soil erosion and landscape stability in southern Iceland: a tephrochronological approach', in K.S.Richards, R.R.Arnett and S.Ellis (eds.) Geomorphology and Soils, George Allen and Unwin, London, 78-95.

Gigon, A. (1983) 'Typology and principles of ecological stability and instability', Mountain Research and Development 3, 95-102.

Grove, J.M. (1988) The Little Ice Age. Methuen, London.

Haraldsson, H. (1981) 'The Markafljót sandur area', Striae 15.

Hooker, W.J. (1813) Journal of a tour in Iceland in the summer of 1809. London.

Jónsson, O. (1974) 'Landslides and mudflows', Jökull 24, 63-76.

Larsen, G. (1982) 'Preliminary report on the historical tephra layer at Gautavík, East Iceland', in T.Capelle (ed.) Untersuchungen auf dem mittelälterlichen Handelspatz Gautavík, Island, Rheinland-Verlag, Cologne, 97-98.

Koch, L. (1945) 'The East Greenland ice', Meddelelser om Grønland 130, no.3.

Ogilvie, A.E.J. (1984) 'The past climate and sea-ice record from Iceland', Climatic Change 6, 131-152.

Rapp, A. and Stromqvist, L. (1988) 'Slope erosion due to extreme rainfall in the Scandinavian Mountains', Geografiska Annaler 58A, 193-200.

Sveinbjarnardóttir, G. (1988) Settlement patterns in medieval and post-medieval Iceland: an interdisciplinary study. Unpublished Ph.D. Thesis, University of Birmingham.

Thórarinsson, S. (1958) 'The Öræfajökull eruption of 1362', Acta Naturalia Islandica 2, No.2.

Thórarinsson, S. (1962) 'L'érosion éolienne en Islande á la lumière des études tephrochronologiques', Revue Geomorphologie Dynamique 13, 107-124.

Thórarinsson, S. (1967) 'The eruptions of Hekla in historical times', in Th.Einarsson, G.Kjartansson and S.Thórarinsson (eds.), The eruption of Hekla 1947-1948, Societas Scientatis Islandica 1, 1-170.

Thórarinsson, S. (1970) 'Tephrochronology and medieval Iceland', in R.Beyer (ed.) Scientific Techniques in Medieval Archaeology, University of California Press, Los Angeles, 295-328.

Thórarinsson, S. (1979) 'Tephrochronology and its application in Iceland', Jökull 29, 33-36.

Thoroddsen, Th. (1916-17) Árferði á Íslandi í þúsand Ár, 1916, 1-192: 1917, 193-342, Hið Íslenska Fræðafjelag, Copenhagen.

Thoroddsen, Th. (1919) Lysing Íslands III, Landbúnaðar á Íslandi, Kaupmannahöfn.

Winiger, M. (1983) 'Stability and instability of mountain ecosystems, United Nations University Workshop 1981', Mountain Research and Development 3, 103-111.

GLACIER FLUCTUATIONS AND ROCK GLACIERS IN TRÖLLASKAGI, NORTHERN ICELAND, WITH SPECIAL REFERENCE TO 1946-1986

H. Elizabeth Martin
152 Tom Lane, Fulwood,
Sheffield, S10 2PG,
UK

W. Brian Whalley
School of Geosciences
The Queen's University of Belfast
Belfast BT7 1NN,
UK

Chris Caseldine
Department of Geography
University of Exeter
Exeter EX4 4RJ,
UK

ABSTRACT. Rock glaciers are variously thought to be of glacial or non-glacial origin. Those in Tröllaskagi, northern Iceland include features that appear to have evolved from debris-covered glaciers and co-exist with glaciers in this marginally-glacierized area. Monitoring of one such rock glacier, in Nautadalur, since 1977 and with aerial photographs since 1946, allows an assessment of its activity and comparison with nearby rock glaciers and glaciers. Aerial photograph analysis reveals that this and other rock glaciers in Tröllaskagi have exhibited little movement or morphological change since 1946. By contrast, neighbouring glaciers have fluctuated significantly over the same period, with some having experienced a net retreat of over 300m. The fluctuations of a valley glacier, Gljúfurárjökull, have been directly linked to temperature changes. A corresponding temperature-induced retreat or glacier core disintegration is not seen on the rock glaciers, either as a contemporaneous or lagged response over this timescale. The stability of rock glaciers is apparently due to the insulation of the relict glacier ice core by the debris cover. Rock glaciers of glacial

255

J. K. Maizels and C. Caseldine (eds.), Environmental Change in Iceland: Past and Present, 255–265.
© 1991 *Kluwer Academic Publishers. Printed in the Netherlands.*

origin in northern Iceland do not respond to temperature fluctuations in the manner of debris-free glaciers but may well show effects of precipitation increases by changes in their flow patterns, albeit over longer time scales. Rock glaciers should be used in paleoenvironmental reconstructions only with care.

1. INTRODUCTION

There has been considerable examination of moraine chronologies in the south of Iceland in recent years (e.g Gordon and Sharp, 1983;Sharp and Dugmore, 1985) but comparatively little has been reported from the north until recent work by Caseldine (1987), Kugelmann (in press) and Stötter (in press). Further, rock glaciers exist in close proximity to debris-covered and debris-free glaciers in the north but not in the south of Iceland. Thus, it is important to discover the environmental reasons for this relationship.

Rock glaciers are generally thought of as either glacial or non-glacial in origin (Washburn, 1979). Those found in northern Iceland have apparently evolved from debris-covered glaciers (Whalley, 1974; Escritt, 1978; Martin and Whalley, 1987a) and, therefore, have a potentially important role to play in understanding the glacial history of this region. Inspection of results published on the movement of one such rock glacier, in Nautadalur, by Martin and Whalley (1987a) and a comparison with those fluctuations of Gljúfurár-jökull, a neighbouring glacier, (Caseldine and Cullingford, 1981; Caseldine, 1983, 1985, 1987, 1988), illustrate very different patterns of activity. As yet, detailed rock glacier observations in Iceland are rather scanty. This paper examines the currently available detail for rock glacier and glacier fluctuations in Tröllaskagi. Analysis of aerial photographs dating from 1946 and results of recent field monitoring involving a number of dating techniques allow comparisons within a small area. Consideration is given to the reasons for the differing patterns of glacier and rock glacier activity in terms of available climatic data.

2. ROCK GLACIER FLUCTUATIONS IN TRÖLLASKAGI, 1946-1986

Rock glaciers are, apparently, rare in Iceland except for the drier areas of the north (Whalley, 1974) although one has been described from the south west (Eyles, 1978). Thus, examination of any relationship between glaciers, rock glaciers and climate comes from north-central Iceland (Fig. 1).

The Tröllaskagi Peninsula (3800 km²), lying between Skagafjörður and Eyjafjörður (Fig. 1), consists of the highest basalt massif in Iceland (rising to 1300-1500m a.s.l.) and is dissected by glacially-modified valleys. Nautadalur rock glacier, the subject of field investigations since 1977, is located in the south-eastern corner of this region, lying in a corrie on the southern side of Skjóldalur (approximately 20km south of Akureyri). Martin and Whalley (1987a) provide a detailed description of Nautadalur rock glacier and evidence for its origin as a debris-covered glacier. Chronological data have been obtained on a purported rock glacier and related corrie glacier at Klængshóll (25 km NW of Akureyri) by Meyer and Venzke (1985). This is however believed to be an ice-cored moraine and has

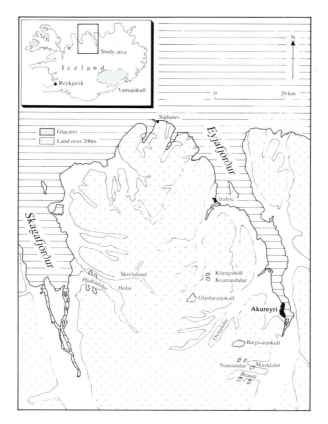

Fig. 1   Iceland, location of Tröllaskagi and map of Tröllaskagi showing the main glacier locations discussed.

few distinct characteristics of a rock glacier. It may well be a transitional form between an active glacier with ice-cored moraine and a rock glacier. This would be expected for rock glaciers formed from heavy surface debris loads on stagnating glaciers (Martin and Whalley, 1987b). There is no evidence that it is permafrost-related however.

   In summary, Nautadalur rock glacier is identified by the purely morphological criteria described by Martin and Whalley (1987b). It possesses all the classic rock glacier criteria, (e.g. a distinct snout and lateral margins and surface relief of ridges and furrows). Located at the head of the rock glacier is a small corrie glacier. The ice of this corrie glacier is contiguous with the rock glacier. Glacier ice exposed in sections along the length of the rock glacier shows that the latter is a debris-covered extension of the corrie glacier. The debris thickness increases from no cover on the corrie glacier ice towards the snout (1.5 - 2m thickness) but glacier ice has been proved at the snout, emerging from beneath the debris cover (Martin and Whalley, 1987a).

In 1977 the Nautadalur rock glacier was surveyed with three transects; boulders were marked to provide the basis for subsequent measurement of surface displacements and assessment of rock glacier activity (Martin and Whalley, 1987a). The snout of the rock glacier is shown to be virtually stationary but with < 1m of 'recession' (Fig. 2). This reflects rock debris displacements as a result of near surface ice melt. As the snout remains steep ('active' in the usual terminology) there still should be some forward ice movement. This too should be <1m over this period, although it has not been possible to measure it directly. The glacier ice core model predicts that, although the ice may be moving, the snout position need not do so if the small amount of movement is matched by melting (Whalley, 1983). This appears to be the case here.

Aerial photographs of Nautadalur rock glacier are available from 1946-1985. Photographs from August 1946, 1960 and 1982 illustrate that, although summer snow is highly variable, the general extent and characteristics have remained remarkably constant. From analysis of aerial photographs of rock glaciers in Tröllaskagi, it is apparent that Nautadalur is not exceptional in its relationship to small corrie glaciers. Further, little, if any, change has occurred in rock glacier activity or morphology since 1946. For two rock glacier complexes in Brandi, the next valley south of Nautadalur, photographic comparison for a similar period also shows that no dramatic changes have occurred. Glacier ice has been observed (1986) at the snout of one of these.

The last forty years at least have seen neither the further development nor destruction of rock glaciers in this area but rather the maintenance of glacier ice bodies which form the substantial core of these glacially-derived landforms.

3. NEOGLACIAL GLACIER EVENTS IN ICELAND

Although outer moraines of Neoglacial age can be identified in several parts of Iceland, reasonable estimates of their age are few, and the pattern of activity is rather unclear and may not necessarily be related to events elsewhere in northen Europe (Grove, 1979, 1988). Recent work in north Iceland (Caseldine, 1985, 1987; Häberle, in press; Stötter, in press) has demonstrated the existence of pre-'Little Ice Age' glacier explansion but only a limited chronology has yet been established.

The glacier systems in Tröllaskagi are rather small compared with Iceland as a whole (Björnsson, 1978). Nevertheless, their response to climatic changes may be a useful indicator of events for this area in the North Atlantic domain (see, e.g. Løken, 1972; Anda et al, 1985).

4. RECENT GLACIER FLUCTUTIONS IN TRÖLLASKAGI

In contrast to the wealth of documentary evidence available for dating moraine sequences in some areas, there is little available in northern Iceland and a variety of techniques such as lichenometry, tephrochronology and comparative rock weathering measurements need

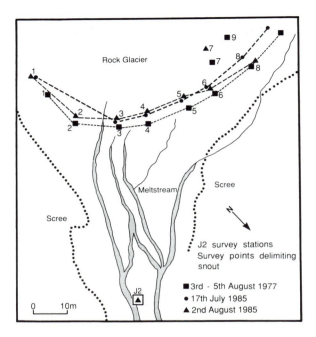

Fig. 2   Nautadalur rock glacier, snout movement, 1977-1986.

to be applied (Caseldine, 1987, in press). However, some data have been derived for the main glaciers in the area west of Akureyri (Fig.1).

Lichenometric results (Caseldine, 1987, in press; Kugelmann, in press) show that the majority of moraines examined in this part of Tröllaskagi are of Late Neuzeitlich/'Little Ice Age' date and that there were five main periods of moraine abandonment. In summary:

i.   ca 1810-1820 In upper Svarfaðardalur there are moraines of this age in Bur-fellsdalur and Vatnsdalur, whilst in western Tröllaskagi Tungnahryggsjökull also has a moraine of this age.

ii. 1845-1875 Kugelmann (in press) argues for a separation of this into two distinct phases with the earliest around 1850 and a later period between 1870-1880. Moraines from the earlier phase can be found at 7 sites and over ten glaciers have moraines from the latter phase.

iii. Late 1880s-1890s There are many moraines which appear to have been abandoned during this period and a combination of the results from Caseldine and Kugelmann show this period to have been equally as important as its immediate predecessor.

iv. 1915-1925 A smaller number of glaciers, particularly at higher altitudes at the head of Skíðadalur, began to retreat at this time.

v. 1930s-1940s Moraines abandoned in the early 1940s occur at several locations, either as recessional terminal features or as fragments of lateral moraine.

Following continued retreat of over 300m since the 1940s, there has also been a recent period of noticeable advance at Gljúfurárjökull, between the mid 1970s and the mid 1980s (Caseldine, 1988). Although this advance appeared to have ceased there has since been some small expansion between 1988 and 1989.

5. GLACIER ACTIVITY AND RECENT CLIMATIC CHANGES IN TRÖLLASKAGI

Over the last century it is possible to relate glacier retreat to an observed increase in temperature in north Iceland by examining temperature records for Akureyri (Fig.3); especially following the main increase in temperature since about A.D.1925 (Jóhannesson, 1986). Surface area depletion for all (115) glaciers in Tröllaskagi 1930-1960 has been about 33% (Björnsson, 1978). Because so few mass balance data are available for Tröllaskagi glaciers (but see Björnsson, 1971) attention must be directed to temperature-precipitation data as a surrogate. Thus, mean summer temperatures and winter precipitation provide the best means for analysis of recent glaciological events (Caseldine, 1985). Caseldine (1987) argues that there is evidence for a lag of about 10 years between a sustained temperature increase and glacier snout response. Despite changes in the chronology of recent deglaciation necessitated by the new lichen growth curve of Kugelmann, this pattern is still valid.

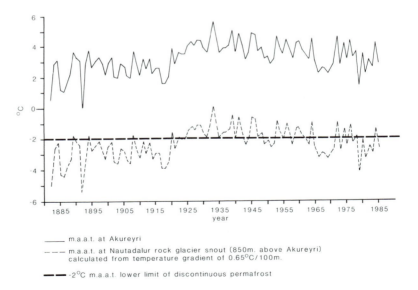

Fig. 3   Temperature records from Akureyri and estimate of mean annual average (m.a.a.t.) temperature at Nautadalur.

For precipitation changes, there have been marked fluctuations (Fig. 4) and it is difficult to relate these to possible mass balance changes, especially as there are also marked spatial variations in precipitation over the peninsula. The recent response of the glaciers has been generally of recession, despite some small advances for some glaciers. For most glaciers in Tröllaskagi, rather smaller than Gljúfurárjökull, (the largest at about 3 km²), the response of the glacier mass has been a marked decline since the 1930s with, especially, rapid snout position recession.

Following Björnsson's (1971) work on Bægisárjökull, Caseldine (1985) suggested that for Gljúfurárjökull, a positive mass balance takes ten years to give an advance (i.e. the response time) but that recession tends to occur more rapidly and is a response to mean summer temperatures above 7.5°C. This critical temperature can be used to measure the response of the larger glaciers in the light of the variable winter precipitation values. It is not yet known whether the smaller glaciers respond in a like manner. However, this is debatable as they receive much of their snow input from 'snowfence' effects beneath the plateau rims. Moreover, it is undesirable to think of an equilibrium line for such glaciers as the receipt of snowfall is so uneven. The ice balance of such glaciers is difficult to assess fully and has not yet been attempted for any of the smallest glaciers in Tröllaskagi.

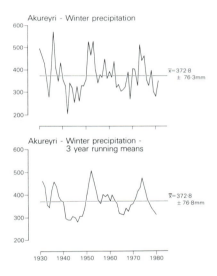

Fig. 4   Winter precipitation, as means and 3 year running means, for Akureyri.

6. EFFECTS OF DEBRIS ON GLACIERS

It is notable that all the largest glaciers in Tröllaskagi are 'conventional' (essentially debris-free) with well-defined moraine sequences: e.g Gljúfurárjökull, Bægisárjökull. Most of the smaller glaciers however have associated moraines and rock glacier complexes. Exceptionally, e.g. Lambarájökull, a larger valley glacier shows rapid retreat as an elongated debris-covered glacier with prominent moraine sequences but looking rather similar to a rock glacier. Because of the close relationship between glacier ice and the rock glaciers we suggest that the debris of the rock glaciers hides glacier ice which has failed to melt during the rapid retreat of at least the last 80 years as a response to climatic warming and (possibly) reduced snow input.

7. DISCUSSION

The valley glacier fluctuations in Tröllaskagi can be directly linked to temperature changes (Björnsson, 1971; Caseldine, 1985). Aside from the largest of the valley glaciers, all the smaller corrie glaciers are receding rapidly and we conclude that Tröllaskagi is marginal for glaciers at the present time. (This is also true of once large glaciers such as Ok and Dranga discussed in Jóhannesson (1986) and Björnsson (1979).) The sustainable ice mass is limited in the case of the corrie glaciers to the extreme limits of corrie headwalls in favourable (generally north but also east-facing) aspects which can collect sufficient snow to maintain a glacier. Altitude alone is not an independent factor, size of site available is significant; see e.g. the corrie to the east of Nautadalur which contains a very diminished ice body compared with that above the neighbouring rock glacier. This remains a rather unexplored problem in the response of an area to changes in glacierization and climate controls. Nevertheless, it is possible that some of the corrie glaciers exist at altitudes below the regionalized equilibrium line altitudes for Tröllaskagi (Ahlmann, 1948).

Given the recent mass balance and snout position history of the large glaciers, especially Gljúfurárjökull, it is possible to predict the response to changes in climate and look at these in the light of the response of the largest Icelandic glaciers (Jóhannesson, 1986). We have not yet been able to assess the loss in ice mass for the glaciers associated with the Nautadalur-Brandi rock glaciers for the period 1946-1985. Although inspection of the aerial photographs shows little change in areal disposition there has been ice lost at the lowest extents of the exposed ice in corrie glaciers. This corresponds to the general wastage seen in Tröllaskagi since the recession of the 1920s. Because the snouts of the glacier/rock glacier systems are hidden, the usual mapping procedures cannot be applied (Whalley et al., 1986). As the rock glacier snouts do not appear to have altered significantly in the last 40 years, the effects of increased summer temperatures have not made substantial differences to their configuration. However, reduced precipitation (or at least some reduced receipt by the glacier) and increased melting has allowed the ice in the corrie head to decline.

A temperature-induced retreat or glacier core disintegration is not seen on the rock glaciers, either as a contemporaneous or even as lagged response. However, downwasting of the exposed glacier ice is seen. Since both rock glaciers and glaciers are subject to the same environmental conditions the difference in activity is best explained by the debris over the ice in the case of the rock glacier. This debris must be sufficient to insulate the glacier ice from the temperature changes which have been crucial to the rapid retreat of exposed glacier ice bodies in this area.

For those glaciers without associated rock glaciers it is posible that major new inputs will be required to bring them back to a 'healthy' state. For those with rock glaciers slightly different conditions may apply. Potter (1972) has discussed some of the possibilities for a glacier/rock glacier system. If there is contiguous ice between the corrie (debris-clear) glacier and the glacier residuum in the rock glacier portion then there could be visible changes on these snouts. Not only might the snout appear to be more 'active' but, particularly, increases in velocity should be determined at an early stage in any swing towards more positive mass balances. This would be the same as changes to ice-cored lateral moraines which have been seen to precede snout advances (e.g Whalley, 1973).

The response time of such glacier-rock glacier systems to climate might be shorter than for the larger valley glaciers such as Gljúfurárjökull. This would apply particularly for increased precipitation because, being high, ablation of the exposed ice is reduced. More significantly however, the rock glacier portion of the system allows a greater volume of ice to be present in the corrie than if it were absent. Debris band inclination as well as velocity measurements on Nautadalur show steeply emergent velocity vectors for the corrie glacier ice. We believe that the flow regime of the corrie glacier is essentially decoupled from that of the rock glacier downstream. Hence, it may be that monitoring velocity vectors of such glaciers may be important in ascertaining climatic changes well before they are evident at the snouts of larger, 'conventional' glacier systems.

Because Tröllaskagi is a marginally glacierized area we would expect to see evidence of increased (winter) precipitation/reduced (summer) temperatures manifest sequentially in several types of glacier system. Despite the lagged response to ice mass reduction which has affected glaciers this century, rock glaciers may well be the first to show signs of climatic events which increase mass of glacier systems. However, we do not yet know the effects of changing dominant wind patterns affecting the 'snowfences' which contribute to the snow accumulation of many of the highest glaciers; this needs more detailed study.

8. CONCLUSIONS

Constraints in addition to climate, such as glacierized area, appear to be important to small glaciers in north Iceland. Rock glaciers of glacial origin illustrate a different and more complex relationship still. The understanding of rock glacier dynamics is complicated by the effect debris can have in protecting and preserving ice masses. Potentially, the 'relict' glaciers with rock glaciers represent an, as yet, untapped record of paleoenvironments as well as providing substantial present-day ice volume in this marginally-glacierized region.

Rock glaciers of glacial origin must be used with care in environmental reconstruction. If debris can protect the ice then there is a possibility that the rock glacier may in fact be preserving ice bodies which may otherwise have melted. Therefore, if we can understand the effects this may produce on lagged responses to climatic change then it may be possible to use these features to help interpret palaeoenvironments as well as identify changes to come.

ACKNOWLEDGEMENTS

We thank the Icelandic Research Council for permission to work in Tröllaskagi and to Landmælingar Íslands for assistance with aerial photographs. Funding has been generously granted from a number of sources and we gratefully acknowledge: The Royal Geographical Society and Mount Everest Foundation, The Royal Society - Dudley Stamp Fund, Gilchrist Educational Fund, The Gino Watkins Fund, William Cadbury Charitable Trust, British Sugar PLC, The Queen's University of Belfast, Exeter University Committee of Deans and the Department of Education of Northern Ireland. We should also like to thank our various field assistants and Hjörtur Þórarinsson at Tjörn and Þorvaldur Hállsson at Ystra Gerði.

REFERENCES

Ahlmann, H.W. (1948) Glaciological research on the North Atlantic coasts. Royal Geographical Society, Research Series No.1. 83 pp.

Anda, E., Orheim, O. and Mangerud, J. (1985) ' Late Holocene glacier variations and climate at Jan Mayen', Polar Research 3, 129-140.

Björnsson, H. (1971) 'Bægisárjökull, North Iceland. Results of glaciological investigations 1967-68. Part I. Mass balance and general meteorology', Jökull 21, 331-348.

Björnsson, H. (1978) 'The surface area of glaciers in Iceland', Jökull 28, 31.

Björnsson, H. (1979) 'Glaciers in Iceland', Jökull 29, 74-80.

Caseldine, C.J. (1983) ' Resurvey of the margins of Gljúfurárjökull and the chronology of recent deglaciation', Jökull 33, 111-118.

Caseldine, C.J. (1985) 'The extent of some glaciers in northern Iceland during the Little Ice Age and the nature of recent deglaciation', Geographical Journal 151, 215-227.

Caseldine, C.J. (1987) 'Neoglacial glacier variations in northern Iceland: examples from the Eyafjörður area', Arctic and Alpine Research 19, 296-304.

Caseldine, C.J. (1988) 'Fluctuations of Gljúfurárjökull, Northern Iceland 1983-1987', Jökull 38, 32-34.

Caseldine, C.J. (1990) 'A review of Holocene dating methods and their application in the developmentof a Holocene glacial chronology for Northern Iceland', Münchener Geographische Abhandlungen, Reihe B, in press.

Caseldine, C.J. and Cullingford, R.A. (1981) 'Recent mapping of Gljúfurárjökull and Gljúfurárdalur', Jökull 31, 11-22.

Escritt, E.A. (1978) 'North Iceland Glacier Inventory - 1977 Season', Jökull 28, 57-58.

Eyles, N. (1978) 'Rock glaciers in Esjufjöll Nunatak area, South-east Iceland', Jökull 28, 53-56.

Gordon, J.E. and Sharp, M. (1983) 'Lichenometry in dating recent glacial landforms and deposits, southeast Iceland', Boreas 12, 191-200.

Grove, J.M. (1979) 'The glacial history of the Holocene', Progress in Physical Geography 3, 1-54.

Grove, J.M. (1988) The Little Ice Age. Methuen, London.

Häberle, T. (1990) Spät- und Postglazial Gletscherschwankungen im Hörgárdalur gebietes, Tröllaskagi, Nordisland, Dissertation an der Universität Zürich, in press.

Jóhannesson, T. (1986) 'The response time of glaciers in Iceland to changes in climate', Annals of Glaciology 8, 100-101.

Kugelmann, O. (1990) ' Datierungen neuzeitlicher Gletscherschwankungen im Svarfaðadalur/Skíðadalur mittels einer neuen Flechten-Eichkurve', Münchener Geographische Abhandlungen, Reihe B, in press.

Løken, O.H. (1972) ' Growth and decay of glaciers as an indicator of long-term environmental changes', International Commission for the Northwest Atlantic Fisheries, Special Publication, 8, 71-87.

Martin, H.E. and Whalley, W.B. (1987a) ' A glacier ice-cored rock glacier in Tröllaskagi, Northern Iceland', Jökull, 37, 49-55.

Martin, H.E. and Whalley, W.B. (1987b) 'Rock glaciers: a review, Part 1', Progress in Physical Geography 11, 260-282.

Meyer, H.H. and Venzke, J.F. (1985) ' Der Klængshóll-Kargletscher in Nordisland', Natur und Museum 115, 29-46.

Müller, H.-N., Stötter, J, Schubert, A, and Betzler, A. (1986) ' Glacial and periglacial investigations in Skíðadalur, Tröllaskagi', Polar Geography and Geology 10, 1-18.

Potter, N. jr., (1972) ' Ice-cored rock glacier, Galena Creek, Northern Absaroka Mountains, Wyoming', Bulletin of the Geological Society of America 83, 3025-3058.

Sharp, M. and Dugmore, A. (1985) 'Holocene glacier fluctuations in eastern Iceland', Zeitschrift für Gletscherkunde und Glazialgeologie 21, 341-349.

Stötter, J. (1990) 'Neue Beobachtungen und überlegungen zum Verlang des Periglazials Islands am Beispiel des Svarfaðar-Skíðadals', Münchener Geographische Abhandlungen, Reihe B, in press.

Thórarinsson, S. (1943) ' Oscillations of the Icelandic glaciers in the last 250 years', Geografiska Annaler 25, 1-54.

Washburn, A.L. (1979) Geocryology. Edward Arnold, London.

Whalley, W.B. (1973) 'An exposure of ice on the distal side of a lateral moraine', Journal of Glaciology 12, 327-329.

Whalley, W.B. (1974) 'Rock glaciers and their formation as part of a glacier debris-transport system', Reading Geographical Papers 24, 60 pp.

Whalley, W.B. (1983) 'Rock glaciers - permafrost features or glacial relics?', Proceedings of the Fourth International Conference on Permafrost, National Academy Press, Washington, Vol.1, 1396-1401.

Whalley, W.B., Martin. E.H. and Gellatly, A.F. (1986) 'The problem of 'hidden' ice in glacier mapping', Annals of Glaciology 8, 181-183.

THE ORIGIN AND EVOLUTION OF HOLOCENE SANDUR DEPOSITS IN AREAS OF JÖKULHLAUP DRAINAGE, ICELAND

Judith Maizels
Department of Geography,
University of Aberdeen
Elphinstone Road,
Aberdeen AB9 2UF
U.K.

ABSTRACT. The impact of jökulhlaup (flood) history on the origin and evolution of the coastal sandar of southern Iceland is explored in this paper through analysis of the morphology, sedimentology and stratigraphy of six different sandur environments. The results indicate that three distinct models of sandur evolution can be identified from the landform and lithofacies assemblages of the deposits. Type I sandur deposits are characterized by repeated, thin, upward-fining cycles of gravels, sands and silts, and are found in areas of seasonal meltwater activity associated with braided river systems. Type II sandur deposits are characterized by thick (>10m), coarsening-upward, clast-supported cobble gravels, overlain by a fining-upward sequence of fine gravels, sands and silts, which are found in areas of prolonged (days and weeks), high magnitude drainage events associated with subglacial geothermal activity or ice-dammed lake drainage. Finally Type III sandur deposits are typified by thick sequences (8m+) of structureless pumice granules, underlain by a crudely bedded basal unit, and capped by several metres of trough cross-bedded and horizontally bedded granules. These deposits are associated with catastrophic drainage events generated by subglacial volcanic eruptions. These floods are characterized by extremely high peak flows (>10^5 m^3 s^{-1}) and sediment concentrations (>35%), and short-lived hydrographs (peak flows reached in hours). Type I sandur plains form an extensive low relief surface marked by abandoned, braided palaeochannel networks, while Type II is normally associated with deep incision of meltwater channels into a pitted sandur surface; Type III sandur plains are characterised by the most complex and varied proglacial morphology and surface landforms. Types II and III appear to dominate the stratigraphy of the major Icelandic sandar of Sólheimasandur and Mýrdalssandur, and Skeiðarársandur, respectively, such that the braided sandur Type I occurs only locally or as a relatively thin surface veneer overlying Type II deposits. The paper concludes that the sandar forming the south coast of Iceland largely owe their origin to the dominating impact of infrequent but catastrophic, jökulhlaup events.

J. K. Maizels and C. Caseldine (eds.), Environmental Change in Iceland: Past and Present, 267–302.

1. INTRODUCTION

The large coastal sandar of south Iceland are normally drained by complex braided meltwater rivers, whose activity has been widely regarded as the dominant agent in sandur development (eg. Hjulström, 1952; Krigström, 1962; Bluck, 1974, 1979, 1982, 1987; Ward et al., 1976; Boothroyd & Nummedal, 1978; Hine & Boothroyd 1978; Rust, 1978). However, many of the sandar have been, and remain, subject to catastrophic flood events, or jökulhaups, produced by a variety of mechanisms. Major floods have resulted from subglacial volcanic eruptions (eg. from Katla and Öræfajökull), by sudden drainage of ice-dammed lakes (eg. Grænalón), or subglacial, geothermal lakes (eg. Grimsvötn), or even from occasional rockfalls and landslides (eg. the Steinholtsá flood of 1967; Kjartansson, 1967), and from floods associated with glacier surging (eg Eyjabakkajökull; Clapperton, pers. comm.). This study explores the role that these catastrophic flood events may have in affecting long-term sandur development. It investigates the possibility that although infrequent, these events may have dominated long-term sandur growth, such that the great extent (eg. Skeiðarársandur covers c 900 km²) and thickness (eg. Mýrdalssandur >120m thick, see Thórarinsson & Guðmundsson, 1979; Skeiðarársandur >200m thick, Boulton, 1989; and Markarfljót sandur >250m thick, see Haraldsson, 1981) of the sandar may largely be attributed to a long history of jökulhlaup drainage, possibly since the end of the Lateglacial stade, rather than to normal, seasonal meltwaters draining across the sandur surface in braided river networks (see discussions in Sigbjarnarson, 1973; Björnsson, 1980; Jónsson, 1982; Smith, 1985; and Nummedal et al, 1986).This paper aims to explore the role of jökulhlaup events in sandur formation by concentrating initially on Sólheimasandur (Fig. 1), through examination of the morphology, sedimentology and stratigraphy of the sandur deposits, and development of a model of sandur evolution in areas of jökulhlaup drainage. The second part of the paper tests the model by examining more briefly some preliminary evidence from several other sandur environments each subject to different meltwater populations, including Mýrdalssandur, Skeiðarársandur and some smaller fan deposits (Klifandi and Kotá fans, see Fig. 1).

2. SÓLHEIMASANDUR

2.1. Rationale: Characteristics of jökulhlaup events and expected impact on sandur development

Sólheimasandur forms a 50 km² area of coastal sand and gravel plains extending southwards from Sólheimajökull, a 15-km long outlet glacier of Mýrdalsjökull. The Mýrdalsjökull ice cap is underlain by the volcano Katla, which has been subject to at least 17 eruptions in recorded history (Thórarinsson, 1957; Björnsson, 1975; Einarsson et al, 1980), with each eruption giving rise to a jökulhlaup event. The characteristics of 'Kötluhlaups' are highly distinctive in terms of both the magnitude and duration of the flood hydrograph, and the sediment concentrations borne by the flood flows (Fig. 2).

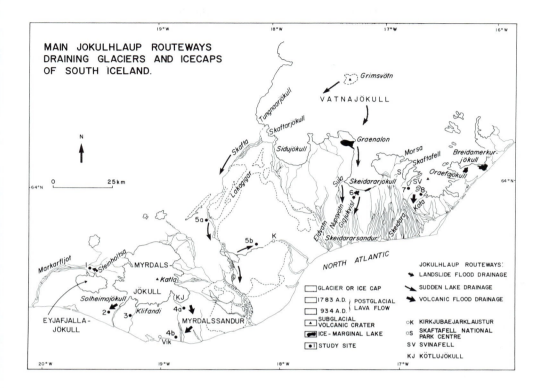

Fig. 1   Major jökulhlaup routeways in southern Iceland, and location of
main study areas. Jökulhlaups are differentiated according to origin and mechanism of
drainage (see legend).

The nature of the flood hydrograph (magnitude and frequency, especially) and
sediment concentrations play a major role in controlling the long-term geomorphic significance of large flood events, particularly through determining the potentials for erosion and
deposition. Jökulhlaups from Katla are typified by an extremely rapid rise to peak flows,
reaching discharges up to $10^5 m^3 s^{-1}$ in a few hours (Björnsson, 1975). In the 1918 jökulhlaup,
for example, flood waters rose to ca $10^5 m^3 s^{-1}$ in only 4-5 hours, and returned to pre-flood
levels within only 24-36 hours (Thórarinsson 1957; Jónsson, 1982). The large volumes of
meltwater appear to be derived not just from ice melt, but by sudden drainage of reservoir
water stored subglacially in the Katla crater (Björnsson, 1975). In addition to high flood
discharges, Katla floods are characterized by extremely high sediment concentrations
associated with abundant supplies of fresh magmatic materials, and particularly of finer-
grained tephra and pumice (Jónsson, 1982). The main flood routeways at Sólheimajökull
follow canyons cut into the palagonite plateau deposits, thereby providing an additional
supply of friable volcanic detritus to the flood flows (Carswell, 1983).

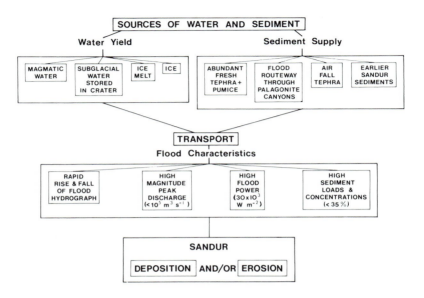

Fig. 2 Simplified model of main hydrologic and sedimentologic
controls on jökulhlaup characteristics at Sólheimajökull.

The high flood powers and abundant sediment supplies associated with Katla floods
are likely to produce a distinctive flood landscape, its long-term preservation depending on
flood recurrence intervals, landscape relaxation times, and the efficacy of non-flood
subaerial processes. At Sólheimasandur, as on most of the Icelandic sandar, two major
meltwater flood populations can be identified (illustrated conceptually in Figure 3 A): first,
normal seasonal ablation flows, and second, infrequent high magnitude jökulhlaup flows.
The relative importance of each flow population in modifying the proglacial landscape
depends as much on the resistance and relaxation time of the initial landscape to imposed
change, as it does on the magnitude of the event. On Sólheimasandur, two contrasting flood
routeways of differing resistance can be identified. Much of the proglacial zone forms an
unconfined coastal plain, rising up to 20m+ above the present meltwater river. Former
jökulhlaup flows were therefore able to expand across an extensive area of unvegetated,
poorly consolidated sands and gravels presenting little resistance to flood flows, and
allowing widespread accumulation of sediments from successive flood events. However,
jökulhlaups occurring since the time of incision into the sandur to form the present river
course (ca 14th-15th centuries AD; Maizels & Dugmore, 1985) have been confined to the
narrow incised channel, causing increased resistance from sandur bluffs, and hence
increased likelihood of removal of successive flood sediments rather than accumulation. In
addition, while the upper plain has been abandoned by meltwater flows since the 14th

century AD, continued meltwater flows along the incised course means that subsequent landscape modification by meltwaters has been possible only in that zone.

This study focuses on the impact of jökulhlaup events on the main sandur, and suggests that while each flood event would have had a major and widespread impact on the sandur surface (Fig. 3A & B), post-flood braided meltwater systems would have had little difficulty in reworking and removing the finer-grained surface materials. Hence, one would expect the sandur to comprise a series of relatively thick, superimposed, fining-upward, waning stage flood deposits, separated (perhaps only locally) by thin, fining-upward braided river gravels (eg see Miall, 1977, 1978; Smith, 1985), incorporating local and volcanically-derived materials to form a heterogeneous mix (Fig. 3C(1) and (2)).

The longer-term impact of the two-population flood regime will reflect not only the history of these events and the progressive seaward progradation of the sandur margin, but also external controls on the system, particularly changes in relative sea-level. It appears, for example, that a rapid fall in relative sea-level, to between -40 and -60m in the early Holocene, associated with rapid isostatic rebound between 11,000 and 9,000 BP, was followed by a less rapid and fluctuating rise to present levels and relatively stable conditions for the remainder of the Holocene (Boulton, pers. comm.; Hansom & Briggs, this volume; Thors, this volume). Over the past 10,000 years one would therefore expect a progressive change in sandur stratigraphy (see Fig. 3C(3)) reflecting the early rise in relative sea-level, followed by the progradation of delta foresets at the sandur edge (stages 2-3 in Fig. 3C(3)), and an overall upward coarsening sediment sequence associated with rising sea-levels and continued progradation (stage 4). The top of the sequence, associated with stable sea-level (stages 6 and 7), but with sandur steepening as proximal aggradation proceeded, would reflect incision of the proximal sandur and washing out of the associated sediments to form units of coarse sediment pulses (stages 8 to 10). The full Holocene sequence would also be expected to be marked by repeated flood depositional sequences, probably with thicker, more frequent glacigenic flood units at the base (associated with rapid melting and short transport distances to the sea), becoming increasingly infrequent, and thinner through the sequence (Fig. 3C(4)). The periodicity of volcanically-generated flood units may have remained relatively constant throughout much of the Holocene (at least until the 14th century AD as the Katla eruption centre migrated to the northeast; Björnsson, 1979), but sediment thickness would be expected to decline as the areal extent of the sandur increased.

2.2 Morphology of Sólheimasandur

Sólheimasandur is composed of a series of spatially distinct outwash surfaces of varying morphological and sedimentological character and age (Figs. 4 and 5). The most distinctive feature of western Sólheimasandur (known as Skógasandur) is a steep, lobate, proximal fan (Skógafan) which extends southwestwards from several former meltwater canyons for ca 2km on to the sandur (Fig. 6). The fan rises ca 15m above the sandur surface, and is dissected by 3 deep channels containing streamlined residual hummocks, mega-ripples, stone lags and scattered, imbricated palagonite boulders (see Figure 6B; Maizels 1987, 1989 a, b).

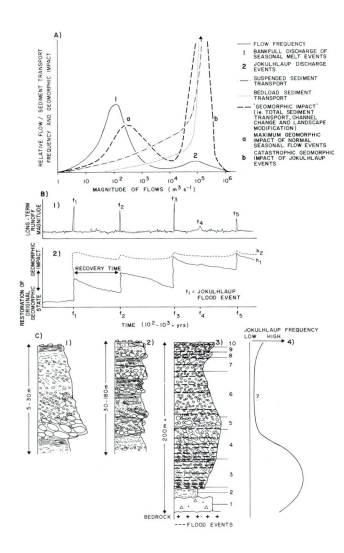

Fig. 3 Simplified models of expected meltwater runoff populations in areas of jökulhlaup drainage, and associated landform and stratigraphic response (see opposite page for detailed annotation).

A. Magnitude-frequency relations of bimodal population of flow events, and likely geomorphic impact (modified from Church, 1988).

B. Long-term meltwater runoff record in areas of repeated jökulhlaup drainage, over a timescale of 102 to 103 years (B1): and likely patterns of geomorphic response to bimodal distribution of flow events (B2).
Figure 3B(2) shows two possible patterns of response:
 h1: Each successive flood event has a major impact on the landscape, although the landscape gradually reverts to its original state during periods between floods. Its final state partly depends on the recovery time (or relaxation time) of the landscape.
 h2: The first flood event has the dominant impact, and subsequent floods and inter-flood flows have little impact on further landscape change.

 C. Expected short-term (C1), intermediate-term (C2) and long-term (C3) stratigraphic record of bimodal population of flow events in areas of jökulhlaup drainage.
 C1: Fining-upward sediment sequence representing waning stage of single flood event.
 C2: Repeated fining-upward sediment cycles representing a series of flood events, with little reworking of sediments during normal flows.
 C3: Holocene sedimentary sequence, reflecting long-term changes in sea-level, ice-marginal position, flood frequency, and progressive sandur progradation and steepening:
Stages - 1: Subglacial till
          2: Glacio-marine sediments
          3: Coarsening-upward, prograding delta foresets
          4: Coarsening-upward delta topsets (sandur surface gravels) associated with sandur progradation
          5: Uniform sediment sequence associated with stable sea-level and ice-margin
          6: Fining-upward sandur gravels associated with progressive ice-marginal retreat
          7: Coarsening-upward sequence associated with progressive proximal aggradation of sandur surface during relatively stable or slightly advancing ice-margin
          8: Influx of coarse sediment associated with incision of steep proximal sandur
          9: As stage 7 (repeated cycle)
         10: As stage 8 (-)
The full stratigraphic sequence is punctuated, and partly composed of, repeated jökulhlaup deposits of varying frequency, thickness and character.

 C4: Long-term variations in jökulhlaup frequency, reaching a primary maximum in the Lateglacial-early Holocene, and a secondary maximum in the recent past (see text for details).

Beyond the edge of the fan lies an extensive area of 'washed' sandur, forming a boulder-strewn lag surface, locally exhibiting transverse and streamlined boulder ridges between networks of incised channels. To the northwest, a relatively abrupt boundary marks the edge of an area of unmodified sandur (Fig. 4), which exhibits low relief, braided palaeochannel systems. The whole of Skógasandur is bounded in the north by the deep (>20m) Hofsá 'Meander Channel', also of probable jökulhlaup origin (Maizels 1989a). To the east of the 'washed' sandur a terrace sequence is found on either side of the Jökulsá river. The most distinctive terraces are the Main Palagonite Terrace, composed almost exclusively of weathered, rounded palagonite cobbles, and the Top Terrace, which forms the most extensive high surface of Sólheimasandur. This top surface descends eastwards to include deposits produced by drainage from the Hólsá and Húsá meltwater routeways (Figure 4), the latter exhibiting fields of highly distinctive, transverse megaripples. The Sólheimasandur complex is bounded on the east by the Klifandi fan, an actively developing, piedmont alluvial fan with no record of Kötluhlaups in historical time.

2.3. Sedimentology of Sólheimasandur

The sedimentology of the individual outwash deposits forming Sólheimasandur is highly complex, and there is scope here only for providing a summary of the dominant sedimentary characteristics and lithofacies sequences observed within the sandur deposits (Table 1). A more detailed discussion of sandur sedimentology is presented elsewhere (see Maizels, 1989a,b; and in prep.).

2.3.1. Lithofacies types

Although Sólheimasandur outwash deposits comprise a wide variety of fines (F), sands (S), gravel (G) and boulder (B) lithofacies types, they are dominated by thick and extensive units of black, pumice granules (GR). The FINES, composed of silts and clays, are relatively uncommon, and are found only locally as laminated units (Fl) up to 20 cm thick, with some units exhibiting small-scale load structures (Fs). The SAND lithofacies types generally comprise thin units (<0.5m) of poorly sorted, small and medium-scale (<0.3m), cross-bedded sands (Sx, St, Sr, Sp) normally exhibiting fining-upward laminae (Suf), and horizontal bedding (Sh). Occasional coarsening-upward laminae, and thin units of massive, structureless sands (Sm) also occur locally. Percentages of silt and clay tend to be consistently low averaging >5%

The GRANULE lithofacies type comprises homogeneous, massive, structureless, uniform, well-sorted, rounded, black pumice/tephra granules (GRm) with D50 averaging between 0.26 and 3.1mm (based on measurements of >200 samples), and constituting only 2.5% silt and clay. Granule units commonly exceed 8m in thickness and spatially form the

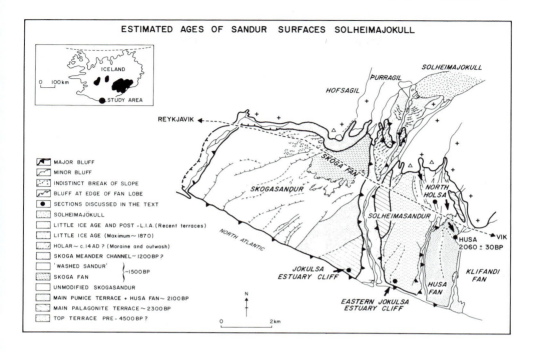

Fig. 4   Morphology of Sólheimasandur and estimated ages of sandur surfaces.

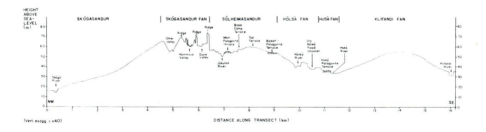

Fig. 5   Topographic profile across Sólheimasandur, from the Skógá river in the west to the Klifandi river in the east.

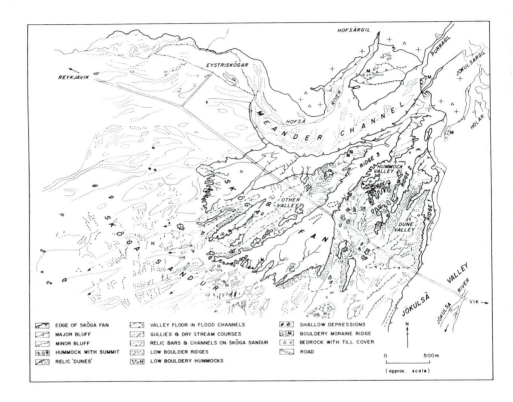

Fig. 6 A : Morphology of the Skóga pumic fan, flood channels and 'washed'
sandur (after Maizels, 1989a).

6B : Flood channel incised into the Skóga pumice fan, exhibiting streamlined residual
hummocks, a scattered palagonite boulder lag, and 'mega-ripples'

Table 1   The dominant lithofacies types present in south
Iceland sandur deposits

FINES (<0.063 mm)

Fl	Laminated silts and clays
Fm	Massive s + c
Fs	S + c with load structure

SANDS (0.063 - 2.0 mm)

Sd	Deformed sand horizons
Sh	Horizontal/plane-bedded sands
Sm	Massive sands
Sp	Plane cross-bedded sands
Sr	Ripple cross-laminated sands
St	Trough cross-bedded sands
Suc	Upward-coarsening sand unit
Suf	Upward-fining sand unit
Sx	Cross bedded sands

GRANULES (2.0 - 8 mm; tephra/pumice)

GRch	Channelled massive granules
GRd	Deformed granule bedding
GRm	Massive, homogeneous granules
GRmb	Massive, pseudo-bedded
GRmc	Massive with isolated clasts
GRmp	Massive with pebble stringers
GRo	Openwork granules
GRruc	Repeated upward-coarsening cycles
GRruf	Repeated upward-fining cycles
GRt	Trough cross-bedded granules
GRuc	Upward-coarsening granule unit
GRuf	Upward-fining granule unit
GRx	Cross-bedded granules

GRAVELS (8 - 256 mm; heterolithic)

Gh	Horizontally bedded gravels
Gm	Massive, clast-supported crudely bedded gravels
Gmi	Massive, clast-supported, imbricated gravels
Gni	Normal-inversely graded gravels
Go	Openwork gravels
Gs	Matrix-supported gravels
Gsi	Matrix-supported, imbricated gravels
Gu	Matrix-supported, unsorted gravels
Guc	Upward-coarsening gravels
Guf	Upward-fining gravels

BOULDERS (>256 mm)

Bi	Imbricated boulders
Blg	Boulder lag
Bsl	Single lithology boulder deposit
Bhl	Heterolithic boulder

most extensive lithofacies group. Locally, some of these homogeneous units may contain isolated larger clasts (GRmc), pebble stringers (GRmp), and a variety of indistinct internal structures including crude 'pseudo'-bedding (GRmb), cross-bedding (GRx and GRt) and channelisation (GRch). Normal (GRuf) and inverse grading (GRuc) is present in some units, and may form repeated graded cycles. In addition, open granular sediments are present locally (GRo), while granular sediments exhibiting deformed bedding (GRd) are found in association with debris flow material.

The GRAVEL lithofacies comprises clasts averaging between 8 and 256 mm, and is dominated by crudely horizontally bedded, massive (Gm), clast-supported, poorly sorted, poorly imbricated (Gmi), subrounded heterogeneous or basalt-rich sandur gravels. The gravel units rarely exceed 1.5m in thickness, and most commonly form a surface lag consisting of a single clast layer. Matrix-supported gravels (Gs) occur locally, and include unsorted (Gu), angular heterogeneous clasts in a clay-rich (<68% silt and clay) or sandy matrix.

BOULDER beds (B) are also found within the sedimentary sequences. Even in distal locations (e.g. at coastal sections; see Fig.4), boulder beds exceed 1m in thickness and comprise strongly imbricated (Bi), clast-supported boulders up to 5m in diameter. Boulder lags (Blg) are also present extensively across the Skóga fan, sandur and terrace surfaces, notably on the Palagonite Terrace (single lithology boulder lag, Bsl), within the flood channels of Skógafan, and across the 'washed' zone of Skógasandur (multiple lithology or 'heterolithic' boulder lag, Bhl).

2.3.2. Dominant lithofacies sequences

The Sólheimasandur outwash deposits are composed of five main vertical lithofacies sequences, each type exhibiting minor variability in lithofacies characteristics and internal structure (Table 2):

LITHOFACIES SEQUENCE A comprises the massive, homogeneous, uniform, well-sorted, fine black pumice granules of lithofacies GRm, GRmc and GRmp, with some evidence of local sub-horizontal 'pseudo-bedding (GRmb). These lithofacies dominate large areas of sandur deposits, exceeding thicknesses of over 4-5m at many sites (e.g. west Holsá site, see Maizels 1989a).

LITHOFACIES SEQUENCE B is composed of thick (>10m) basal sediment units similar to those of sequence A, but capped by an erosional contact overlain by thin (<1m) cross-bedded (GRx and GRt) or channelled (GRch) granule units (sequence B1); horizon-tally bedded (GRh) (Sequence B2); or cross-bedded units overlain by horizontally bedded granular sheets (Sequence B3). Large areas of sandur are dominated by lithofacies Sequence Type B1, which in turn is commonly capped by a gravel or boulder lag. At the eastern Jökulksá estuary site for example (see Fig.4 for location; and Fig.7), over 7m of Types A1 and A2 granular lithofacies (Unit 1 in Fig.7) are overlain by ca 2m of horizontally bedded, repeated upward-fining cycles (GRh ruf - unit 2), scoured by a series of channel structures (GRch - unit 3) up to 3.2m wide and 1.4m deep, infilled with sandur gravels (GRm and GRmi). Unit 4 at this section comprises a 2m thick capping of horizontally bedded, alternating coarse and fine granular laminae (GRh), in which the laminae form 'nested' fining-upward cycles, with secondary fining-upward cycles (ca 10m thick) within each of 4 primary cycles (ca 0.5m thick). None of the granular lithofacies over the 10m thickness of outwash deposits contains more than 4.5% silt and clay. Granules and isolated clasts up to 14cm in diameter, both from units 1 and 2 (Fig. 7A), exhibit some marked upstream imbrication of both a and b axes.

Table 2
The dominant lithofacies sequences present in sandur deposits of south Iceland.

LITHOFACIES SEQUENCE	CHARACTERISTIC STRUCTURES	DOMINANT LITHOFACIES TYPE	MINOR LITHOFACIES TYPE
A	Massive, homogeneous pumice granules	GRm	GRmc, GRmp, Blg
B	Massive, homogeneous, pumice granules, with:		
B1	Cross-bedding at top	GRm→ GRx, GRt	GRch, Blg
B2	Horizontal bedding at top	GRm→ GRh	GRo, Sh, Fl, Fs
B3	Cross-bedding and horizontal bedding at top	GRm→ GRx, GRt→ GRh	GRch, GRo, Sh, Fl, Fs
B4	Basal bedded gravels; and cross-bedding and horizontal bedding at top	GRh→ GRm→ GRx, GRt → GRh	GRch, GRo, Sh, Fl, Fs, Gs, Blg
C	Graded pumice granules or gravel		
C1	Normal Sh, Fl		Gmi→ GRuf→ Suf
C2	Cyclic normal Suf		GRruf
C3	Normal - inverse	Gmi→ GRuf→ GRuc	Suf, Sh
C4	Inverse Suc, Gm		GRuc
C5	Composite inverse-normal Sh, Fl	GRuc→ GRuf	Gh, GRh, Suf, Suc, GRo,
D	Deformed bedding	GRd, Sd	Fs, Gs, Gu
E	Poorly sorted, matrix-supported, heterogeneous mix	Gs, Gu	Gsi
F	Boulder deposits	Blg	Bsl, Bhl, Bi
G	Poorly sorted, clast-supported, rounded heterogeneous gravels	Gmi	Gm, Guf, Sh

See Table 1 for key to lithofacies codes

Fig. 7 Sedimentologic sequence exposed at the eastern Jökulsá estuary site (see Figure 4 for location). Jökulhlaup flow was from left (north) to right.
7A: Section ca 12m high exhibiting Type B vertical lithofacies sequence, with basal GRm lithofacies (>4m thick), overlain by GRuf and GRch units, capped by GRh bedding.

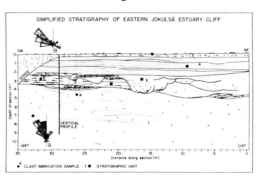

7B: Diagrammatic interpretation of vertical section, indicating main sedimentologic units and imbrication directions.

LITHOFACIES SEQUENCE C comprises a range of graded sediments, again dominated by pumice granule lithofacies, but with only minor proportions of homogeneous, structureless Type A deposits. Type C1 comprises normally graded pumice sand and granules, with a basal layer of basalt clasts, and fining upwards over depths of <1m. Type C2 deposits are

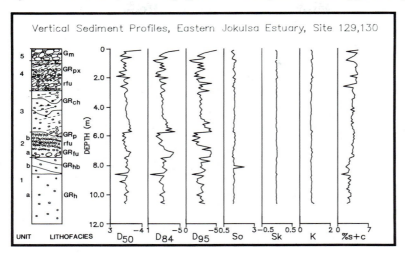

7C: Sedimentologic characteristics of the vertical section at the eastern Jökulsá estuary site. D_{50}, D_{84} and D_{95} = 50th, 84th and 95th percentiles of the sediment size distributions, respectively; So = sorting index; Sk = skewness index; K = kurtosis index; %s+c = percent silt and clay (<0.063 mm) in sample. Stratigraphy is shown in left-hand column. Key to lithofacies codes is given in Table 1.

represented by thicker sequences of cyclic, upward-fining sub-units of C1 sediments. Up to 4 such cycles have been identified within a single unit at the Jökulsá Bridge site (see Fig. 4 for location). Type C3 lithofacies comprises normal-inversely graded granular sediments. At the North Hólsá site, for example (see Fig. 1 for location), coarse, clast-supported, poorly sorted, and moderately well imbricated basal gravel (see unit 2E, Fig. 8) grades upwards into a more homogeneous, matrix-supported granular sub-unit containing isolated clasts exhibiting some degree of imbrication, but with the modal class deviating from that of the lower sub-unit by ca 90 degrees. The uppermost sub-unit (Unit 3G, Fig. 8) forms a second clast-supported sub-unit, comprising moderately well imbricated clasts exhibiting similar modal imbrication direction to that of the basal coarse layer. Finally, Type C4 lithofacies are represented by inversely graded granular sediments, grading upwards from basal fine-grained, well sorted granules either into coarser openwork granules, or into massive, heterogeneous pebble gravels (Gm).

LITHOFACIES SEQUENCE D comprises deformed granule beds, intermixed with disturbed sand and silt beds, and lumps and stringers of diamicton and/or gravels (Sequences E and G, respectively, see below). Deformation structures include steeply dipping shear planes, low angle folds, and evidence of contortion, compaction, compression, thrusting and over-turning. Severely disturbed bedding is most clearly demonstrated at the Húsá Cavity site (see Figs. 1 and 9), where intermixing GRm and GRh lithofacies are folded and contorted around a debris flow deposit (see below).

Fig. 8 Sedimentology of the North Hólsá site (see Fig. 4 for location). Flow was from right (north) to left.

8A: Section ca 4.5 m high, exhibiting basal pre-surge gravels overlain by type C4 lithofacies (normal-inverse grading) of flood surge deposit.

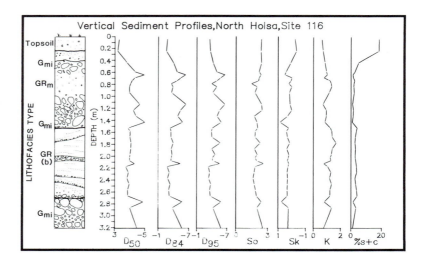

8B: Diagrammatic interpretation of vertical section, indication main sedimentologic units and imbrication directions.

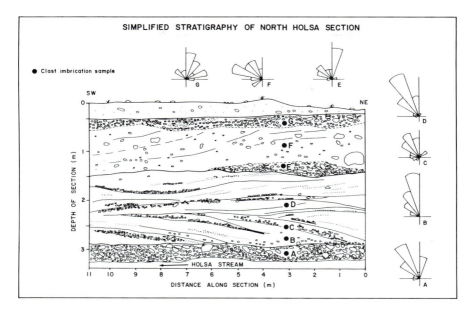

8C: Sedimentologic characteristics of the vertical section at the North Hólsá site (keys given in Table 1 and in caption for Fig. 7C)

LITHOFACIES SEQUENCE E comprises a poorly sorted, matrix-supported, heterogeneous, yellow, clay-rich mix up to 1.9m thick, containing angular to rounded clasts up to 13 cm in diameter (eg. Húsá cavity site; Fig. 9). This lithofacies only occurs locally as discrete lenses or irregular inclusions and lumps within adjacent granular (or deformed granular) lithofacies. Both the a and b clast axes exhibit some crude imbrication.

LITHOFACIES SEQUENCE F is dominated by cobble and boulder deposits, comprising either single lithology clasts (Bsl, normally palagonite) or multiple lithology clasts, (Bhl); it either forms a surface lag (Blg) or is included within the sedimentary sequence. Single palagonite boulders and cobbles up to 2m in diameter are scattered across the floor of the flood channels cutting through the Skóga pumice fan (Maizels, 1987), and large clusters and spreads of weathered palagonites (boulders >7m diameter) form the exhumed Main Palagonite Terrace (Maizels and Dugmore, 1985). At the Skóga sandur coastal cliff, by contrast, a basal horizon contains boulders <0.5m in diameter, of which only 20 per cent are palagonites, overlain by over 6m of shallow, gravel and granule cross-beds.

LITHOFACIES SEQUENCE G is characterized by massive, clast-supported, heterogeneous, crudely imbricated gravels (Gmi), containing rounded to subrounded clasts, and less than 2.5% silt and clay, and exhibiting crude sub-horizontal bedding. Some local normal grading is present, with gravels grading upwards into plane-bedded, or massive coarse/medium sands (Sh and Sm, respectively). This sequence occurs only locally within the Sólheimasandur outwash deposits, and rarely exceeds 1.5m in thickness.

Fig. 9 Sedimentology of the Húsá Cavity site (see Figure 4 for location).
Flow was from right (north) to left.
9A Section ca 5m high, exhibiting basal outwash gravels overlain by interbedded debris
flow and massive granular surge deposits

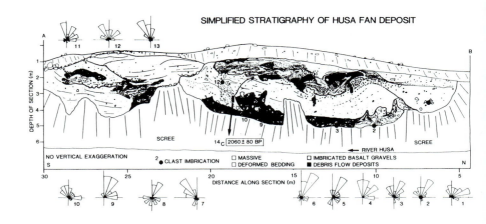

9B Diagrammatic interpretation of the Húsá Cavity section, indicating the main sedimentologic units, imbrication directions, and position of radiocarbon dated birchwood
fragment

2.4 Stratigraphy

The stratigraphy of Sólheimasandur reflects the complex spatial and vertical relationships that exist both between successive depositional units and within individual units whose characteristics show marked lateral variations. The stratigraphy has been determined from morphologic and stratigraphic relationships, together with dating of surfaces and deposits using tephrochronology, lichenometry and ^{14}C dating (summarized in Maizels and Dugmore, 1985; Maizels, 1989a). The evidence available indicates that at least 8 major pumice-producing events can be identified which pre-date the 'Little Ice Age' (Table 3). The oldest

Table 3   Major stratigraphic units and approximate chronology of
Sólheimasandur deposits (pre-'Little Ice Age')

Stratigraphic Unit	Deposit or Event	Approximate date of formation
13	Low Jökulsá terraces (T2 to T5)	Post-1840 AD
12	Main distal terrace (T1, SW Hólar)	ca late C14 AD
11	Hólar/Eystrihólar	ca mid C14 AD
10	Main period of terrace incision	ca C10 AD
9	Hófsá Meander Channel	ca 1200 yrs BP
8b	Skóga Fan flood channels	}
		} ca 1500 yrs BP
8a	Skóga Pumice Fan	}
7	High Jökulsá terraces	}
		} ca 1500-2100 yrs BP
6	Skógasandur	}
5	Palagonite Terrace	2100-2300 yrs BP
4	Balsalt Gravel Sheet Sandur	
3	Pre-Palagonite Channelled Gravels	ca 3500 yrs BP
2	Basalt Cobble Sandur	
1	Lower Pumice Gravels	ca 4500 yrs BP

Chronology based on Dugmore (1989), Jónsson (1982), Maizels and Dugmore (1985),and Maizels (1989a, b)

deposit appears to pre-date ca 4,500 BP; the Palagonite Terrace dates from ca 2,300 BP, the Skóga pumice fan from ca 1,500 BP, the Jökulsá terrace sequence from the 14th century AD, and the Hólar deposit from between ca 1400 and 1700 AD (Fig. 4; Table 3). A recently acquired radiocarbon date from the Húsá fan deposits to the east, (derived from birchwood incorporated within the base of a debris flow deposit; indicates an age of 2,060±80 BP for the Húsá pumice event (see Fig.9) (Maizels, in press). This event may represent the Katla eruption of ca 2,300 BP, which was associated with the formation of tephra layer 'L' (Dugmore, 1989), and hence correlates with the Palagonite Terrace at Sólheimasandur (Maizels, 1989a).

Sólheimasandur therefore comprises deposits of a variety of ages, spanning at least the past 2,100 years, and possibly the past 4,500 years (Maizels and Dugmore, 1985). In addition, only three non-pumice events, comprising heterogeneous pebble gravels of lithofacies Type G, have been identified in the stratigraphic sequence at Sólheimasandur. These gravel units are relatively thin, representing only 15 per cent of the total observed sandur thickness.

2.5. Palaeohydrologic and palaeohydraulic interpretation of Sólheimasandur sediment sequences

2.5.1. Palaeoflow interpretation of lithofacies sequences

The dominant lithofacies sequences at Sólheimasandur, comprising thick units of granular pumice sediments, are not characteristic of normal braided meltwater facies. They are interpreted as deposits produced by high magnitude flows containing high concentrations of sediments in suspension, fully dispersed within the fluid-sediment mix. Such flows are typical of hyperconcentrated, cohesionless grain flows which occur during flood surges (eg. Fisher, 1971; Lowe, 1976; Costa, 1984, 1988; Nemec & Steel, 1984). The vertical lithofacies sequences at Sólheimasandur also provide a valuable key to understanding not only the nature of the flow event (i.e. type of flow and sediment concentrations), but also the nature of the flood hydrograph. The dominant lithofacies sequence is Type B, which is interpreted as representing a 2-stage flood event. The thick basal unit of massive sediments is considered to represent the main flood surge, containing high concentrations (40-60%) of newly erupted, fine-grained pumice (cf. Pierson, 1981; Smith, 1987; Costa, 1988). The deposits containing 'pseudo'-bedding and pebble stringers suggest that flow was character-ised by successive flow pulses, with indistinct shear planes forming between each pulse (where events may have been pulsed either temporally or spatially during the event; cf. Larsen & Steel, 1978; Cossey & Ehrlich, 1979; Pierson, 1980,1981; Postma, 1986). The pebble stringers may have formed in two ways: at the top of a local surge, being supported by dispersive and buoyant forces (Bagnold, 1954; Enos, 1977), or at the base of a unit where shear forces caused imbrication to develop, but dispersive stresses were insufficient to raise clasts into the flow. The uppermost units of the sequence, comprising a basal erosion surface capped by smaller-scale bedform structures, appear to represent the later, more fluid stages or 'runout flows' (Scott, 1988) of the flood event. These developed when sediment supply had become relatively depleted, when dewatering was occurring from sediments deposited further up-sandur, and as flow depths and velocities were progressively declining (cf. Pierson, 1980). Thus, the dominant lithofacies sequence occurring on Sólheimasandur represents the record of sedimentation and scour during both the rising and falling limits of the flood hydrograph.

The more minor vertical lithofacies sequences (Type A, and Types C to G) represent local variations in flow conditions, sediment concentrations, flood pathways, and rates of dewatering. Type A sediments, for example, appear to represent a truncated form of a Type B sequence, suggesting that late stage flood flows were sufficiently powerful to prevent formation of (or preservation of) migrating bedforms, and removed all sediment to more distal reaches. The graded and cyclically graded sediments of sequence Type C also reflect local variations and pulsed flows, with normal grading associated with lower sediment concentrations and internal dispersive stresses, compared with more hyperconcentrated, poorly sorted flows producing sequences exhibiting inverse and normal-inverse grading (cf Carter, 1975; Pierson, 1986; Scott, 1988).

Type D deposits include both massive and bedded jökulhlaup pumice granules which have become disturbed through intermixing with a contemporaneous debris flow (Type E deposits). At the Husá Cavity site, for example (see Fig. 9), the upward shearing of a basal debris flow, and associated overfolding, deformation and inclusions in the adjacent granular sediments, suggest that a relatively slow debris flow (generated from local slopes) was overridden by the more rapidly flowing, hyperconcentrated, main flood surge (cf. Pierson, 1980) (Fig. 9). Type F sediments, where composed of single lithologies, are of volcanically-generated jökulhlaup origin, and represent the clastic fraction of the flood sediment load, scoured from canyon walls along the flood routeway, and transported on the surface of the flow as palagonite boulders held up by buoyant and dispersive forces. The heterolithic boulder deposits are also likely to be of flood origin, although forming in more distal zones and therefore incorporating more heterogeneous materials from large areas across the sandur.

Finally, lithofacies sequence G, which comprises clast-supported, heterogeneous, crudely imbricated, subrounded gravels, locally interbedded with sand lenses, is interpreted as a braided river deposit, following the criteria for braided river facies described by Bluck (1974, 1974, 1982, 1987), Miall (1977, 1978), Boothroyd and Nummedal (1978), Rust (1978), and Smith (1985). These deposits are likely to represent the downstream migration, vertical accretion, and local preservation of longitudinal gravel bars, channel-lag gravels and minor channel fills. The detailed palaeohydraulic significance of individual sedimentary sequences will be discussed more fully in a separate paper (and see Maizels, 1989a, b).

2.5.2. Palaeohydrology of Sólheimasandur flood events

The sedimentologic and stratigraphic evidence indicates clearly that there are two distinctive sedimentary facies assemblages, representing the two dominant flow populations:

facies assemblage 1, which comprises the wide range of pumice lithofacies types (Types A to F) produced by high magnitude jökulhlaup events; and facies assemblage 2, which consists of the heterogeneous, rounded, clast-supported gravels (lithofacies Type G), representing the 'normal', seasonal ablation meltwater events.

Modelling of palaeodischarge amounts for nineteen of the sandur deposits (see Maizels, 1987, 1989a, b) indicates the clear hydrologic distinction between the estimated peak flows of the jökulhlaup flood population associated with Holocene volcanic eruptions of Katla (with discharge ranging from 25 to 330 x 10³ m³ s⁻¹), and those of the post-1840, 'Little Ice Age' floods (with discharges of <2.5 x 10³ m³ s⁻¹) (Fig. 10A and 10B; and cf. with Fig. 3A). The post-1840 floods are associated with the low terrace surfaces bounding the

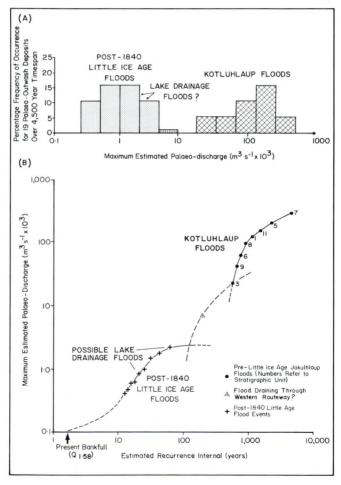

Fig. 10 Magnitude-frequency relation for Holocene flood populations at Sólheimajökull (based on palaeoflow calculations in Maizels, 1989a).

present Jökulsá river, are characterized by abandoned braided channel systems, and are unrelated to any known volcanically-generated flood events. However, it is possible that some of the earlier terrace deposits may have been associated with sudden drainage of a small, marginal lake. This lake would have been dammed up in a northern tributary valley (the Jökulsárgil) during a period of ice advance, with lake drainage floods possibly continuing until the late 1930's (Björnsson, 1976).The maximum volume of the Jökulsárgil lake could have been between ca 50 and 150 x 10^6 m³ of water, such that maximum instantaneous flood discharges during sudden lake drainage is likely to have ranged between ca 1,000 and 2,200 m³ s^{-1} (after Clague, 1973). Such lake drainage events would therefore represent the upper range of the post-1840 flood events recorded in the Sólheima-jökull terrace sequence (Fig.10). These terraces are likely to be composed of coarse-grained, locally-derived cobble and boulder gravels. A third type of flood event of intermediate magnitude (discharges of approx. 4 to 7.5 m³ s^{-1}; Fig. 10A & 10B) may represent a separate population of floods associated with drainage through a more westerly routeway through the plateau during a period of ice advance in the eighteenth century. The peak discharges estimated for the jökulhlaup and lake drainage floods are between ca 10^6 and 10 times as great, respectively, as the normal bankfull flows of ca 100 m³ s^{-1} in the present Jökulsá á Sólheimasandi.

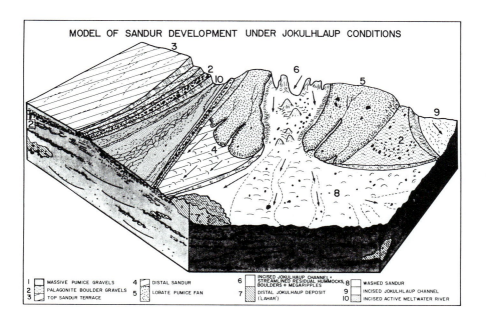

Fig. 11 Model of sandur development in an area of jökulhlaup drainage, with recent glacier entrenchment and meltwater incision.

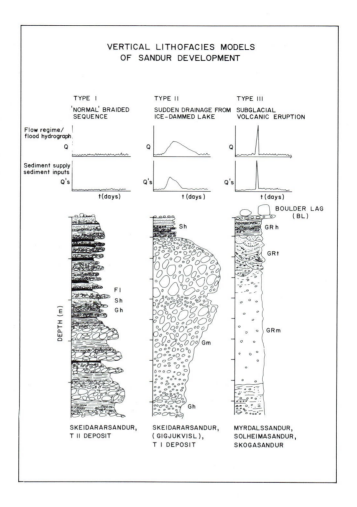

Fig.12 Vertical lithofacies models of sandur development in areas of different characteristic meltwater flow regimes:

TYPE I SANDUR: sandur subject to normal, ablation-controlled, seasonal meltwater regime,

TYPE II SANDUR: sandur subject to drainage from an ice-dammed or subglacial lake, exhibiting a flood hydrograph extending over periods of days to weeks,

TYPE III SANDUR: sandur subject to sudden 'catastrophic' floods generated by subglacial volcanic eruptions, exhibiting highly peaked hydrograph, high sediment concentrations, and flood duration of a few hours.

2.5.3. Geomorphic significance of different flood populations

Although the high magnitude jökulhlaup events occur only rarely, exhibiting recurrence intervals of between 550 and 4,500 years, compared with ca 10-130 years for the most recent events (Fig. 10B), it is the former which have dominated the morphological, sedimentological and stratigraphic evolution of Sólheimasandur (cf. Fig. 3C). At least 85 per cent of the observed sandur thickness (<20m in proximal zones), and 82 per cent of the sandur surface area, appear to be composed of volcanically-generated jökulhlaup sediments. The highly significant impact of jökulhlaup events on the origin and evolution of Sólheimasandur suggests the need for an alternative model to the traditional 'braided river models' of sandur development (eg. Miall, 1977, 1978; Bluck 1974; Rust, 1978; Boothroyd and Nummedal, 1978). The models invoking repeated, upward-fining cycles of Gm, Gt, Sp and Fl lithofacies sequences, for example (see Fig. 3A and 3B), do not describe the Sólheimasandur deposits. A simple schematic landform model for Sólheimasandur is proposed here (Fig. 11), representing a sandur subject to repeated, volcanically-generated jökulhlaup events in an area of glacier entrenchment and river incision. A generalised vertical lithofacies model is also proposed, representing accumulation of fine-grained sediment from hyperconcentrated fluid flows during successive stages of each jökulhlaup event (Fig.12).

As a means of testing the wider applications of this new model of sandur origin and evolution, the remainder of this paper presents some brief and preliminary results of landform and lithofacies sequences found at five other sandur sites in south Iceland (sites 2-7 from west to east on Fig. 1). Details of the geomorphology and palaeohydrology of these other sites will be presented in a separate publication.

3. TESTING THE SÓLHEIMASANDUR MODEL: MORPHOLOGY AND LITHO-FACIES SEQUENCES AT OTHER SANDUR SITES

3.1 Klifandi Fan

The Klifandi fan abuts, and has partially buried, the eastern margin of the Húsá flood deposits (Figs. 1 and 5). The catchment of Klifurárjökull lies south of the Katla crater rim, such that there are no records of large volcanically-generated jökulhlaups affecting the hydrologic regime of the Klifandi meltwater stream during at least the past 6,800 years (Dugmore, 1989). The Klifandi fan, unlike the Sólheimasandur outwash, forms a low relief, undissected alluvial fan, in which the proximal terrace sequence, comprising 2 main terraces, extends over an elevation range of only 3.75m. The surface of the fan is marked by extensive networks of steeply graded (with a mean slope of 0.021), dry, braided channels, abandoned only since ca 1940 AD (i.e. since construction of artificial embankments on the present river) (Maizels et al, 1987). The distribution of palaeochannels indicates that, prior to the 1940's, the Klifandi River had regularly migrated across the whole fan surface, and that this pattern of river behaviour had apparently remained uninterrupted by any significant erosional episodes, resulting in long-term, progressive aggradation of the fan.

Analysis of sandur sedimentology indicates that the sandur material is not significantly different from that found in the present river channel. Boulder sizes rarely exceed 30 cm diameter either on the sandur surface or in the channel (Maizels et al, 1987), while stratigraphic sections exhibit 2-3 m of poorly bedded, crudely imbricated, basalt-rich, rounded, clast-supported gravels in a coarse sandy matrix (Fig. 13). The Klifandi sandur sediments are interpreted as Type G lithofacies, and hence characteristic of 'normal' seasonal drainage in a braided river environment.

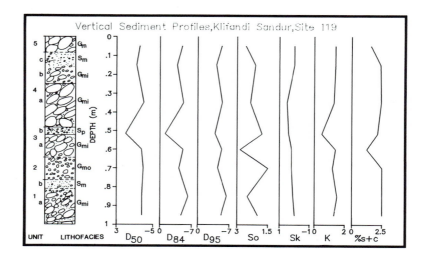

Fig. 13 Sedimentology of distal deposits on Klifandi alluvial fan. The sedimentological characteristics represent a vertical section ca 2m high, exhibiting alternating sand and gravel beds (Sm, Gm especially). The gravels comprise heterogeneous, sub-rounded, clast-supported materials characteristic of lithofacies facies Type G.

3.2 Mýrdalssandur

Mýrdalssandur represents the major jökulhlaup routeway for floods generated by eruptions of Katla (see Fig.1). The last major jökulhlaup occurred in 1918, while a minor event occurred in 1955 (Thórarinsson, 1957; Jónsson, 1980, 1982). The 1918 flood appears to have reached a peak discharge of c $1.5m \times 10^6 m^3 s^{-1}$ within 4 to 5 hours of the start of the flood, declining to pre-flood levels within 24 hours. Peak flow depths ranged from c 70 m in proximal areas to c 10 m near the south coast, with flows having travelled c.15-20 km from the ice-margin (Jónsson, 1980). Large volumes of freshly generated pumice and ice were transported during this event, such that >8m of new sandur material accumulated in distal zones, while the coastline was extended by up to 4 km offshore. Jónsson (1982) considered this Kötluhlaup to have been a 'volcano-glacial debris flow' or 'lahar' event, characterised particularly by hyperconcentrated flows, and dispersive stresses acting to transport ice and boulders along the flow surface. The 1955 event reached a measured peak discharge of ca 3000 $m^3 s^{-1}$ within one hour of its visible start, and was largely confined to the present channel zone (subsequently modified by bridge construction).

Mýrdalssandur exhibits a wide variety of landforms and sediments. The sandur surface is locally dissected by active meltwater streams, both flowing across the sandur surface and emerging as distal gully systems. In addition to terrace bluffs up to ca 18 m high, the surface is marked by extensive (<3km long), streamlined, remnant boulder bars associated with ice block 'crater-kettles', and fields of remnant, streamlined pendant bars downstream of bedrock obstacles (lava knob outcrops). Sandur sediments exposed in a 8m high, distal gully section forms a highly distinctive sequence. The distal sandur at this site is composed entirely of fine-grained, black pumice granules (Fig. 14). The basal 1.5m comprises crudely bedded, poorly sorted, granular gravels, overlain by 3.7m of massive, structureless, well-sorted pumice granules, exhibiting indistinct subhorizontal laminations towards the base and top of the unit. This massive unit is truncated at the top by an erosional contact, capped by 1.6m of trough cross-bedded granules and finally, 1m of horizontally bedded, pumice sands and granules (Fig. 14). This site represents a clear example of lithofacies Sequence B, although in this case a crudely bedded, basal unit can also be identified. This sequence is interpreted as a 3-stage jökulhlaup event comprising pre-surge, surge, and post-surge stages (cf. Scott, 1988), the vertical sequence being closely matched to the known flow characteristics of the 1918 jökulhlaup. The sequence therefore also provides valuable confirmation of the interpretation of the Sólheimasandur model in terms of successive jökulhlaup events generated by subglacial eruptions of Katla.

3.3 Skeiðarársandur (Gigjúkvísl)

Gigjúkvísl forms a major drainage outlet from the southern margin of Vatnajökull (Fig. 1), and therefore acts as one of the routeways used by catastrophic floods from the subglacial, geothermally active area of Grimsvötn (Tómasson, 1974). Grimsvötn has a well-recorded history of sudden drainage since the 12th century AD, with about one flood per decade occurring over the period 1600 to 1934 AD (Björnsson, 1988). The largest of the more recent of these floods produced a peak discharge of ca 40,000 $m^3 s^{-1}$. Flood peaks have been declining over the past 30-40 years (since the late 1930's), partly in association with more frequent flooding (every 3 to 6 years), and partly as a result of increased geothermal activity in the local Grimsvötn area (Björnsson, 1988). Since 1954, at least 16 'Skaftahlaups' (draining the west lobe of Vatnajökull; see Fig.1) have been recorded, the maximum reaching a discharge of only ca 1,500 $m^3 s^{-1}$ (Rist, 1983). Grimsvötn floods are characterized by a relatively prolonged jökulhlaup hydrograph, with rising flows extending over a period of 7-10 days followed by rapid recession to normal flows (Thórarinsson, 1957). In addition,

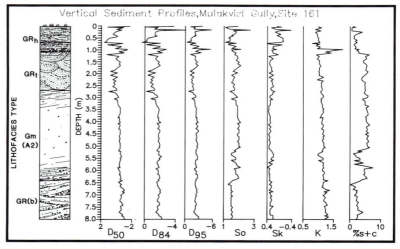

Fig. 14 Sedimentology of distal deposits at Mýrdalssandur (Múlakvísl, eastern gully).
A. Vertical section ca 8m high, exhibiting lithofacies sequence Type
B3, with basal bedded, pre-surge granules (unit 1), overlain by thick, massive Type A
granules (unit 2), capped by trough cross-bedded (unit 3) and horizontally-bedded (unit
4) granules.
B. Sedimentologic characteristics of the vertical section at
Múlakvísl eastern gully on Mýrdalssandur.

Grimsvötn hydrographs often reveal two discharge peaks, an earlier one related to melting of the basal ice and ice tunnel, and a later, short-lived one related to the main eruptive event (Tómasson, 1974). Jökulhlaup flows on Skeiðarársandur contain high suspended sediment loads, with values of up to 7 x 10⁴ kg of sediment being recorded in flows of 2 x 10³ m³ s⁻¹, representing a peak sediment concentration of ca 35%, during the 1972 Skeiðará flood (Tómasson, 1974; Tómasson et al, 1980).

The sandur surface adjacent to the Gigjúkvísl gap, where the river cuts through the 'Little Ice Age' moraine complex, lies some 30m above the present river (Galon, 1973). The proximal sandur forms a composite surface, with an upper surface clearly extending beneath the outer 'Little Ice Age' moraine ridge (see Galon, 1973; Klimek, 1973), forming the TI surface. The lower surface, TII, is cut into the TI deposit, and forms the main upper sandur. It is mantled by extensive, braided palaeochannels that originate from a more easterly gap in the 'Little Ice Age' moraine ridge.

Fig. 15 Sedimentology of proximal deposits at Skeiðarársandur. Gully in foreground exhibits thick, coarsening-upward cobbel gravels, fining towards the top, capped by horizontally laminated sands and silts (Upper TI surface). In the distance, the Lower TII surface is cut into the edge of the TI deposit, and exhibits >4.5 m of repeated upward-fining cycles of gravel-sand-silt. The TII surface is marked by an extensive braided palaeochannel network.

Sedimentology of the older, TI deposit is quite distinctive from that described earlier from the Kötluhlaup and non-jökulhlaup sandar. The deposit comprises over 10m of coarsening-upward, heterogeneous, subrounded, clast-supported, pebble and cobble gravels, marked by a thinner, coarsening-upward unit towards the top. The uppermost 2-3 m of the sequence comprises a fining-upward sequence of cobble and pebble gravels (Gm) (Fig.15). The gravels are overlain by up to 10m of alternating fine- and coarse-grained silts, sands and fine gravel. This sequence is interpreted as a 'type C5' jökulhlaup deposit (composite inverse-normal grading; see Table 2) in which, as at Sólheimasandur and Mýrdalssandur, the sedimentary record of both the rising and falling limbs has been preserved. The main coarsening-upward cobble-gravel sequence is interpreted as the main rising limb sequence associated with the first discharge peak (ice melt stage); the secondary, upper, coarsening-upward sequence as the second discharge peak (eruptive stage); and the topmost fines as the waning stage flows. Thick, coarsening-upward cobble-pebble gravel sequences have also been interpreted as representing rising-stage jökulhaup flows in Pleistocene deposits (eg the Wabash valley, see Fraser and Bleuer, 1988).

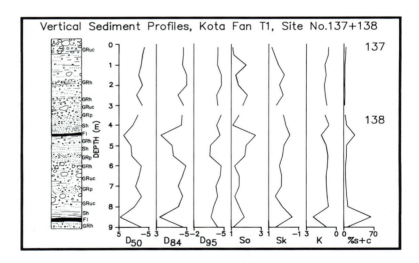

Fig. 16 Sedimentology of the top-most jökulhlaup deposits in the proximal zone of the Kotá fan, Öræfajökull.

Sedimentology of the lower, TII sandur deposit contrasts significantly with that of TI. The upper 4.5 m of this deposit comprises repeated, upward-fining cycles of gravels, sands and silts (Fig. 16), similar in character (i.e. Gm, Sh, Fl) to the 'normal' braided river facies (see summaries in Miall, 1977, 1978; Smith, 1985). The close association between these Type G gravels and the braided palaeochannel systems on the sandur surface suggest that the uppermost part of the TII surface at least, is derived from waning stage aggradation, probably representing numerous diurnal periodic meltwater events. The stratigraphic aggradation, probably representing numerous diurnal periodic meltwater events. The stratigraphic evidence from the Gigjúkvísl sediment sequences suggests that Skeiðarársandur is largely composed of thick Type C5 jökulhlaup sediments, capped by a veneer of non-flood, braided river gravels (lithofacies Types G and C).

3.4 Svínafellssandur, Öræfajökull

Svínafellsjökull is an outlet glacier of Öræfajökull which has erupted twice during historic time (1362 and 1727 AD), although Svínafellssandur is believed to have escaped the effects of any jökulhlaup drainage. The outwash from Svínafellsjökull forms a series of gravel and boulder terraces (Thompson and Jones, 1986), comprising heterogeneous, clast-supported, rounded to subrounded, moderately imbricated cobble and boulder gravels, containing <2.5% silt and clay, locally interbedded with plane-bedded and structureless sands. This lithofacies sequence closely resembles that of Type G, interpreted as characterizing a braided, non-jökulhlaup, river system.

3.5 Kotá fan

Kotájökull descends from the slopes of the subglacial volcano Öræfajökull, which generated catastrophic jökulhlaups during the two historic eruptions (1362 and 1727 AD). The proglacial zone of the Kotá is marked in the south by a high and distinctive sequence of up to 8 terraces, the highest extending to c.30m above the present river in the proximal reach. The northern flood route is marked by an extensive boulder field, comprising large kettle holes, boulder ridges and bars, and dissected, irregular topography. The sediments composing the terrace sequence comprise a thick, basal, poorly sorted, matrix-supported mix, described by Thompson and Jones (1986), and apparently similar to sequence Type E (i.e. a debris flow deposit). The uppermost 9m is composed of crudely horizontally bedded, alternating coarse and fine laminae (with silt and clay content ranging from 2 to 64%), capped by a coarse-grained gravel unit <30cm thick (Fig. 16), containing isolated palagonite clasts. This sequence is interpreted as a composite Type E/B sequence, with the pre-surge/early surge event represented by the basal debris flow deposit (Type E), overlain by deposits produced by more fluidal, later stage flows of the jökulhlaup.

4. CONCLUSION

The evidence available suggests that the sandur of south Iceland do not owe their origin exclusively to the regular seasonal activity of braided meltwater rivers. Instead many sandar are the product of infrequent, catastrophic jökulhlaup events, caused by subglacial volcanic eruptions (eg Katla, Öræfajökull), and/or by the emptying of ice-contact lakes (eg. Grimsvötn). The contrasts in flood hydrology, sediment supply and sediment concentration associated with these different populations of flow events have produced distinctive contrasts in sandur sedimentology. Hence, three generalised models of sandur sedimentology can be recognized, representing the contrasting response of the sandur environment to different jökulhlaup conditions: Type I Sandur, associated with normal braided river activity, Type II Sandur with prolonged, high magnitude drainage from ice-contact lakes, and Type III Sandur with sudden, catastrophic, sediment-rich drainage during a subglacial volcanic eruption (Fig. 12, Table 4).

TABLE 4
Dominant lithofacies sequences of three characteristic sandur types, south Iceland.

SANDUR TYPE	DOMINANT LITHOFACIES SEQUENCE	SECONDARY LITHOFACIES SEQUENCE	DOMINANT LITHOFACIES TYPE	SECONDARY LITHOFACIES TYPE	DOMINANT MELTWATER REGIME AND DEPOSITIONAL ENVIRONMENT
TYPE I	G	E, F	Gmi	Sh, Sm, Sx, Guf, Fl	Normal, seasonal meltwater flows in braided river environment
TYPE II	C5	E, F, G	Guc, Guf	Gh, Sh, Suf, Fl	Jökulhlaup drainage from ice-dammed or sub-glacial lake ('Limno-glacial' flood)
TYPE III	B3, B4	A, B1, B2, C1 - C4, D, E, F	GRm→ GRx→ GRh	GRch, GRo, GRni, GRd, GRuc, GRmb, GRmp, GRmc	Jökulhlaup drainage during subglacial volcanic eruption ('Volcano-glacial' flood)

See Tables 1 and 2 for key to lithofacies codes

Jökulhlaup events therefore appear to have played a significant role in controlling the formation of many Icelandic sandar. As the oldest jökulhlaup deposits on Sólheimasandur have been dated to 4,500 BP, and those of Skeiðarársandur to pre-18th century (Galon, 1973; Klimek, 1973), although possibly dating from earlier in the Holocene, it appears that jökulhlaup events have at least dominated the later Holocene evolution of the Icelandic sandar. The extent to which jökulhlaup events have affected long-term sandur development since the Lateglacial can best be assessed by a programme of seismic stratigraphic work on the sandar combined with analysis of the off-shore sedimentary record. At present it does appear that the sandar of southern Iceland largely represent the outcome of a long history of repeated, catastrophic, sediment-rich jökulhlaup events.

REFERENCES

Bagnold, R.A. (1954) ' Experiments on a gravity-free dispersion of large solid spheres in a Newtonian fluid under shear,' Proceedings of the Royal Society of London Series A 225, 49-63.

Björnsson, H. (1975) ' Subglacial water reservoirs, jökulhlaups and volcanic eruptions', Jökull 25, 1-14.

Björnsson, H. (1976) ' Marginal and supraglacial lakes in Iceland', Jökull 26, 40-50.

Björnsson, H. (1980) 'Glaciers in Iceland', in Geology of the European Countries, I.G.C., 203-209.

Björnsson, H. (1988) 'Hydrology of Ice Caps in Volcanic Regions', Visindafélag Íslendinga, Societas Scientarium Islandica Rit. XLV, 139pp.

Bluck, B. (1974) 'Structure and directional properties of some valley sandur deposits in southern Iceland', Sedimentology 21, 533-554.

Bluck, B. (1979) ' Structure of coarse - grained braided alluvium', Transactions of the Royal Society of Edinburgh 70, 181-221.

Bluck, B. (1982) 'Texture of gravel bars in braided stream', in R.D.Hey, J.C.Bathurst and C.R.Thorne (eds.) Gravel-bed Rivers, Wiley & Sons Ltd, Chichester, 339-355.

Bluck, B. (1987) 'Bed forms and clast size changes in gravel-bed rivers', in K.S.Richards (ed.) River Channels, Environment and Process, Blackwell, Oxford, 159-178.

Boothroyd, J.C. and Nummedal, D. (1978) 'Proglacial braided outwash: A model for humid alluvial-fan deposits', in A.Miall (ed.) Fluvial Sedimentology, Canadian Society of Petroleum Geologists Memoir 5, 641-668.

Carswell, D.A. (1983) 'The volcanic rocks of the Sólheimajökull area, southern Iceland', Jökull 33, 61-71.

Carter, R.M. (1975) 'A discussion and classification of subaqueous mass-transport with particular attention to grain-flow, slurry-flow and fluxoturbidites', Earth Science Reviews 11, 145-177.

Church, M. (1988) 'Floods in cold climates', in V.R. Baker, R.C. Kochel and P.C.Patton (eds.) Flood Geomorphology, Wiley, Chichester, 205-229.

Clague, J. (1973) 'Sedimentology and paleohydrology of Late Wisconsin outwash, Rocky Mountain Trench, South-eastern British Columbia', in A.V.Jopling and B.C. McDonald (eds.) Society of Economic Paleontologists and Mineralogists Special Publication No. 23, Tulsa, 223-237.

Cossey, S.P.J. and Ehrlich, R. (1979) 'A conglomerate, carbonate flow deposit, northern Tunisia: a link in the genesis of pebbly mud-stones', Journal of Sedimentary Petrology 49, 11-22.

Costa, J.E. (1984) 'Physical geomorphology of debris flows', in J.E.Costa and P.J.Fleisher (eds.) Developments and Applications of Geomorphology, Springer-Verlag, New York, 268-317.

Costa, J.E. (1988) 'Rheologic, geomorphic, and sedimentologic differentiation of water flood, hyperconcentrated flows, and debris flows', in V.R.Baker, R.C.Kochel & P.C.Patton (eds.) Flood Geomorphology, Wiley, Chichester, 113-122.

Dugmore, A.J. (1989) 'Tephrochronological studies of Holocene glacier fluctuations in south Iceland', in J.Oerlemans (ed.) Glacier Fluctuations and Climatic Change, Kluwer Academic Publishers, Dordrecht, 37-55.

Einarsson, E.H., Larsen, G., and Thórarinsson, S. (1980) 'The Sólheimar tephra layer and the Katla eruption of 1357', Acta Naturalia Islandica 28, 24pp.

Enos, P. (1977) 'Flow regime in debris flow', Sedimentology 24, 133-142.

Fisher, R.V. (1971) 'Features of coarse-grained, high-concentration fluids and their deposits', Journal of Sedimentary Petrology 41, 916-927.

Fraser, G.S. and Bleuer, N.K (1988) 'Sedimentological consequences of two floods of extreme magnitude in the late Wisconsonian Wabash Valley', in H.E.Clifton (ed.) Sedimentologic Consequences of Convulsive Geologic Events, Geological Society of America Special Paper 229, 111-125.

Galon, R.(1973) 'Geomorphological and geological analysis of the proglacial area of Skeiðarárjökull, central section', Geographica Polonica 26, 15-56.

Haraldsson, H. (1981) 'The Markarfljót sandur area, S. Iceland: Sedimentological, petrographical and stratigraphic studies', Striae 15, 1-65.

Hine, A.C. and Boothroyd, J.C. (1978) 'Morphology, processes and recent sedimentary history of a glacial-outwash plain shoreline, southern Iceland', Journal of Sedimentary Petrology 48, 901-920.

Hjulström, F. (1952) 'The geomorphology of the alluvial outwash plains (sandurs) of Iceland and the mechanics of braided rivers', Proceedings of the 17th International Geographical Congress, Washington, 337-342,

Jónsson, J. (1982) 'Notes on the Katla volcanoglacial debris flows', Jökull 32, 61-68.

Kjartansson, G. (1967) 'The Steinholtshlaup, central-south Iceland on January 15th, 1967', Jökull 17, 249-262.

Klimek, K. (1973) 'Geomorphological and geological analysis of the proglacial area of Skeiðarárjökull. Extreme eastern and extreme western sections', Geographica Polonica 26, 89-113.

Krigström, A. (1962) 'Geomorphological studies of sandur plains and their braided rivers in Iceland', Geografiska Annaler 44, 328-346.

Larsen, V. and Steel, R.J. (1978) 'The sedimentary history of a debris-flow dominated, Devonian alluvial fan - a study of textural inversion', Sedimentology 25, 37-59.

Lowe, D.R. (1976) 'Grain flow and grain flow deposits', Journal of Sedimentary Petrology 46, 188-199.

Maizels, J.K. (1983) 'Lichenometry and paleohydrology of terraced sandur deposits, Sólheimajökull', in Geomorphological and Environmental Studies in Southern Iceland, Report of Aberdeen University Iceland Expedition 1983, Department of Geography, University of Aberdeen, 40-50.

Maizels, J.K. (1987a) 'Modeling of paleohydrological change during deglaciation', Geographie Physique et Quaternaire 40, 263-277.

Maizels, J.K. (1987b) 'Large-scale flood deposits associated with the formation of coarse-grained, braided terrace sequences', in F.G.Ethridge, A.M.Flores and M.D.Harvey (eds.), Recent Developments in Fluvial Sedimentology, Society of Economic Paleontologists and Mineralogists, Tulsa, 135-148.

Maizels, J.K. (1989a) 'Sedimentology, paleoflow dynamics and flood history of jökulhlaup deposits: palaeohydrology of Holocene sediment sequences in southern Iceland sandur deposits', Journal of Sedimentary Petrology 59, 204-223.

Maizels, J.K. (1989b) 'Sedimentology and palaeohydrology of Holocene flood deposits in front of a jökulhlaup glacier, south Iceland', in K.Bevan and P.Carling (eds.) Floods: Hydrological, Sedimentological and Geomorphological Implications, Wiley and Sons, Chichester, 239-251.

Maizels, J.K. (in press) 'A radiocarbon date from Holocene sandur deposits in areas of jökulhlaup drainage, southern Iceland', Radiocarbon.

Maizels, J.K. and Dugmore, A.J. (1985) 'Lichenometric dating and tephrochronology of sandur deposits, Sólheimajökull, southern Iceland', Jökull 35, 69-78.

Maizels, J.K., Fyfe, G. and Dugmore, A.J. (1987) 'Geomorphic criteria for distinguishing jökulhlaup and non-jökulhlaup sandur deposits, south Iceland', Unpublished field report, Aberdeen University, Department of Geography, 25pp.

Miall, A. (1977) 'A review of the braided river environment', Earth-Science Reviews 13, 1-62.

Miall, A. (1978) 'Lithofacies types and vertical profile models in braided river deposits: a summary', in A.D.Miall (ed.) Fluvial Sedimentology, Canadian Society of Petroleum Geologists Memoir 5, Calgary, 597-604.

Nemec, W., and Steel, R.J. (1984) 'Alluvial and coastal conglomerates: their significant features and some comments on gravelly mass-flow deposits', in E.H.Koster and R.J.Steel (eds.) Sedimentology of gravels and conglomerates, Canadian Society of Petroleum Geologists Memoir 10, 1-31.

Pierson, T.C. (1980) 'Erosion and deposition by debris flows at Mt. Thomas, north Canterbury, New Zealand', Earth Surface Processes 5, 227-247.

Pierson, T.C. (1981) 'Dominant particle support mechanisms in debris flows at Mt. Thomas, New Zealand, and implications for flow mobility', Sedimentology 28, 49-60.

Pierson, T.C. (1985) 'Initiation and flow behaviour of the 1980 Pine Creek and Muddy River lahars, Mount St. Helens, Washington', Bulletin of the Geological Society of America 96, 1056-1096

Postma, G. (1986) 'Classification for sediment gravity-flow deposits based on flow conditions during sedimentation', Geology 14, 291-294.

Rist, S. (1983) 'Floods and flood danger in Iceland', Jökull 33, 119-132.

Rust, B. (1978) 'Depositional models for braided alluvium', in A.D. Miall (ed.) Fluvial Sedimentology, Canadian Society of Petroleum Geologists Memoir 5, Calgary, 605-625.

Scott, K.M. (1988) 'Origin, behaviour and sedimentology of prehistoric catastrophic lahars at Mount St. Helens, Washington', in H.E.Clifton (ed.) Sedimentologic consequences of convulsive geologic events, Geological Society of America Special Paper 229, 23-35.

Smith, G.A. (1987) 'The influence of explosive volcanism on fluvial sedimentation: the Deschutes Formation (Neogene) in Central Oregon', Journal of Sedimentary Petrology 57, 613-629.

Smith, N.D. (1985) 'Proglacial fluvial environments', in G.M. Ashley, J.Shaw and N.D.Smith (eds.) Society of Economic and Petrological Mineralogists, Short Course Lecture Notes No.16, 85-134.

Thompson, A. and Jones, A. (1986) 'Rates and causes of proglacial river terrace formation in southeast Iceland: an application of lichenometric dating techniques', Boreas 15, 231-246.

Thórarinsson, S. (1957) 'The jökulhlaup from the Katla area in 1955 compared with other jökulhlaups in Iceland', Jökull 7, 21-25.

Thórarinsson, S. and Guðmundsson, H. (1979) 'Mýrdalssandur. A geophysical survey', Jarðefnaiðnaður h/f, OS79022JKD06, Reykjavík, 34pp.

Tómasson, H. (1974) 'Grimsvatnahlaup 1972, mechanism and sediment discharge', Jökull 24, 27-39.

Tómasson, H., Palsson, S. and Ingólfsson, P. (1980) 'Comparison of sediment load transport in the Skeiðará jökulhlaups in 1972 and 1976', Jökull 30, 21-33.

Ward, L.G., Stephen, M.F., and Nummedal, D. (1976) 'Hydraulics and morphology of glacial outwash distributaries, Skeiðarársandur, Iceland', Jökull 46, 770-777.

Þórarinsson, S. (1975) 'Katla og annall Kötlugosa', Árbok Ferðafélags Íslands, 1975, 124-149.

SEDIMENT AND SOLUTE YIELD FROM THE JÖKULSÁ Á SÓLHEIMASANDI GLACIERIZED RIVER BASIN, SOUTHERN ICELAND

Damian Lawler
School of Geography
The University of Birmingham
Edgbaston
Birmingham B15 2TT
UK

ABSTRACT. This chapter discusses change and variability in suspended sediment and solute concentration, load and yield for the Jökulsá á Sólheimasandi glacial river system in southern Iceland. Using a combination of low-frequency Icelandic Hydrological Survey data for the 1973 - 1988 period and the author's results from intensive monitoring of water quality in the 1988 melt season, variations are examined at three timescales: inter-annual, seasonal and diurnal. The combined yield of suspended sediment and dissolved material is 14,500 t $km^{-2}a^{-1}$, equivalent to an erosion rate of 5.4 mm a^{-1}, which is one of the highest values recorded anywhere. Slight restrictions on suspended sediment concentration, however, possibly associated with longer-term exhaustion effects, may now exist. One reason for caution in the interpretation of these estimates, however, is the pronounced variability, at all timescales, of runoff and sediment output from this basin, particularly in relation to the low-frequency data available. High seasonality of flow and sediment output typical of glaciofluvial systems occurs here, and a clear peak in total flux from the system emerges in the July-September period, when discharge, suspended sediment loads and solute loads all reach their maxima.

1. INTRODUCTION

Glaciofluvial suspended sediment transport has attracted renewed attention recently because of its demonstrable value in shedding light on processes acting within the glacier 'black-box', including subglacial erosion rates (e.g. Collins, 1977, 1979; Fenn, 1983; Gurnell and Clark, 1987; Gurnell and Fenn, 1984; Richards, 1984). Direct observation of

303

J. K. Maizels and C. Caseldine (eds.), Environmental Change in Iceland: Past and Present, 303–332.
© 1991 Kluwer Academic Publishers. Printed in the Netherlands.

many glaciological processes is extremely difficult and hazardous, but by monitoring meltwater quality variations it is often possible to gain valuable insights into the operation of parts of the system. Subarctic environments have received less attention than alpine regions in this respect. Also, studies on the 'larger' glaciofluvial systems, which may display significant inertia and storage effects (but cf. Church and Slaymaker, 1989), are relatively few.

This paper helps to redress this imbalance by examining, at a variety of timescales, suspended sediment and solute variations in the Jökulsá á Sólheimasandi glacial river system of southern Iceland. Largely drawing upon a database of approximately monthly observations made over the 1973 - 1988 period by the Icelandic National Energy Authority, this chapter sets provisional estimates of suspended sediment and solute export and erosion rate for this catchment in a wider Icelandic and global context, and offers tentative inferences about temporal change in glacial sediment transfer patterns through an exploration of fluctuations in suspended sediment concentrations over annual, seasonal and, to a lesser extent, diurnal timescales. It is the first in a series of papers that spring from a more comprehensive and long-term investigation into the influence of meteorological and geothermal events on glacial runoff and the dynamics of sediment transfer. The detailed results of intensive field monitoring of meltwater quality variations carried out in the Jökulsá á Sólheimasandi in the melt seasons of 1986, 1988 and 1989 are being worked up and presented elsewhere (e.g. Dolan, 1990; Lawler, 1989).

2. STUDY AREA AND SEDIMENT SOURCES

All field measurements and sampling reported here were carried out at a road-bridge, about 4 km from the snout of Sólheimajökull and approximately 6 km east of the village of Skógar in southern Iceland (Fig. 1), where Route 1 crosses the Jökulsá á Sólheimasandi glacial river (19°25'W, 63°30'N). The sampling site is the nearest to a glacier snout of any of those operated by the Hydrological Survey division of the Icelandic National Energy Authority (Orkustofnun). Sólheimajökull is a southward-flowing lobate outlet glacier which protrudes approximately 8 km from between two domes (of altitude 1450 m and 1493 m) of the ice-cap Mýrdalsjökull which overlies the active geothermal area of Katla (Fig. 1). The terminus of the glacier is around 100 m a.s.l (Maizels and Dugmore, 1985) and its Equilibrium Line Altitude is around 1100 m (Dugmore, 1989). Sólheimajökull has been advancing down its dog-legged valley at about 20 m a^{-1} since 1968 (Björnsson, 1979). It is an example of a reasonably large, subarctic, glaciofluvial system, in which approximately 60 per cent of the catchment is ice-covered. Drainage basin area at the bridge is estimated to be 78 km^2 (H. Tómasson, 1989, pers. comm.), where bankfull discharge is estimated to be 100 m^3 s^{-1} (Lawler, 1989). Catchment length to the bridge is approximately 17 km. Annual precipitation (1931 - 1960) is estimated to rise from at least 1600 mm at the bridge gauging site to over 4000 mm near the highest boundary of the catchment (Eythorsson and Sigtryggsson, 1971 (cited in Björnsson (1979)). Because of a spatially and temporally coarse precipitation sampling network, however, as well as the inhospitable terrain around

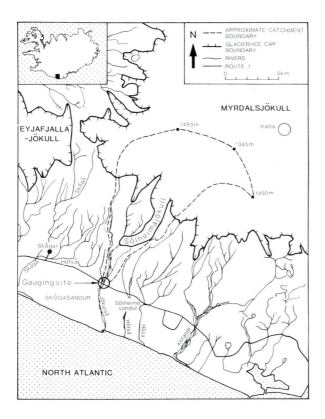

Fig. 1 Location of the Jökulsá á Sólheimasandi glaciofluvial system in southern Iceland.

the catchment head in the snow accumulation areas, these values are subject to considerable uncertainty. For example, Rist (1957, cited in Rist (1967)) found at the top of the Sólheimajökull basin that the snow layer marking accumulation during the winter of 1954/55 was 9200 mm thick and had an average specific gravity of 0.63, corresponding to a liquid precipitation receipt of 5800 mm.

Results of detailed geological mapping of the area have recently been published by Carswell (1983). Hydroclastic and acid volcanic rocks dominate. Yellow-brown palagonite tuffs and breccias are common. Subglacial lithologies, though, are essentially unknown. Soils are generally very thin on the terrace surfaces and hillslopes, but can thicken considerably in depressions. The contemporary human impact in the Jökulsá á Sólheimasandi catchment is minimal, and is dwarfed by the natural dynamism of sediment-delivery and landscape-forming processes.

No detailed evaluation of runoff and sediment dynamics has yet been attempted in this area, so it is difficult to suggest the dominant sources and pathways of meltwater and clastic material output from the system. Sólheimajökull is highly-charged with distinct layers of englacial debris,   (not unlike the snout area of Glacier de Tsidjiore Nouve in Switzerland (see Gurnell, 1987, her Fig. 12.10)), which become clearly visible at the glacier margin near the terminal region. During periods of high solar radiation in summer, particularly, the glacier surface becomes increasingly plastered in black melt-out detritus from these layers which is washed off in subsequent rainstorms. Also, for the lowest three kilometres of the glacier tongue, there are large quantities of material exposed at the surface of Sólheimajökull. Additionally, there is probably a liberal supply of sediment at the glacier bed particularly if subglacial channels, like their proglacial counterparts, are prone to rapid course-switching. Subglacial channels may also liberate englacial debris exposed in their ice-walls. The amount of material produced at the sole of this actively-advancing glacier by plucking, crushing, abrasion, shearing and fluvial erosion (see, for example, Fenn (1987)) is unknown, but is likely to be considerable.

Waters draining subglacially and englacially, therefore, and supraglacially in the terminal region, are likely to have large stores of sediment at their disposal. Given the proximity of the Sólheimajökull basin to active volcanic zones, the englacial and supraglacial debris is likely to have a high ash content (possibly of low specific gravity), and may not be strictly erosional in origin. Moreover, it may well have been produced outside the basin. These points should be borne in mind when erosion rates are estimated from sediment yield data later in this paper.

The hillslopes either side of Sólheimajökull and in front of the southern edge of Mýrdalsjökull are deeply gullied and sparsely vegetated, and it would be surprising - especially given the very high precipitation of the region - if substantial amounts of sand- and silt-sized material were not frequently washed out into the subglacial and proglacial drainage channels. Other sediment sources include the numerous frost-shattered free-faces with sizeable talus accumulations at their bases which surround the glacier. Also, around the glacial margin and in the proglacial zone lie extensive moraines (Dugmore, 1989). These are characterised by a very wide range of particle sizes and are occasionally unvegetated. Marginal streams can undercut these morainic deposits which then become available for fluvial transport. Proglacial streams and the main Jökulsá á Sólheimasandi channel itself are also free to tap the large quantities of valley-fill material, including colluvial and outwash deposits. Their entrainment is probably maximized at high flows and/or during channel-change episodes. The sandur and valley-fill materials in front of Sólheimajökull have been extensively mapped, dated and interpreted by Dugmore (1989), Maizels (1989) and Maizels and Dugmore (1985). Although this paper is not intended to identify dominant sediment sources and production and delivery processes, these aspects are reviewed in Gurnell and Clark (1987) for alpine areas.

3. FIELD AND LABORATORY METHODS

The estimation of sediment yield from a catchment ideally demands reliable, long-term, continuous records of the discharge of water, bedload, suspended sediment and solutes. Only in rare instances, however, are such records available, and it is often necessary to work with short-term data of low sampling frequency. Bedload measurements are rarely included as part of the routine monitoring programmes of water agencies, and no bedload data are available for the Jökulsá á Sólheimasandi. All these limitations particularly apply to remote and/or hostile environments - especially high-latitude regions. This hampers attempts to obtain fully representative global comparisons of sediment yield and erosion rate, and can frustrate efforts to detect environmental change in the sediment-flux record.

Orkustofnun has operated for a number of years a sampling network for river discharge, suspended sediment and total dissolved solids concentrations (TDS) on many of the glacial rivers of Iceland. The Jökulsá á Sólheimasandi has been 'spot' sampled on a regular, approximately monthly, basis since at least 1973, though no continuous flow measurement or automatic water/sediment sampling instrumentation has operated here for any length of time. At the time of each manual sample, river discharge is estimated at the gauging site (Fig. 1) using a float technique. A weighted container on a known length of graduated cable (usually around 30m) is lowered from the bridge (which is approximately 6m above river level) to the water surface where it is allowed to float downstream over a recorded and timed distance to give a measure of surface velocity. The float is then hauled in and the measurement repeated at about 6 different points across the section. The geometry of the river section is obtained by plumbing with a weighted suspended sediment sampler (see below), again at about 6 points across the profile. Discharge is then calculated using the continuity equation with unadjusted surface velocities. This may cause a slight overestimation of discharge although, because of presumed relative uniformity in the velocity field of the near-surface water (associated with very high turbulence, steep water surface slopes of around 0.01 and maximum midstream velocities up to around 4 m s^{-1} (Lawler, 1989)), this may not be too serious.

Suspended sediment samples are withdrawn from the bridge using a US D-49 depth-integrating cable-and-reel sampler - a device of considerable stability when immersed in rivers of high velocity and turbulence and one which allows sampling to within 0.1 m of the bed (Guy and Norman, 1970). It is suspended with a heavy weight by a cable and winch built into the side of the Orkustofnun field vehicle (Fig. 2). Samples are taken from about six verticals in the section. Suspended sediment concentrations are subsequently determined in the Reykjavík Orkustofnun laboratories using a sedimentation column. The values used in this paper pertain to the data set on file at Orkustofnun headquarters, and represent mean-section suspended sediment concentrations (SSC) derived from averaging separate 'verticals'. Although no bedload transport data or hydraulic and sedimentological information on which to base bedload estimates are yet available for this glacial river, rates are probably substantial. Note that although in temperate environments bedload may be much less than 10 per cent of total load (Richards, 1982, p.106), in glacierized catchments

Fig. 2 Orkustofnun field vehicle for hydrometric measurement, with US D-49 sampler ready for use.

this fraction may be much higher (but cf. the 2-10 per cent figure of Ferguson (1984)). For instance, Gurnell's (1987a) tabulation for catchments in Switzerland, Norway and Canada shows that melt-season bedload accounted for between 30 and 59 per cent of total sediment flux (ignoring solute load). Gurnell et al. (1988) demonstrate for Bas Arolla, Switzerland, that this proportion can rise to 64 per cent.

In the summer melt seasons of 1986 and 1988, the author measured suspended sediment concentration, load and yield variability in the Jökulsá á Sólheimasandi main channel at the minute, hourly and daily timescales. This formed part of a more comprehensive investigation into the controls on sediment and solute transport in subarctic glacierized catchments. A datalogging system was installed at the Orkustofnun gauging site (Fig. 1) in 1988 to obtain, for 2.5 months, quasi-continuous data on river level, turbidity and electrical conductivity (the latter a surrogate of TDS), as well as river temperature and meteorological variables. A preliminary flow rating relationship, based on the plumbed cross-section of 21 July 1988 and salt-velocity tracings, was also established for the Jökulsá á Sólheimasandi, which agreed fairly well with Orkustofnun float-gauging data (Fig. 3). This allowed the 1988 stage values measured by a pressure transducer to be converted to discharges to allow the computation of short-term fluctuations in sediment load. Also, for

12 days (28 July - 9 August 1988), an automatic water sampler was programmed to withdraw, at hourly intervals, a ca 700-ml sample of Jökulsá á Sólheimasandi water from near the right bank. These samples provided information on the temporal variability of suspended sediment concentrations, albeit from one point in the section (not anticipated to be a serious problem in this highly turbulent reach where vertical and lateral mixing are probably substantial). For 127 of these samples, a volume of around 300 ml was filtered, within 24 hours of collection, through 11-cm diameter Whatman No. 52 filter papers (initial penetration 4.5 microns) at a nearby base-camp laboratory. The filter-paper residues were retained to allow (a) future sediment-source analysis and (b) weighing on return to Birmingham to provide values of suspended sediment concentration (SSC), which could also be used to calibrate a custom-built turbidity meter deployed on the Jökulsá á Sólheimasandi to detect short-term (<1 h timescale) sediment pulsing (Lawler, 1989; Lawler and Brown, in prep.).

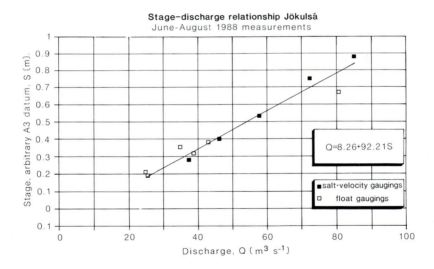

Fig. 3 Preliminary discharge rating curve for the Jökulsá á Sólheimasandi at the bridge site (SDG = author's salt-velocity gaugings; fg = Orkustofnun float gaugings).

4. SEDIMENT YIELDS

4.1 Error and uncertainty in sediment yield estimation

The calculation of sediment loads, sediment yields and erosion rates from river discharge and sediment concentration data is not an exact science. In fact, Olive et al. (1980) called it a 'geomorphic guessing game'! A number of averaging and interpolation procedures have to be used depending on the nature of the available data. Walling and Webb (1981) cite examples where workers on the same river, often using the same basic data, have calculated widely different sediment loads because of procedural variations. A considerable amount of methodological work, warning of pitfalls and suggesting degrees of accuracy (proximity to absolute truth, or absence of systematic bias) and precision (repeatability of a measurement or sample-to-sample variability), has been accomplished recently (e.g. Dickinson, 1981; Ferguson, 1986, 1987; Gurnell, 1987b; Walling, 1978; Walling and Webb, 1981). The potential sources of error and uncertainty, which apply more to suspended sediment than solutes, are tabulated below:

1. Errors due to sampling instrumentation. With most manual samplers, there is an unsampled zone near the bed which may lead to underestimation of true mean-vertical suspended sediment concentration, especially in shallow streams. For example, for two of the most common samplers, the US DH-48 and the US D-49, this unsampled region is the lowermost 9 and 10 centimetres respectively. Too large an intake nozzle may result in preferential trapping of the larger particles. Automatic samplers generally sample from one point in the cross-section only which may not be representative (see below). The intake will probably be fixed at this point, too, and its relative position with respect to a continuously-fluctuating water level, therefore, will change. Its filter may screen out the larger suspended particles. There may be sample-to-sample contamination if a common withdrawal tube is used for all samples in a sequence. If left unattended for long periods, some suspended sediment may go into solution, causing inaccuracy in both SSC and TDS determinations.

2. Errors due to spatial variability in SSC. This can affect the representativeness of specific sampling schemes. Suspended sediment concentrations vary across natural river cross-sections, usually increasing towards the channel centre and towards the bed. Such stratification decreases with turbulence, and thus is less troublesome in steep, rough mountain or glacial rivers. Downstream variations in suspended sediment concentration can also occur. Conceivably, too, differences at the bend or reach scale could emerge, although less research has been conducted into this area.

3. Errors due to temporal variability in SSC. Suspended sediment concentrations vary at a range of timescales. Short-term sediment pulsing (e.g. at the sub-hourly timebase) may be cumulatively important for gross sediment export from a catchment yet will be missed in all but the most detailed investigations. Similarly, individual storm-events are often missed by the typical weekly or monthly sampling systems operated by water agencies, and many workers (e.g. Dickinson, 1981; Ferguson, 1987; Walling, and Webb, 1981) cite this as the main reason why many values of suspended sediment loads are seriously underestimated. Moreover, suspended sediment yield estimates must remain tentative unless available records are of sufficiently lengthy to embrace significant year-to-year variability due to fluctuations in sediment supply and/or runoff production.

4. Laboratory errors. These should be small in comparison to the above, but can become significant when dealing with small absolute SSC or TDS values. The main frustration is the lack of consistency in procedure - especially in the choice of the filter paper

pore-size (see Gurnell, 1987b).

5. Sediment-load estimation procedures. At least six methods are in use for the calculation of suspended sediment yields (Walling and Webb, 1981) - each appropriate for different data-records and each with different levels of accuracy and precision (see Dickinson (1981) and Ferguson (1987)). Even for a given method, variations in the time-periods over which loads are integrated can make large differences. Data are commonly skewed, too, making some measures of central tendency misleading.

6. The sediment delivery problem. Only a small proportion of the material eroded from catchment surfaces is exported 'immediately'. Much goes into storage in, for example, fans, valley-fills, colluvial deposits and in-channel bars. This can often explain the lack of synchroneity between, for example, anthropogenic disturbance or climatic change and apparent erosional response deduced from sedimentary records downstream. Thus, depending on the catchment sediment delivery ratio (Roehl, 1962), which itself is subject to temporal change and variability, contemporary sediment-yield values may convey confusing messages regarding upstream geomorphic activity.

7. Uncorrected material inputs and outputs. Not all material exported from a basin has arisen from denudation processes within the catchment. Ideally, inputs of, say, chloride in precipitation need to be isolated in basin geochemical budgets before reliable estimates of solute yield can be made. Similarly, dry fallout of clastic material (e.g. of loessic or volcanic origin) may be difficult to correct for. Part of a river's exported load may be organic, it may be derived from geothermal activity or, especially in heavily cultivated or urbanised areas, it may be composed of significant proportions of agrochemicals or water-treatment products.

8. Assumptions about unmeasured properties. Estimates of sediment and solute yield are normalised for basin area to facilitate spatial comparisons, and it is usual to assume that the topographically-defined catchment boundary defines the hydrological catchment. This may be highly questionable where substantial natural or artificial inter-basin transfer of water takes place. In glacierized catchments, poorly-known subglacial topography and water routing patterns should inspire extra caution. When yields are converted to erosion rates, the assumption of a single lumped value of rock specific gravity (e.g. 2.65 t m^{-3}) for the mosaic of catchment surface materials is also subject to error. Finally, surface-lowering is assumed to be spatially uniform, although most erosion is likely to be concentrated along fluvial (and glacial) 'corridors', unstable hillslopes and free faces.

The important point is that all sediment yield and erosion rate values, unless based on continuous and long-term flow and sediment transport/turbidity data in basins where many of the above assumptions can be checked and corrected for, should be interpreted with caution. All workers should declare their assumptions, areas of uncertainty and calculation procedures. A set of standardised procedures would greatly ease the problem of inter-study comparability.

4.2 The database

The Orkustofnun flow and sediment database for the Jökulsá consists of 199 'instantaneous' measurements of mean-section suspended sediment concentration and total dissolved solids concentration. Discharge values are available for 189 of these measurements (Table 1). The record starts on 26 June 1973 and, although sampling has continued throughout

Table 1   Summary of flow and water quality data for the Jökulsá over the 1973-1988 period

Variable	Discharge (m^3s^{-1})	Suspended sediment concentration $(mg\,l^{-1})$	Instantaneous suspended sediment load $(kg\,s^{-1})$	Total dissolved solids concentration $(mg\,l^{-1})$	Instantaneous solute load $(kg\,s^{-1})$
Maximum	103	6358	305.2	139	5.25
Minimum	1	34	0.09	35	0.09
Mean	23.82	979.4	34.10	78.86	1.72
Sample standard deviation	17.47	985.4	51.51	20.66	1.14
Coefficient of variation (%)	73.3	100.6	151.1	26.2	66.2
Median	23	795	15.72	77	1.55
Lower quartile	10	288	2.22	63	0.78
Upper quartile	32	1318	38.43	91	2.19
Interquartile range	22	1030	36.21	28	1.41
n	189	199	189	199	199

1989, for the purposes of this paper finishes on 30 November 1988. This gives around 12 samples per annum to work with, distributed rather unevenly throughout the year with a peak sampling frequency in June, July and August (Fig. 4) - the season of peak flows and suspended sediment transport (see below). The diurnal timing of sampling is important because cyclic discharge variation, in response to fluctuations in radiation loading of the glacier surface, is particularly important in summer here with peak stage around 1800 h GMT (= Icelandic local time) and low flow around 0600 h (Lawler, 1989). Fig. 5 shows that most samples have been taken between 1100 h and 2200 h with a distinct peak at 1800 h. On a seasonal and daily basis, therefore, a slight bias towards higher flow and higher sediment transport conditions exists. This is likely to be outweighed, however, by the high probability that the sampling scheme misses short-lived, but highly significant, peak flows. Indeed, Ferguson (1987, p.103) argues that 'the disproportionate importance for sediment transport of high discharges suggests it is reasonable to bias sampling towards higher flows, though not at the expense of a representative low-flow fit as well'.

4.3 Computations and assumptions

One way of obtaining an approximate indication of contemporary erosion rates within catchments is to combine, in a variety of ways, data on suspended sediment concentration

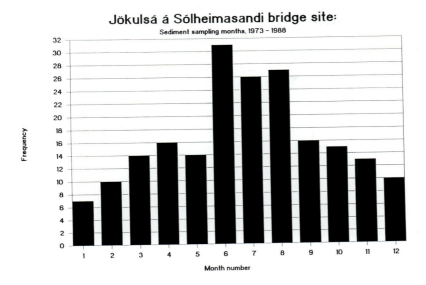

Fig. 4 Seasonal bias of suspended sediment samples towards June, July and August (high-flow months).

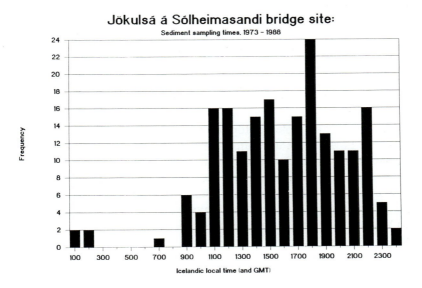

Fig. 5 Daily timing of suspended sediment samples, showing slight bias towards higher flows between 1100 hours and 2200 hours.

(usually expressed in mg l^{-1}) with river discharge values (in m^3 s^{-1}) to produce a sediment load (in kg s^{-1} or t a^{-1}). Loads may then be divided by catchment area to define the specific sediment yield of a basin (usually expressed in t km^{-2} a^{-1}) - a measure which facilitates comparison with other basins. Finally, if the specific gravity of the catchment surface material is known then an erosion rate (in mm a^{-1}) for the basin can be estimated.

Preliminary sediment loads here have been estimated using two different procedures (Methods 1 and 2 of Walling and Webb (1981)):

$$L_1 = \overline{Q} \cdot \overline{C} = \left(\frac{\sum_{i=1}^{n} Q_i}{n}\right)\left(\frac{\sum_{i=1}^{n} C_i}{n}\right) \tag{1}$$

and

$$L_2 = \overline{Q_i C_i} = \frac{\sum_{i=1}^{n}(Q_i C_i)}{n} \tag{2}$$

where:

L   = estimated sediment (or solute) load (g s^{-1})
Q_i = river discharge at time of sampling (m^3 s^{-1})
$\overline{Q}$ = arithmetic mean discharge for period of record (m^3 s^{-1})
C_i = suspended sediment (or solute) concentration at time of sampling (g m^{-3} i.e. mg l^{-1})
$\overline{C}$ = arithmetic mean suspended sediment (or solute) concentration for period of record
     (g m^{-3})
$\overline{Q_i C_i}$ = arithmetic mean instantaneous sediment (or solute) load (g s^{-1})
n   = number of samples

These two methods are the only ones appropriate here, given the lack of continuous or quasi-continuous data on flow and water quality variations for the Jökulsá á Sólheimasandi. Method 2 (equation (2)) is a discharge-weighted estimate and is the approved technique of the UK Harmonized Monitoring Programme (Department of Environment, 1979 (cited in Walling and Webb, 1981)), and these results are considered the more realistic. Method 1 is considered to result in gross underestimation of true sediment loads (Ferguson, 1987; Walling and Webb, 1981). In fact, in one of the few published tests of the accuracy and precision of the various techniques of load estimation, Walling and Webb (1981) found that Method 1 was very precise but inaccurate, producing estimates of around 20 per cent of true load, while Method 2 produced reasonably accurate results but of very low precision.

  Some comment is appropriate on the specific limitations of the analysis to be presented here, in relation to the general and potential sources of error listed in section 4.1. The depth-integrated samples taken by Orkustofnun, using a vertical transit method in a highly turbulent and presumably well-mixed river, are considered to be representative. The sampling section lies on a straight reach in a single channel in an otherwise highly-braided river. The biggest problems are the lack of continuous flow data and a low frequency of water-sampling; other workers have found that this situation almost invariably results in underestimation of true suspended sediment loads and a slight overestimation of solute loads. There is no reason to suspect any systematic bias in the laboratory determinations of sediment and solute concentrations. No data on bedload discharge here are yet available.

In this high-energy environment, few opportunities for storage of fine sediment between glacier terminus and gauging station are apparent, and a fairly high sediment delivery ratio is probable. Thus sediment yield estimates are not thought to be seriously underestimating contemporary rates of upstream erosion and material reworking. River water samples withdrawn in the 1986 and 1988 melt seasons showed negligible amounts of organic matter, and so no allowance has been made for this in the calculations. Nor has any allowance been made for the proportion of exported fine sediment that is not strictly erosional in origin but eruptive material (e.g. volcanic ash layers in Sólheimajökull itself), and possibly derived from outside the basin. Chloride deposition on the catchment is considered to be of minimal quantitative importance in this system (despite its proximity to the North Atlantic coast (Fig. 1)), and no adjustments have been made to computed solute yields. Furthermore, a substantial geothermal component in solute loads exists (Dolan, 1990), although this has not been subtracted from the solute yield calculations that follow, as further geochemical work is necessary. With minimal human impact on the catchment, no adjustments of loads need be made for anthropogenically-introduced materials.

Limited mapping of glacier surfaces and echo-sounding of subglacial topography here, a complete absence of water tracing, and the possibility of substantial inter-basin fluxes of (geothermal) groundwater means that the assumed drainage basin area of 78 km^2 (H. Tómasson, pers. comm.) is subject to greater error than in non-glacierized catchments. Also, it is thought that the ice/drainage divide of the Jökulsá á Sólheimasandi basin has migrated over time (Dugmore, 1989), making long-term extrapolation of specific yields hazardous. To facilitate comparisons with Tómasson's (1976, 1986) work, the same rock specific gravity value of 2.7 t m^{-3} (H. Tómasson, pers. comm.) has been used to convert yields to notional erosion rates.

To summarise, those factors tending to encourage overestimation of sediment loads, sediment yields and erosion rates include (a) the slight biasing of sampling towards conditions of higher flow and sediment transport; (b) the use of uncorrected surface float velocities in discharge estimation; and (c) not subtracting non-denudational components of the river's load (e.g. atmospheric chloride and geothermal compounds, organic matter, and dry fallout). Factors which lead to underestimation include (a) the ignoring of bedload; (b) an infrequent flow- and sediment-sampling programme which probably misses many of the high-magnitude events; (c) the presence of the unsampled zone beneath the US D-49 sampler used; and (d) the possibility of net losses of fine sediment to storage between entrainment sites and gauging station. The net effect of these counteracting sets of factors is almost certainly an underestimation of true loads and erosion rates.

4.4 Results

The combination of high suspended sediment concentrations (e.g. above 6000 mg l⁻¹ (Table 1)) and discharges leads to very high sediment loads and exceptionally high suspended sediment yields of 13,787 t km⁻² a⁻¹ and total yields (excluding bedload) of almost 14,500 t km⁻² a⁻¹ (Table 2). Mean annual sediment yield is 12674 t km⁻² a⁻¹, with a high standard error of 1785 t km⁻² a⁻¹. These are amongst the highest published values anywhere in the world (Table 3), and apparently the highest recorded rates for a glacial river (see Gurnell, 1987b). Specific sediment yields here are almost three times larger than those published for the Hunza River in the Karakoram Himalayas by Ferguson (1984), and almost 100 times greater than the global average value cited by Walling and Webb (1983).

Table 2   Provisional estimates of sediment and solute yields and erosion rates for Jökulsá á Sólheimasandi catchment. Estimates assume a rock specific gravity of 2.7t m⁻³ and a drainage basin area of 78km².

Method (see Walling, 1981)	Suspended sediment only			Solutes only			Suspended sediment and solutes		
	Mean load	Specific suspended sediment yield	Erosion rate	Mean solute load	Specific solute yield	Erosion rate	Sediment yield	Specific sediment yield	Erosion rate
	(kg s⁻¹)	(t km⁻² a⁻¹)	(mm a⁻¹)	(kg s⁻¹)	(t km⁻² a⁻¹)	(mm a⁻¹)	(Mt a⁻¹)	(t km⁻² a⁻¹)	(mm a⁻¹)
1	23.3	9432	3.5	1.88	760	0.28	0.795	10192	3.8
2	34.1	13787	5.1	1.72	695	0.26	1.130	14482	5.4

These exceptionally high suspended sediment loads are probably due to (a) the high debris content of Sólheimajökull which forms a readily-available sediment supply to be tapped by supraglacial, englacial and subglacial drainage; (b) the rapid advance of the glacier which encourages sediment production by abrasion, and the introduction of fresh subglacial sediment stores for entrainment by the glacial drainage system; (c) an active stream system, with frequent course changes, partly related to glacial advance; (d) a maintenance of moderate discharges all year round (note that winter discharges average around 10 m³ s⁻¹ (about 25 per cent of mean summer rates (Fig. 8)), and preliminary analysis suggests that the winter (October to April) proportion of runoff here is very high for glacierized catchments, at approximately 38 per cent; compare this to, for example, the 5 per cent figure of Bezinge (1987) for the Ferpècle basin, Switzerland); (e) a very high precipitation and specific runoff for the catchment (although sediment yields per cumec for the Jökulsá still plot out at the upper end of the distribution of Gurnell (1987b, her Fig. 12.2)); and (f) few opportunities for significant sediment storage to take place in the immediate proglacial area because of high stream powers and relatively efficient channels.

The Jökulsá loads and yields are considered typical of rivers draining glaciers in southern Iceland, with similarly high values having been determined by Tómasson (1976,

1986) for other rivers draining parts of Mýrdalsjökull. Global maps of sediment yield (e.g. Clark, 1987; Walling and Webb, 1983) commonly exclude Iceland or ascribe the country a very low value and these may need amending to incorporate these new data from a high-precipitation sub-arctic environment, and those of Tómasson (1976, 1986).

Table 3   The world's highest fluvial sediment yields
(after Walling and Webb (1983) in which source references are cited)

Country	River	Drainage basin area (km^2)	Mean annual sediment yield (t km^{-2} a^{-1})	Rank	Source
Taiwan	?	?	31,700	1	Li (1976)
China	Dali	96	25,600	2	Mou and Meng (1980)
China	Dali	187	21,700	3	Mou and Meng (1980)
New Zealand	Waiapu	1378	19,970	4	Griffiths (1982)
Kenya	Perkerra	1310	19,520	5	Dunne (1974)
New Zealand	Waingaromia	175	17,340	6	Griffiths (1982)
New Zealand	Hokitika	352	17,070	7	Griffiths (1982)
China	Dali	3893	16,300	8	Mou and Meng (1980)
Iceland	Jökulsá á Sólheimasandi	78	14,482*	9	Lawler (this paper)

* mean over whole 16-year period

5. VARIABILITY IN SEDIMENT CONCENTRATIONS AND LOADS

5.1 Year-to-year variability

Table 1 gives some idea of the range and variability of 'average' flow and water quality levels for the Jökulsá á Sólheimasandi over the whole sampling period. Measured discharge and suspended sediment concentration has ranged over at least two orders of magnitude, while instantaneous suspended sediment loads have varied by more than three orders of magnitude (Table 1). Solutes, as in other systems, are the more conservative variables but even here there is a four-fold difference between maximum and minimum concentrations, and peak instantaneous solute loads are more than 50 times greater than minimum load (Table 1). Actual variability will be considerably greater than measured variability inferred from a data series derived from coarse-interval sampling. Separate calculations for individual years on suspended sediment outputs alone reflects this great variability; mean annual specific suspended sediment yield is 12674 t km^{-2} a^{-1} (about 8% less than the value computed from the aggregated 1973 - 1988 data (Table 2)) but the standard error of 1785 t km^{-2} a^{-1} is high. Even assuming that each year's sampling programme accurately and representatively captures information on true yields, then, we can only be 95% confident that long-term

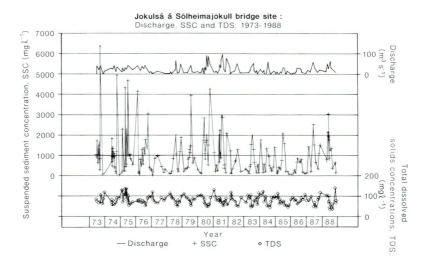

Fig. 6 Time series of discharge, suspended sediment concentration (SSC) and total dissolved solids (TDS) concentration at Jökulsá á Sólheimasandi bridge site, June 1973 - December 1988.

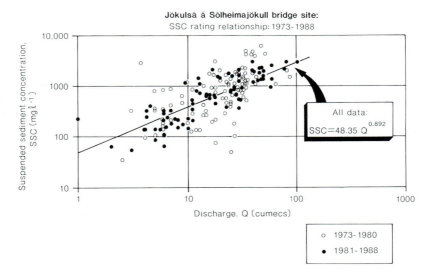

Fig. 7 Suspended sediment rating relationship (n = 189).Suspended sediment yield lies between 9104 and 16243 t km⁻² a⁻¹.

suspended sediment yield lies between 9104 - 16243 t km^{-2} a^{-1}. Figure 6 shows the temporal distribution of flow, suspended sediment concentration and solute concentration for the sampled period 1973 - 1988. Although possibly nothing more than a sampling artifact, there is a tendency for instantaneous suspended sediment concentrations to be highest in the 1973 - 1976 period (Fig. 6). A subsidiary peak is evident in the late 1970s/early 1980s, while the last seven years are characterised by relatively subdued values. The four highest suspended sediment concentration values recorded all took place in the 1973-76 period, despite only moderately high flows of between 35 and 55 m^3 s^{-1}. In fact, when all the suspended sediment values are normalised for river discharge differences, sediment concentration still tends to be greater in the first half of the record than in the second. No substantial differences emerged, however, between suspended-sediment rating equations derived for earlier and later parts of the record, although the 1973 - 1980 period is characterised by much greater scatter (r^2 = 34.3%) than the 1981 - 1988 data (r^2 = 70.8%) (Fig. 7).

This (albeit weak) tendency for a decline in suspended sediment concentration over time may partly reflect the timing of the data record relative to glacial readvance. After retreating for much of this century, Sólheimajökull has been advancing since 1968 by around 20 m a^{-1} (Björnsson, 1979; Rist, 1984). It may be that one factor encouraging high suspended sediment concentrations in the early part of the record relates to the initial stages of advance of the glacier over fresh and largely unvegetated melt-out and outwash deposits; these would be likely to provide a copious (but also a highly spatially-variable) sediment supply for the glacial drainage system to transport. Although new sediment supplies associated with an advancing glacier will continue to become available, the net sediment store may progressively be depleted by successive flows. This point will be returned to below.

5.2 Seasonality of flow and sediment export

As with many rivers, there is a clear positive relationship between discharge and suspended sediment concentration for the Jökulsá á Sólheimasandi (Fig. 7). The scatter of the points is not unusual, and represents an order-of-magnitude variation in suspended sediment concentration for a given flow level. Fig. 7 is a rating plot to which the following power-function regression equation can be fitted:

$$SSC = 48.35 \ Q^{0.892} \qquad (3)$$

where SSC is suspended sediment concentration in mg l^{-1}, and Q is discharge in m^3 s^{-1}. Discharge accounts for 50.8% of the variance in suspended sediment concentration and is significant at p = 0.0001.

Within individual years, however, there is a pronounced but expected summer increase in meltwater discharge, in response to increases in temperature, day length, and solar radiation receipts on the glacier surface during June, July and August. Peak flow tends to arrive around day 200 of the year i.e. the latter half of July (Fig. 8), a lag after peak

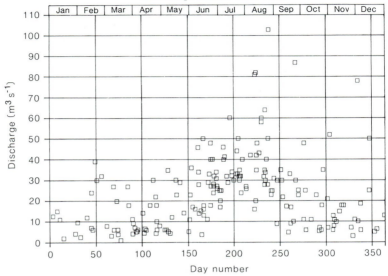

Fig. 8 Aggregated seasonal distribution of instantaneous flows for each year of record
1973 - 1988.

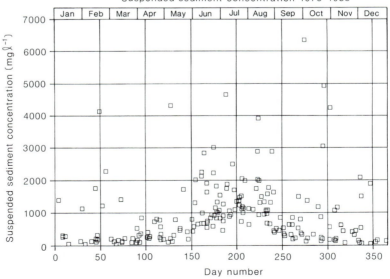

Fig. 9 Aggregated seasonal distribution of instantaneous suspended sediment
concentration values for each year of record 1973 - 1988.

insolation that is consistent with the proportion of the basin that is glacierized, according to results for Cascade Glacier, Washington (Fountain and Tangborn, 1985, their Fig. 4). No discharge greater than 40 m³ s⁻¹ has ever been measured before 7 June. A further aspect of this asymmetry, though, is that the really high flows have only ever been observed in the August - October period (Fig. 8). Work is continuing at Birmingham University to produce a simple model of seasonal and diurnal flow response to meteorological inputs, which can then be used to refine sediment transport estimates for the system. Being in part flow-related, suspended sediment concentrations also start to rise around the beginning of May, and reach a peak towards the end of July (Fig. 9).

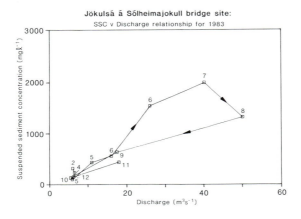

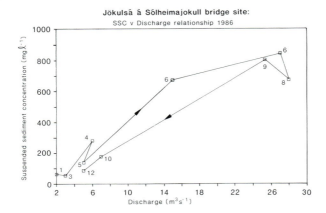

Fig. 10 Seasonal pattern of suspended sediment concentration - discharge relationships for example years. Numerals on plotted points refer to month number (note that in some months more than one sample was taken). Recent years tend to show signs of late-season declines in sediment concentration.

However, other workers on glacial rivers (e.g. Ferguson, 1984) have noted a tendency for higher suspended sediment concentrations for a given discharge to occur early in the melt-season. This produces a positive hysteretic loop when discharge is plotted against suspended sediment concentration for any given year. The condition is usually explained by reference to large amounts of sediment available for transport during the first few flood events of the melt period - supplies which become depleted as successive flow events take place. Hence autumn flows tend to carry less suspended sediment than equivalent events in spring.

As far as the limited sampling can reveal, a degree of positive hysteresis (i.e. seasonal recession-flow sediment shortage) is present for the Jökulsá system, but only in the more recent period. Example trends are shown in Figure 10. An apparent increase in seasonal hysteresis through time implies ample sediment stores soon after glacier advance had recommenced in 1968, but a slightly restricted supply in the 1980s leading to reductions in the suspended sediment concentration of late-season flows. This is consistent with the evidence discussed above of higher and more variable absolute, and discharge-normalised, suspended sediment concentrations in the first half of the record (e.g. Fig. 6). This area is worthy of further investigation, although ideally data are needed which span the onset of glacial advance.

5.3 Diurnal sediment yield variability

The detailed water quality measurement programme mounted by the author in the 1988 melt season provided turbidity data at two- and ten-minute frequencies. The results show substantial flow and suspended sediment pulsing at the <1 hour timescale, probably in response to temporary damming of meltwaters by ice-falls. Details are presented elsewhere (e.g. Lawler, 1989). Space permits only the brief presentation here of results of a concurrent hourly water-sampling programme mounted for the 11-day period 28 July - 9 August 1988. These reveal a substantial diurnal variability in flow and sediment delivery that the longer-term, coarse-interval, sampling programme inevitably misses. The 1988 study period included a wide range of discharge, from low flows during dry intervals to almost bankfull conditions brought about by persistent rain in the first week of August (Fig. 11). The oft-cited distinctive daily pulsing of discharge and suspended sediment output as a result of diurnal variations in glacier melt-rate can clearly be seen (Fig. 11). The peak suspended sediment concentrations (up to 3809 mg l^{-1}), associated with the highest flows around 8-9 August, are almost 25 times greater than those measured during low flows a week earlier (Fig. 11). Table 4 summarises for this melt-season period average and extreme levels of stage, discharge and suspended sediment transport, and provides some measure of the degree of short-term variability. Long-term, all-year-round, monitoring schemes such as those carried out by Orkustofnun are vitally important, but there is a need to complement these with specific studies based on automatic instrumentation, at least in the important melt-seasons when most water and sediment flux is enabled, to detect short-lived pulsing which is often crucial to the inference of meltwater- and sediment-production processes.

High-frequency data also suggest levels of confidence that we may place in longer-term sediment yield estimates.

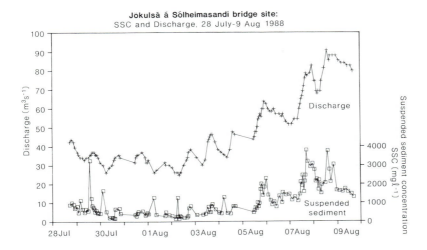

Fig. 11 Variability at the diurnal timescale of discharge and suspended sediment concentration (SSC) for part of the melt season of 1988.

Table 4 Summary statistics for 127 samples processed from Automatic Water Sampler programme, 28th July - 9th August 1988

Variable	Stage above arbitrary datum (m)	Discharge (m^3s^{-1})	Suspended sediment concentration ($mg\ l^{-1}$)	Suspended sediment load ($kg\ s^{-1}$)
Maximum	0.894	90.7	3809	343.4
Minimum	0.179	24.7	153	4.0
Mean	0.439	48.8	1120	67.2
Sample standard deviation	0.202	18.63	838	71.1
Coefficient of variation (%)	46.0	38.2	74.8	105.9
Median	0.368	42.23	802.6	36.51
Lower quartile	0.281	34.15	467.4	15.24
Upper quartile	0.558	59.73	1599.4	105.09
Interquartile range	0.277	25.58	1132	89.85
n	127	127	127	127

6. SOLUTE LOADS AND YIELDS

Calculation of erosion rates using river loads should not exclude the dissolved component. Total dissolved solids (TDS) concentrations in glacier-fed rivers are generally much lower and more stable than suspended sediment concentrations, and the Jökulsá á Sólheimasandi is no exception (Table 1, Fig. 6). However, the 1973 - 1988 data set shows that, in comparison with other proglacial river systems, this reach carries high solute concentrations, up to 139 mg l^{-1}, with a mean of almost 79 mg l^{-1} (Table 1). Ferguson (1984), for example, quotes a range of 13 - 61 mg l^{-1} for the Batura proglacial torrent in the Karakoram. Similarly, summer melt-season electrical conductivity values in the Jökulsá (Lawler, 1989) are much higher than those presented by Collins (1981) for the Gornera or by Gurnell and Fenn (1985) for the Tsidjiore Nouve system, both in Switzerland.

Solute concentrations in Icelandic glacial rivers tend to be high partly because of 'the abundance of reactive basaltic rocks' (Gíslason and Arnórsson, 1988, p.197; Gíslason, 1989). For the Jökulsá á Sólheimasandi in particular, it may also indicate a tortuous routing of meltwater through the heavily debris-laden Sólheimajökull, particularly in the lower and marginal regions of the outlet glacier, and a strong geothermal component in glacial melt processes. (The local name for Jökulá á Sólheimasandi is 'Fúlilækur' (The Foul River)

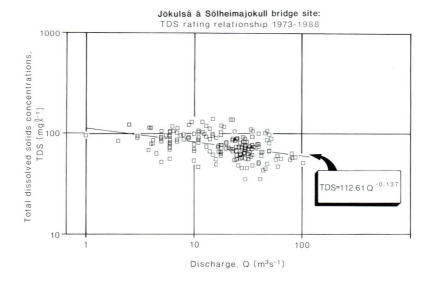

Fig. 12 Relationship between total dissolved solids concentration (T.D.S.) and discharge (n = 189).

derived from its characteristic 'bad eggs' smell of H_2S, and which Sigvaldason (1963) found in river water samples in quantities up to 1.2 ppm.) It is tempting to deduce from the high solute values that rates of chemical denudation in this subarctic catchment are as unexpectedly high as those recently identified from solute load measurements in alpine glacierized basins by Souchez and Lemmens (1987, p.301) and Collins (1983) but, first, the possibly substantial non-denudational component must be isolated, and work is continuing on this problem (Dolan, 1990).

In the classic way, total dissolved solids concentrations tend to decrease with increasing discharge (Fig. 12) because of dilution effects. For example, low flows (characterised by higher TDS values) are associated with rain-free periods and/or reduced insolation at the glacier surface during winter, or in cloudy conditions in summer. Runoff at these times would tend to be composed of slowly-moving waters following tortuous englacial or subglacial routes where availability of materials for dissolution or ion exchange would enhance opportunities for solute acquisition. This should be especially important during winter when the glacial conduit network has been partially closed due to plastic deformation. Higher flows on the other hand tend to be associated with (a) substantial fluxes of incoming radiation in summer which release large volumes of relatively dilute meltwater into the glacial drainage system, or (b) periods of intense or extended rainfall which directly input dilute water to the glacier along with any meltwater. In either case, the result is the dilution of chemically-enriched slowly-moving drainage by 'newer' parcels of solute-poor water which are routed quickly through the available conduits. Figure 12, of course, simply reflects a generalised and aggregated relationship the specific patterns of solute concentration in relation to melt-periods, rainfall events and recession flows may be very different due to complex storage and flushing effects within the glacier.

The TDS rating relationship for the Jökulsá á Sólheimasandi bridge site is:

$$TDS = 112.61 \ Q^{-0.137} \qquad (4)$$

where TDS is in mg l^{-1}, and Q is in $m^3 \ s^{-1}$, but in which discharge only explains 18.8% of the variation in TDS. The low correlation coefficient (r=0.434), though significantly different from zero at p = 0.0001, confirms that the export of dissolved material here cannot be interpreted solely in terms of flow variability. It is perhaps indicative of the subglacial geothermal influence here, which results in an essentially random injection of solutes into meltwaters (e.g. see Dolan, 1990; Sigvaldason, 1963). The lowest and most stable concentrations (between 50 and 70 mg l^{-1}) are found in the summer period when melt rates are maximized, and residence times of meltwater low (Fig. 13). Minima tend to occur in the six-week period beginning mid-July (Fig. 13). However, summer increases in water discharge are more than enough to offset lower solute concentrations at that time, resulting in a tendency (amidst great variability) for solute loads to peak in mid- to late-summer (Fig. 14). This is opposite to the tentative conclusion of Sigvaldason (1963) - reached in the absence of discharge data for the Jökulsá á Sólheimasandi - but is consistent with the results of Collins (1983) for the Gornera basin in Switzerland.

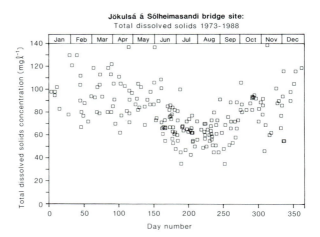

Fig. 13 Aggregated seasonal distribution of instantaneous total dissolved solids concentration values for each year of record 1973 - 1988.

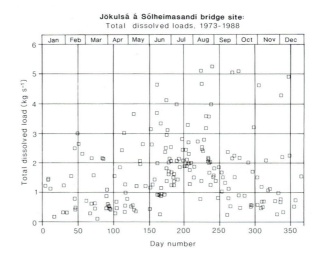

Fig. 14 Aggregated seasonal distribution of instantaneous total dissolved loads for each year of record 1973 - 1988.

Solute loads were calculated using total dissolved solids concentrations in equations (1) and (2), and are summarised in Table 2. They are more than an order of magnitude lower than the instantaneous suspended sediment loads calculated previously (Tables 1-2). Specific solute yields of around 700 t km^{-2} a^{-1} are high by global standards for glacierized catchments, however (e.g. see Ferguson, 1984), but represent only 4.8% - 7.5% of total yield from the basin, depending on the estimation method used (Table 2). July, August and September can be seen as the most important period for total flux from the Jökulsá á Sólheimasandi basin: discharge, suspended sediment loads, and solute loads all peak at this time.

7. EROSION RATES

It is interesting in the wider context of environmental change in Iceland to estimate notional erosion rates here from river load data. Table 2 displays estimated values of surface lowering, separately-calculated for suspended sediment loads only, solute loads only, and sediments and solutes together. By world standards a rate of 5.4 mm a^{-1} is very high indeed (Table 2), and is probably a conservative estimate. Separate calculations on the suspended sediment data alone for the individual years reveals a mean annual erosion rate of 4.69 mm a^{-1} (cf. 5.1 mm a^{-1} for the whole-period computation, Table 2) but with a high standard error of 0.66 mm a^{-1}, giving 95% confidence limits of 3.4 and 6.0 mm a^{-1}. Despite an apparent small reduction in suspended sediment concentration over time, no long-term change in sediment yield is discernible.

The values above agree closely with those computed from river load data by Tómasson (1976, 1986) for the Mýrdalsjökull ice-cap as a whole (Table 5). The estimates are also consistent with the suggestion that Mýrdalsjökull, including its valley glaciers, is the most erosive of all Icelandic ice-caps (Table 5). Furthermore, these new results offer tentative support to the suggestion of Humlum (1981, p.71) that 'the glacier bed of Mýrdalsjökull is apparently among the most intensively eroded surfaces on Earth'.

8. CONCLUSIONS

The Jökulsá á Sólheimasandi basin is apparently one of the most eroded surfaces in the world. Provisional estimates of specific sediment yield (almost 14,500 t km^{-2} a^{-1}) seem to be the highest ever recorded for a glacierized catchment, and amongst the highest for all rivers, regardless of environment. Problems of high temporal variability in material export, a low sampling frequency, and questions of sediment origin, however, suggest that these should only be regarded as preliminary estimates. Suspended sediment export completely dominates solute transfer: dissolved material load is only 4.8 per cent of the combined suspended sediment and solute yield. The July-September period is the dominant season of water, suspended sediment and solute flux.

It may be possible in the future to refine the tentative sediment-yield and erosion-rate estimates presented here by (a) incorporating a synthesized continuous flow series for the

Table 5   Erosion rates of glacierized catchments of Iceland and elsewhere
estimated from river sediment load data

Glacier/Ice-cap basin	Country	Erosion rate (mm a^{-1})	Period of data	Source
Sólheimajökull	Iceland	5.4	1973-1988	Lawler (this paper)
Mýrdalsjökull	Iceland	4.5	?	Tómasson (1976)
Mýrdalsjökull	Iceland	5.0	?	Tómasson (1986)
Vatnajökull	Iceland	3.2	?	Tómasson (1976)
Vatnajökull	Iceland	3.0	?	Tómasson (1986)
Hofsjökull	Iceland	0.9	?	Tómasson (1976)
Hofsjökull	Iceland	1.0	?	Tómasson(1986)
Langjökull	Iceland	0.4	?	Tómasson (1976)
Langjökull	Iceland	0.2	?	Tómasson (1986)
Tsidjiore Nouve	Switzerland	1.55-2.29	1981-1982	Small (1987)
Hunza	Pakistan	1.8	1966-1975	Ferguson (1984)

Jökulsá á Sólheimasandi, simulated statistically from overlapping continuous discharge data for the nearby Skógá river (Fig. 1) and various melt and precipitation variables; (b) establishing a permanent gauging station on the Jökulsá á Sólheimasandi to detect the crucial flow peaks and the true seasonality of the runoff regime; and (c) using continuously-observed melt-season water quality and quantity data that are now becoming available for this system (Dolan, 1990; Lawler, 1989).

Other future work aims to (i) relate amounts of material currently being exported from this and other basins in southern Iceland to the huge volumes of sediment recently discovered in trenches offshore on the south Iceland continental shelf (Boulton et al., 1988); (ii) explore the dynamics of the sediment and solute transfer system, including sources, pathways and sinks, to account fully for the magnitudes and variability encountered here, and to provide a platform for the predictions of responses to future climatic and environmental change; and (iii) investigate the possibility of using water quality perturbations to predict the imminent occurrence of subglacial geothermal activity (Sigvaldason, 1963; Bransdóttir, 1984; Einarsson and Bransdóttir, 1984; Dolan, 1990). It would also be interesting to test the hypothesis that, given the very large fluxes of sediment noted here, more material is exported from the Jökulsá á Sólheimasandi system during 'normal' flows of the annual regime than by catastrophic, infrequent, jökulhlaups (e.g. see Maizels, 1989), even though the latter may be more important, through sandur and terrace construction, for landform development in the proglacial area.

ACKNOWLEDGEMENTS

The members of the 1988 expedition (approved by the Royal Geographical Society) are extremely grateful for the funding or grants-in-kind from the following organisations: Birmingham University Field Research and Expeditions Fund; Royal Meteorological Society; Scott Polar Research Institute (Gino Watkins Memorial Fund); Royal Society (20th IGC Fund); British Geomorphological Research Group; Grant Instruments (Cambridge) Ltd; Philips Scientific; Camden Contract Hire Ltd; Shelley Signs Ltd; Rabone Chesterman Ltd; W. & T. Avery Ltd; Walden Precision Apparatus Ltd; Julius Sax & Co. Ltd (Automatic Liquid Samplers Ltd); Grattan Plc; Premier Brands Ltd; Rowntree Mackintosh Confectionery Ltd; Seabrook Potato Crisps Ltd; Warburton's (Yorkshire) Ltd.

Enormous thanks are due to Dr Arni Snorrason, Snorri Zóphóníasson and Haukur Tómasson of Orkustofnun for their advice, logistic support and data provision, and Þordur Tómasson of the Museum in Skógar for local arrangements. Dr Judith Maizels (Aberdeen University), Jason Phillips and Jim Berridge (both of Birmingham University) and Helgi Gunnarsson (Orkustofnun) helped with equipment transfer to, from, and within, Iceland. I greatly appreciate the research permit from the National Research Council of Iceland. Finally, I owe a huge debt to Dr Heather Lawler, and Mary and Robert Brown, for all their enthusiastic assistance throughout the 1988 field season.

REFERENCES

Bezinge, A. (1987) 'Glacial meltwater streams, hydrology and sediment transport: the case of the Grande Dixence hydroelectricity scheme', in A.M.Gurnell and M.J.Clark, M (Eds), Glacio-Fluvial Sediment Transfer: an Alpine Perspective, Wiley, 473-498.

Björnsson, H. (1979) 'Glaciers in Iceland', Jökull 29, 74-80.

Boulton, G.S., Thors, K. and Jarvis, J. (1988) 'Dispersal of glacially derived sediment over part of the continental shelf of south Iceland and the geometry of the resultant sediment bodies', Marine Geology 83, 193-223.

Bransdóttir, B. (1984) 'Seismic activity in Vatnajökull in 1900 - 1982 with special reference to Skeiðarárhlaups, Skaftárhlaups and Vatnajökull eruptions', Jökull 34, 141-150.

Carswell, D.A. (1983) 'The volcanic rocks of the Sólheimajökull area, south Iceland,' Jökull 33, 61-71.

Church, M. and Slaymaker, O. (1989) 'Disequilibrium of Holocene sediment yield in glaciated British Columbia', Nature 337, 452-454.

Clark, M.J. (1987) 'The alpine sediment system: a context for glacio-fluvial processes', in A.M.Gurnell and M.J.Clark (Eds), Glacio-Fluvial Sediment Transfer: an Alpine Perspective, Wiley, 9-31.

Collins, D.N. (1977) 'Hydrology of an alpine glacier as indicated by the chemical composition of meltwater', Zeitschrift für Gletscherkunde und Glazialgeologie 13, 219-238.

Collins, D.N. (1979) 'Sediment concentration in meltwaters as an indicator of erosion processes beneath an alpine glacier', Journal of Glaciology 23, 247-257.

Collins, D.N. (1981) 'Seasonal variation of solute concentration in melt waters draining from an Alpine glacier', Annals of Glaciology 2, 11 - 16.

Collins, D.N. (1983) 'Solute yield from a glacierized high mountain basin', in Dissolved Loads of Rivers and Surface Water Quantity/Quality Relationships, IAHS Publication No. 141, 41-49.

Dickinson, W.T. (1981) 'Accuracy and precision of suspended sediment loads', in Erosion and Sediment Transport Measurement, IAHS Publication No. 133, 195 - 202.

Dolan, M. (1990) The effects of subglacial geothermal activity on solute and sediment concentrations in the Jökulsá á Sólheimasandi melt river, south Iceland. Unpublished M.Sc. (Qual.) Thesis, University of Birmingham, 238pp.

Drewry, D. (1986) Glacial Geologic Processes. Edward Arnold, London.

Dugmore, A.J. (1989) 'Tephrochronological studies of Holocene glacier fluctuations in south Iceland', In J.Oerlemans (Ed.), Glacier Fluctuations and Climatic Change, Kluwer Academic Publishers, 37-55.

Einarsson, P. and Bransdóttir, B. (1984) 'Seismic activity preceding and during the 1983 volcanic eruption in Grímsvötn, Iceland', Jökull 34, 13-24.

Fenn, C.R. (1983) Proglacial streamflow series: measurement, analysis and interpretation. Unpublished Ph.D. Thesis, University of Southampton.

Fenn, C.R. (1987) 'Sediment transfer processes in alpine glacier basins', in A.M.Gurnell and M.J.Clark (Eds), Glacio-Fluvial Sediment Transfer: an Alpine Perspective, Wiley, 59-85.

Ferguson, R.I. (1984) 'Sediment load of the Hunza River', in K.J.Miller (Ed.) The International Karakoram Project, Cambridge University Press, Vol. 2, 581-598.

Ferguson, R.I. (1986) 'River loads underestimated by rating curves', Water Resources Research 22, 74-76.

Ferguson, R.I. (1987) 'Accuracy and precision of methods for estimating river loads', Earth Surface Processes and Landforms 12, 95-104.

Fountain, A.G. and Tangborn, W.V. (1985) 'The effect of glaciers on streamflow variations', Water Resources Research 21, 579-586.

Gislason, S.R. (1989) 'Kinetics of water-air interactions in rivers: A field study in Iceland', in Miles (Ed.) Water-Rock Interaction, Balkema, Rotterdam, 263-266.

Gislason, S.R. and Arnórsson, S. (1988) 'Chemistry of rivers in Iceland and the rate of chemical denudation', Náttúrufræðingurinn 58 (4), 183-197 (in Icelandic with English summary).

Gurnell, A.M. (1987a) 'Fluvial sediment yield in alpine, glacierized catchments', in A.M.Gurnell and M.J.Clark (Eds), Glacio-Fluvial Sediment Transfer: an Alpine Perspective, Wiley, 415-420.

Gurnell, A.M. (1987b) 'Suspended sediment', in A.M.Gurnell and M.J.Clark (eds), Glacio-Fluvial Sediment Transfer: an Alpine Perspective. Wiley, 305-354.

Gurnell, A.M. and Clark, M.J. (eds) (1987) Glacio-Fluvial Sediment Transfer: an Alpine Perspective, Wiley, Chichester.

Gurnell, A.M. and Fenn, C.R. (1984) 'Flow separation, sediment source areas and suspended sediment transport in a proglacial stream', Catena Supplement 5, 109-119.

Gurnell, A.M. and Fenn, C.R. (1985) 'Spatial and temporal variations in electrical conductivity in a pro-glacial stream system', Journal of Glaciology 31, 108-114.

Gurnell, A.M., Warburton, J. and Clark, M.J. (1988) 'A comparison of the sediment transport and yield characteristics of two adjacent glacier basins, Val d'Hérens, Switzerland', in M.P.Bordas and D.E.Walling (eds), Sediment Budgets, Proceedings of the Porto Alegre Symposium, December 1988, IAHS Publication No. 174, 431-441.

Guy, H.P. and Norman, V.W. (1970) 'Field methods for measurement of fluvial sediment', Techniques of Water-Resources Investigations of the U.S. Geological Survey, Chapter C2, Book 3, 59pp.

Humlum, O. (1981) 'Observations on debris in the basal transport zone of Mýrdalsjökull, Iceland', Annals of Glaciology 2, 71-77.

Lawler, D.M. (1989) Glaciofluvial sediment transport in southern Iceland, Report to the Royal Geographical Society on the University of Birmingham School of Geography expedition to Sólheimajökull, S. Iceland, July - October 1988, 29pp.

Maizels, J. K. (1989) 'Sedimentology, paleoflow dynamics and flood history of jökulhlaup deposits: paleohydrology of Holocene sediment sequences in southern Iceland sandur deposits', Journal of Sedimentary Petrology 59, 204-223.

Maizels, J.K. and Dugmore, A.J. (1985) 'Lichenometric dating and tephrochronology of sandur deposits, Sólheimajökull area, southern Iceland', Jökull, 35, 69-77.

Olive, L.J., Rieger, W.A. and Burgess, J.S. (1980) 'Estimation of sediment yields in small catchments: a geomorphic guessing game?', Proceedings of the Conference of the Institute of Australian Geographers, Newcastle, NSW.

Richards, K.S. (1982) Rivers : Form and Process in Alluvial Channels. Methuen, London.

Richards, K.S. (1984) 'Some observations on suspended sediment dynamics in Storbregrova, Jotunheimen', Earth Surface Processes and Landforms 9, 101-112.

Rist, S. (1967) 'The thickness of the ice cover of Mýrdalsjökull, southern Iceland', Jökull 17, 237-242.

Rist, S. (1984) 'Glacier variations', Jökull 34, 173-179.

Roehl, J.W. (1962) 'Sediment source areas, delivery ratios and influencing morphological factors', in Proceedings of the Bari Symposium, IASH Publication No. 59, 202-213.

Sigvaldason, G.E. (1963) 'Influence of geothermal activity on the chemistry of three glacier rivers in southern Iceland', Jökull 13, 10-17.

Small, R.J. (1987) 'Moraine sediment budgets', in A.M.Gurnell and M.J.Clark (eds), Glacio-Fluvial Sediment Transfer: an Alpine Perspective, Wiley, 165-197.

Souchez, R.A. and Lemmens, M.M. 1987 'Solutes', in Gurnell, A.M. and Clark, M.J. (Eds), Glacio-fluvial sediment transfer an Alpine perspective, Wiley, 285-303.

Tómasson, H. (1976) 'The sediment load of Icelandic rivers', Nordic Hydrological Conference, V-1 - V-16.

Tómasson, H. (1986) 'Glacial and volcanic shore interaction Part I: on land,' in G.Sigbjarnarson (ed.) Proceedings of the Iceland Coastal and River Symposium, Reykjavík, 2-3 September 1985, Icelandic National Energy Authority, Reykjavík, 7-16.

Walling, D.E. (1978) 'Reliability considerations in the evaluation and analysis of river loads', Zeitschrift für Geomorphologie, Supplement Band 29, 29-42.

Walling, D.E. and Webb, B.W. (1981) 'The reliability of suspended sediment load data', in Erosion and Sediment Transport Measurement, IAHS Publication No. 133, 177-194.

Walling, D.E. and Webb, B.W. (1983) 'Patterns of sediment yield,' in K.J.Gregory (ed.) Background to Palaeohydrology, Wiley, 69-100.

Glaciology and Quaternary Geology

1. V. V. Bogorodsky, C. R. Bentley and P. E. Gudmandsen: *Radioglaciology.* 1985   ISBN 90-277-1893-8

2. I. A. Zotikov: *The Thermophysics of Glaciers.* 1986   ISBN 90-277-2163-7

3. V. V. Bogorodsky, V. P. Gavrilo and O. A. Nedoshivin: *Ice Destruction.* Methods and Technology. 1987   ISBN 90-277-2229-3

4. C. J. van der Veen and J. Oerlemans (eds.): *Dynamics of the West Antarctic Ice Sheet.* 1987   ISBN 90-277-2370-2

5. J. S. Aber, D. G. Croot and M. M. Fenton: *Glaciotectonic Landforms and Structures.* 1989   ISBN 0-7923-0100-5

6. J. Oerlemans (ed.): *Glacier Fluctuations and Climatic Change.* 1989
                                                                                          ISBN 0-7923-0110-2

7. J. K. Maizels and C. Caseldine (eds.): *Environmental Change in Iceland: Past and Present.* 1991   ISBN 0-7923-1209-0

Kluwer Academic Publishers – Dordrecht / Boston / London